# Large Antennas
## of the
## Deep Space Network

# Large Antennas
## of the
## Deep Space Network

William A. Imbriale

WILEY-
INTERSCIENCE

A JOHN WILEY & SONS, INC., PUBLICATION

*Library of Congress Cataloging-in-Publication Data is available.*

ISBN 0-471-44537-1

Printed in the United States of America.

10 9 8 7 6 5 4 3 2 1

# Table of Contents

# Foreword

The Deep Space Communications and Navigation Systems Center of Excellence (DESCANSO) was recently established for the National Aeronautics and Space Administration (NASA) at the California Institute of Technology's Jet Propulsion Laboratory (JPL). DESCANSO is chartered to harness and promote excellence and innovation to meet the communications and navigation needs of future deep-space exploration.

DESCANSO's vision is to achieve continuous communications and precise navigation—any time, anywhere. In support of that vision, DESCANSO aims to seek out and advocate new concepts, systems, and technologies; foster key scientific and technical talents; and sponsor seminars, workshops, and symposia to facilitate interaction and idea exchange.

The Deep Space Communications and Navigation Series, authored by scientists and engineers with many years of experience in their respective fields, lays a foundation for innovation by communicating state-of-the-art knowledge in key technologies. The series also captures fundamental principles and practices developed during decades of deep-space exploration at JPL. In addition, it celebrates successes and imparts lessons learned. Finally, the series will serve to guide a new generation of scientists and engineers.

Joseph H. Yuen
DESCANSO Leader

# Preface

*Large Antennas of the Deep Space Network* traces the development of the antennas of NASA's Deep Space Network (DSN) from the network's inception in 1958 to the present. This monograph deals primarily with the radio-frequency design and performance of the DSN antennas and associated front-end equipment. It describes all the new designs and technological innovations introduced throughout the evolution of the DSN. There is also a thorough treatment of all the analytical and measurement techniques used in design and performance assessment.

This monograph is meant to serve as an introduction to newcomers in the field as well as a reference for the advanced practitioner. The technical terms in the text assume that the reader is familiar with basic engineering and mathematical concepts as well as material typically found in a senior-level course in electromagnetics.

Portions of this monograph (in Chapters 2 through 6) were originally written in 1990 to be a chapter in a proposed update to Joseph H. Yuen's book, *Deep Space Telecommunications Systems Engineering* (published in 1983 by Plenum Press). However, the update was never completed, and some of the material was subsequently published elsewhere. After the formation at the Jet Propulsion Laboratory (JPL) of the Deep-Space Communications and Navigation Systems Center of Excellence, a decision was made to publish a monograph series to capture the fundamental principles and practices developed during decades of deep-space exploration at JPL. Since the many technological innovations implemented in the DSN antennas helped make significant improvements in deep-space telecommunications over the decades, there was a real desire to capture these contributions in a comprehensive reference.

Chapters 2 through 6 originally covered the antennas that existed in the DSN prior to 1990. These chapters have been updated to include additions since 1990, such as X-band uplink and Ka-band downlink on the 70-meter antenna. Chapters 7 through 9 cover the beam-waveguide antennas that were introduced during the 1990s. Chapter 10 discusses possible future directions for the DSN. Chapter 1, the last to be written, presents the mathematical principles used to design and analyze all the antennas. Chapter 1 also includes a section on the design of beam-waveguide antennas as well as the measurement techniques used to assess antenna performance

The content of this monograph has been drawn primarily from the work of JPL colleagues, past and present, who have supported the development of the DSN antennas. In many cases, the text merely serves as a summary, the complete story being told by in the works referenced at the end of each chapter.

William A. Imbriale
February 2002

# Acknowledgments

I would like to express my appreciation to Joseph H. Yuen for his constant prodding, encouragement, and support during the writing of this manuscript. I am also deeply indebted to Cynthia D. Copeland for her typing and Diana M. Meyers for her thorough editing of the manuscript. I would like to thank Mark Gatti for supplying much of the documentation used for describing the Ka-band systems, and Daniel J. Hoppe, Vahraz Jamnejad, and David J. Rochblatt for their careful reading and helpful suggestions on several of the chapters. I am especially grateful to the following individuals for technical information, advice, and helpful suggestions during the preparation of the manuscript: Michael J. Britcliffe, Robert J. Cesarone, Paul W. Cramer Jr., Charles D. Edwards, Manuel S. Esquivel, Manuel M. Franco, Farzin Manshadi, Tommy Y. Otoshi, Samuel M. Petty, Harry F. Reilly, Charles T. Stelzried, John B. Sosnowski, Lawrence Teitelbaum, Watt Veruttipong, Victor A. Vilnrotter, and Sander Weinreb.

# Chapter 1
# Introduction

The U.S. National Aeronautics and Space Administration (NASA) Deep Space Network (DSN) is the largest and most sensitive scientific telecommunications and radio navigation network in the world. Its principal responsibilities are to support radio and radar astronomy observations in the exploration of the solar system and the universe. The network consists of three deep-space communications complexes (DSCCs), which are located on three continents: at Goldstone, in Southern California's Mojave Desert; near Madrid, Spain; and near Canberra, Australia. Each of the three complexes consists of multiple deep-space stations equipped with ultrasensitive receiving systems and large parabolic dish antennas. At each complex, there are multiple 34-m-diameter antennas, one 26-m antenna, one 11-m antenna, and one 70-m antenna. A centralized signal processing center (SPC) remotely controls the 34- and 70-m antennas, generates and transmits spacecraft commands, and receives and processes spacecraft telemetry.

The main features of the complex are the large parabolic dish antennas and their support structures. Although their diameters and mountings differ, all antennas employ a Cassegrain-type feed system that is essentially the same as that of a Cassegrain telescope, used in optical astronomy. Each antenna dish surface is constructed of precision-shaped perforated aluminum panels of specified surface accuracy, secured to an open steel framework.

This monograph details the evolution of the large parabolic dish antennas from the initial 26-m operation at L-band (960 MHz) in 1958 through the present Ka-band (32-GHz) operation on the 70-m antenna and provides a rather complete history of the DSN antennas. Design, performance analysis, and measurement techniques are highlighted. Sufficient information is provided to allow the sophisticated antenna professional to replicate the radio frequency optics designs.

The first DSN antenna began operating in November 1958 and was initially used to support Pioneers 3 and 4. The antenna site was subsequently named the Pioneer Deep Space Station. It began operation with a focal-point feed system but was quickly converted to a Cassegrain configuration, which remained until the antenna was decommissioned in 1981. In 1985, the U.S. National Park Service declared the site a national historic landmark.

The present Echo antenna, originally 26 m in diameter, was erected in late 1962 and was extended to 34 m in 1979. Both the Pioneer and Echo antennas were patterned after radio-astronomy antennas then in use at the Carnegie Institution of Washington (D.C.) and the University of Michigan. The main features borrowed from the radio-astronomy antenna design were the mount and the celestial-coordinate pointing system.

The Venus site began operation in 1962 as the Deep Space Network research and development (R&D) station and has remained the primary R&D site for DSN antennas. Almost all technological innovations in DSN antennas have first been tested at this site. The original 26-m antenna was equipped with an azimuth–elevation-type mount that could operate at 2 deg/s in azimuth and elevation. This antenna is no longer in use, replaced in 1990 by the first beam-waveguide (BWG) antenna.

The first 64-m antenna was completed in 1966 and was a physical scale-up of the 26-m antenna. Initial operation was at S-band (2.110–2.120 GHz transmit and 2.27–2.30 GHz receive). In 1968, an X-band feed cone was installed. A feed cone is a sheet-metal conical shell that houses the feed horn, low-noise amplifiers, and ancillary equipment. As the need developed for changing feed cones rapidly, the standard feed-cone support structure was replaced with a structure capable of supporting three fixed feed cones. In addition, the subreflector was modified to permit rapid changing of feed cones by rotating an asymmetrically truncated subreflector about its symmetric axis and pointing it toward the feed. Simultaneous multifrequency was provided by a reflex–dichroic feed system. In support of the NASA Voyager spacecraft encounter at Neptune in 1989, additional performance was realized by increasing the size of the antenna to 70 m, using dual-shaped optics, and improving the main-reflector surface accuracy through the use of high-precision panels. During the 1990s, the potential use of Ka-band on the 70-m antenna was demonstrated through R&D experiments, and in June 2000, X-band (7.145–7.190 GHz) uplink capability was added.

In 1982, a new design for a 34-m antenna with a common-aperture feed horn as well as dual-shaped optics designed for optimum gain over noise temperature ($G/T$) was introduced into the DSN and designated the high-efficiency (HEF) antenna. There is now one HEF at every DSCC.

Early in 1990, a new R&D antenna was fabricated and tested as a precursor to introducing BWG antennas and Ka-band frequencies into the DSN. The R&D antenna was initially designed for X- and Ka-bands but was subsequently upgraded to include S-band. Based upon the knowledge gained with this antenna, operational BWG antennas were included in the network. There are three 34-m BWG antennas at Goldstone, one in Spain, and one in Australia, with one under construction (as of 2002) in Spain and a second planned for Australia.

Also in the 1990s, two BWG antennas capable of supporting X-band high power were built and tested. As they were only a technology demonstration, their original design was never introduced into the network. Rather, one of the two was retrofitted for S-band support and is still in use in that capacity.

This monograph will trace the history of DSN large-antenna technology through the development and implementation of the antennas at the Goldstone complex, since, with the exception of the 64-to-70-m upgrade, the technology improvements were first introduced there. Chapter 1 presents the methods of analysis, with supporting mathematical details, and measurement and design techniques for reflector antennas. Chapters 2 through 9 cover each type of antenna, and Chapter 10 explores current thinking.

## 1.1  Technology Drivers

The prime mission of the DSN is to receive extremely weak signals over vast interplanetary distances. A key element of the telecommunications-link performance is the received power signal-to-noise ratio (SNR), which is given by

$$S/N = \frac{P_T G_T A_R}{4\pi R^2 N} = \frac{P_T G_T G_R}{kBT_S (4\pi R/\lambda)^2} \qquad (1.1\text{-}1)$$

where

$$
\begin{aligned}
P_T &= \text{spacecraft transmit power} \\
G_T &= \text{transmit gain} \\
G_R &= \text{receive gain} \\
R &= \text{distance to the spacecraft} \\
N &= \text{total noise} \\
A_T &= \text{the effective area of the transmit (spacecraft) antenna} \\
A_R &= \text{the effective areas of the receive ground antennas} \\
T_s &= \text{receive system-noise temperature} \\
\lambda &= \text{wavelength}
\end{aligned}
$$

$k$   =   Boltzman's constant

$B$   =   bandwidth.

To do its part effectively, the ground antenna system must maximize the ratio of received signal to the receiving system noise power, which is measured by an antenna figure of merit (FM), defined as the ratio of antenna effective area (or equivalently gain) to system-noise temperature.

The receive temperature consists primarily of the antenna feed system and amplifier contributions. For assessing the antenna FM, it is desirable to draw an imaginary reference plane between the receiver system and the antenna system, thus placing all noise contributions in one of these categories. If the receiver contribution (including the feed-system losses) is given by $T_R$ and the antenna noise contribution by $T_A$, then the FM will be given by

$$FM = \frac{G_R}{T_R + T_A} \qquad (1.1\text{-}2)$$

It can be readily seen that the antenna noise-temperature properties are very significant contributors to the FM, especially for the cases of low receiver noise-temperature systems. Also, the significant effect of total system-noise temperature on the FM is apparent. Thus, to maximize the FM for a given antenna size and frequency of operation, it is necessary to both maximize the antenna gain and minimize the total system-noise temperature. Since the individual contributions to noise temperature are additive and essentially independent of each other, it is necessary to individually minimize each contribution. If any one of the noise contributions is large, minimizing the others only marginally improves the FM. However, by using cryogenic amplifiers, the receiver noise-temperature contribution can be small (as low as 2 to 3 K for the highest-performing masers), and it then becomes imperative to minimize both the antenna and feed-system contributions. For ambient conditions, it should be noted that feed-system losses contribute to noise temperature at the rate of 7 K per 0.1-dB loss.

Techniques for maximizing the antenna gain and aperture efficiency of reflector antennas involve control of the illumination function. By definition, a feed system, which uniformly illuminates the antenna aperture with the proper polarization and has no spillover energy or other losses, possesses 100 percent aperture efficiency. In practice, however, some spillover is present due to the finite size of the feed illuminating a single reflector system or the subreflector illuminating the main reflector of a dual reflector system. One key result of this spillover is a thermal-noise contribution from the physically hot ground. Clearly, then, an ideal feed system has a very rapid energy cutoff at the reflector edge, thereby maintaining relatively uniform reflector illumination while at the same time minimizing spillover energy.

Many design principles for large reflector antennas are borrowed from those successfully used in designing optical telescopes. For example, the earliest antenna design used a two-reflector Cassegrain system in which the main reflector was a paraboloid and the subreflector a hyperboloid. Subsequent designs incorporated dual-shaped optics, which offer higher efficiency for the same-size aperture. Although not as yet used in the DSN (as of 2002), clear-aperture designs (in which the radiating aperture is not blocked by either the subreflector or subreflector supports) offer the potential for even higher efficiencies.

Surface accuracy is also an important parameter in determining effective aperture area, as is demonstrated in Fig. 1-1, which plots the reduction in gain versus root-mean-square (rms) surface error. It demonstrates that the rms surface error must be an extremely small fraction of the reflector diameter.

Surface errors fall into two main categories: (a) time-invariant panel mechanical manufacturing errors and panel-setting errors at the rigging angle and (b) time-varying errors induced by gravity, wind, and thermal effects. Improvements in panel manufacturing and in panel setting using microwave holography have left gravity distortions as a function of elevation angles as the major error source in DSN antennas for higher-frequency operation.

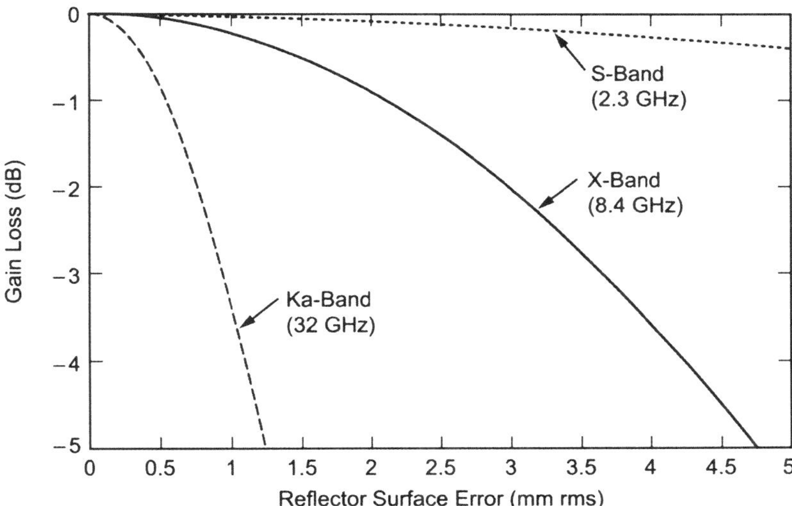


**Fig. 1-1. Gain loss due to reflector surface error.**

### 1.1.1  Frequency Bands Allocated to the Deep Space Network

Frequency ranges have been allocated by the International Telecommunication Union (ITU) for use in deep-space and near-Earth research. These ranges are listed in Table 1-1.

**Table 1-1. Allocated frequency bands (GHz).**

| Band | Deep-Space Bands for Spacecraft Farther Than 2 Million km from Earth | | Near-Earth Bands for Spacecraft Closer Than 2 Million km from Earth | |
|---|---|---|---|---|
| | Uplink[a] | Downlink[b] | Uplink[a] | Downlink[b] |
| S | 2.110–2.120 | 2.290–2.300 | 2.025–2.110 | 2.200–2.290 |
| X | 7.145–7.190 | 8.400–8.450 | 7.190–7.235 | 8.450–8.500 |
| Ka | 34.200–34.700 | 31.800–32.300 | Not applicable | Not applicable |

[a] Earth to space.
[b] Space to Earth.

## 1.2  Analysis Techniques for Designing Reflector Antennas

Reflector antennas have existed since the days of Heinrich Hertz. They are one of the best solutions to requirements for cost-effective, high-dB/kg, high-performance antenna systems.

For a large ground antenna, it is not feasible to build a scale model to verify a new design. Hence, extremely accurate design-analysis tools are required. Fortunately, physical optics (PO) provides the required performance-estimate accuracy. All of the DSN antennas were designed and analyzed using PO, and the measured performance is within a few percent of the calculated values.

In addition to PO, many other techniques are required to completely design and characterize an antenna system. For example, accurate programs to design and analyze the feed horn and transform far-field patterns to near-field patterns for use in the PO analysis are essential. Programs to determine the reflector shape for maximum gain, or $G/T$, are also used. And quasioptical techniques such as geometric optics (GO), Gaussian-beam analysis, and or ray tracing are useful for a quick characterization of BWG systems. Tools to design and analyze frequency-selective surfaces are needed for use in multifrequency systems. And since the FM for these antennas is $G/T$, accurate programs to predict noise performance are required.

The basic mathematical details of each of these techniques are given in this section, with examples of their use sprinkled throughout this monograph.

## 1.2.1 Radiation-Pattern Analysis

By far the most important analytical tool is physical optics, which is used to calculate the scattered field from a metallic reflecting surface—in this case, a reflector antenna. Electric currents, which excite the scattered field, are induced on the conducting surface by an incident wave assumed to be of a known amplitude, phase, and polarization everywhere in space (from a feed or other reflecting surface, for example). The PO approximations to the induced surface currents are valid when the reflector is smooth and the transverse dimensions are large in terms of wavelengths. The closed reflecting surface is divided into a region $S_1$, which is illuminated by direct rays from the source ("illuminated region") and a region $S_2$, which is geometrically shadowed ("shadowed region") from direct rays from the source (Fig. 1-2). The PO approximations for the induced surface current distribution are

$$
\begin{aligned}
J_s &= 2(\hat{n} \times H_{inc}) \quad \text{on } S_1 \\
J_s &= 0 \quad\quad\quad\quad\;\; \text{on } S_2
\end{aligned}
\tag{1.2-1}
$$

where $\hat{n}$ is the surface normal and $H_{inc}$ the incident field. The expressions are then inserted into the radiation integral [1] to compute the scattered field.

Rusch and Potter [2] provide a good introduction to the early techniques used for analyzing DSN reflector antennas. More recently, due primarily to the orders-of-magnitude improvements in computer speed and memory, a very simple but extremely robust algorithm has emerged as the computer program of choice for computing the PO radiation integral. Its initial limitation to small reflectors was primarily due to the speed and memory limitations of the then-existing computers. The algorithm is documented in [3] and [4], but because of its extreme importance and to provide a fairly complete reference, it is repeated here.

One of the simplest possible reflector-antenna computer programs is based on a discrete approximation of the radiation integral. This calculation replaces the actual reflector surface with a triangular facet representation so that the

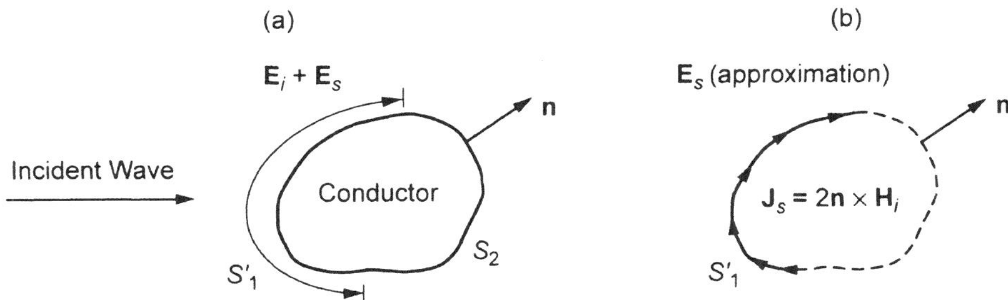


**Fig. 1-2. The physical optics approximation: (a) original problem and (b) approximation.**

reflector resembles a geodesic dome. The PO current is assumed to be constant in magnitude and phase over each facet, so the radiation integral is reduced to a simple summation. This program was originally developed in 1970 and has proven to be surprisingly robust and useful for the analysis of reflectors, particularly when the near field is desired and the surface derivatives are not known.

Two improvements significantly enhanced the usefulness of the computer programs: the first was the orders-of-magnitude increase, over time, in computer speed and memory and the second was the development of a more sophisticated approximation of the PO surface current, which permitted the use of larger facets. The latter improvement is due to the use of a linear-phase approximation of the surface current. Within each triangular region, the resulting integral is the 2-D Fourier transform of the projected triangle. This triangular-shape function integral can be computed in closed form. The complete PO integral is then a summation of these transforms.

**1.2.1.1 Mathematical Details.** The PO radiation integral over the reflector surface, $\Sigma$, can be expressed as [5]

$$\mathbf{H}(\mathbf{r}) = -\frac{1}{4\pi} \int_{\Sigma} \left( jk + \frac{1}{R} \right) \hat{\mathbf{R}} \times \mathbf{J}(\mathbf{r}')\frac{e^{-jkR}}{R} ds' \qquad (1.2\text{-}2)$$

in which

$\quad \mathbf{r} \quad$ = the field point

$\quad \mathbf{r}' \quad$ = source point

$\quad R \quad$ = $|\mathbf{r} - \mathbf{r}'|$ is the distance between them

$\quad \hat{\mathbf{R}} \quad$ = $(\mathbf{r} - \mathbf{r}')/R$ is a unit vector

$\quad \mathbf{J}(\mathbf{r}')$ = the surface current

$\quad \lambda \quad$ = wavelength

$\quad k \quad$ = $2\pi/\lambda$.

For the purpose of analysis, the true surface, $\Sigma$, is replaced by a contiguous set of triangular facets. These facets, denoted $\Delta_i$, are chosen to be roughly equal in size with their vertices on the surface, $\Sigma$. Fig. 1-3 shows a typical facet and its projection onto the $x$, $y$ plane. Let $(x_i, y_i, z_i)$ represent the centroid of each triangle where the subscript $i = 1, \cdots, N$ is associated with a triangle. Then, the field obtained by replacing the true surface, $\Sigma$, by the triangular facet approximation is

$$\mathbf{H}(\mathbf{r}) = -\frac{1}{4\pi} \sum_{i=1}^{N} \int_{\Delta_i} \left( jk + \frac{1}{R} \right) \hat{\mathbf{R}} \times \mathbf{J}(\mathbf{r}')\frac{e^{-jkR}}{R} ds' \qquad (1.2\text{-}3)$$

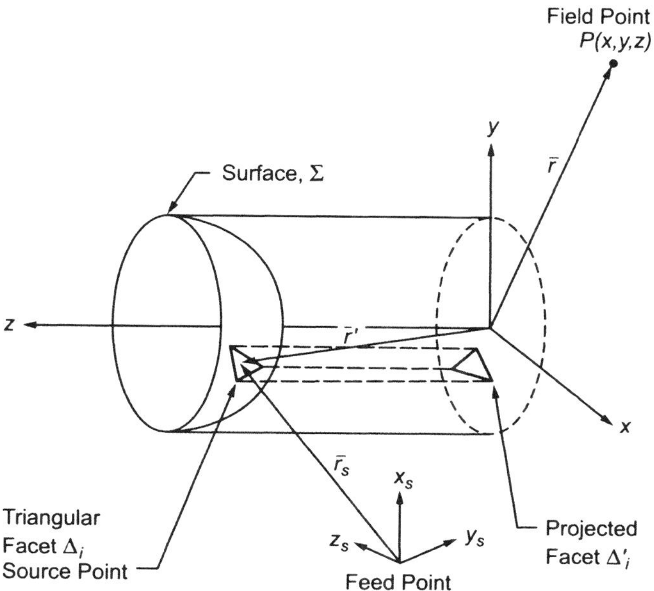



**Fig. 1-3. Reflector-analysis coordinate systems and a typical triangular facet.**

In Eq. (1.2-3), $\mathbf{J}$ is now the equivalent surface current evaluated on the triangular facets. Since the triangles are small, it is expected that $\hat{\mathbf{R}}$ and $R$ do not vary appreciably over the area of a given facet. Thus, let $\hat{\mathbf{R}}_i$ and $R_i$ be the value obtained at the centroid $(x_i, y_i, z_i)$ of each facet and approximate Eq. (1.2-3) by

$$\mathbf{H}(\mathbf{r}) = -\frac{1}{4\pi} \sum_{i=1}^{N} \left[ jk + \frac{1}{R_i} \right] \hat{\mathbf{R}}_i \times \mathbf{T}_i(\mathbf{r}) \qquad (1.2\text{-}4)$$

$$\mathbf{T}_i(\mathbf{r}) = \int_{\Delta_i} \mathbf{J}_i(\mathbf{r}') \frac{e^{-jkR}}{R_i} ds' \qquad (1.2\text{-}5)$$

Assume that the necessary transformations have been performed so that the incident field, $\mathbf{H}_s$, is given in terms of the reflector coordinate system. Then

$$\mathbf{J}_i(\mathbf{r}') = 2\hat{\mathbf{n}}_i \times \mathbf{H}_s(\mathbf{r}') \qquad (1.2\text{-}6)$$

Next, assume that the incident field can be represented by a function of the form

$$\mathbf{H}_s = \mathbf{h}_s(\mathbf{r}_i) \frac{e^{-jkr_s}}{4\pi r_{si}} \qquad (1.2\text{-}7)$$

where $r_s$ is the distance to the source point and $r_{si}$ is the distance from the triangle centroid to the source point. Then, Eq. (1.2-5) can be written

$$\mathbf{T}_i(\mathbf{r}) = \frac{\hat{n}_i \times \mathbf{h}_s(\mathbf{r}_i)}{2\pi R_i r_{si}} \int_{\Delta_i} e^{-jk(R+r_s)} ds' \qquad (1.2\text{-}8)$$

Making use of the Jacobian and approximating

$$R(x,y) + r_s(x,y) = \frac{1}{k}\left(a_i - u_i x - v_i y\right) \qquad (1.2\text{-}9)$$

in which $a_i$, $u_i$, and $v_i$ are constants, the expression can be rewritten as

$$\mathbf{T}_i(\mathbf{r}) = \frac{\hat{\mathbf{n}}_i \times \mathbf{h}_s(\mathbf{r}_i)}{2\pi R_i r_{si}} J_{\Delta_i} e^{-ja_i} \int_{\Delta_i'} e^{j(u_i x' + v_i y')} dx' dy' \qquad (1.2\text{-}10)$$

where the surface normal to the surface $z = f(x,y)$ is

$$\mathbf{N}_i = -\hat{\mathbf{x}} f_{xi} - \hat{\mathbf{y}} f_{yi} + \hat{\mathbf{z}} \qquad (1.2\text{-}11)$$

where $f_x = \dfrac{\partial f}{\partial x}$ and the Jacobian is

$$J_{\Delta_i} = |\mathbf{N}_i| = \left[ f_{xi}^2 + f_{yi}^2 + 1 \right]^{1/2} \qquad (1.2\text{-}12)$$

It may now be observed that this integral is the 2-D Fourier transform of the $i^{\text{th}}$ projected triangle $\Delta_i'$, expressed as

$$S(u,v) = \int_{\Delta_i'} e^{j(ux' + vy')} dx' dy' \qquad (1.2\text{-}13)$$

which can be computed in closed form as described in [6]. The full radiation integral is then the sum of all the transforms of the individual triangles.

**1.2.1.2  Application to Dual-Reflector Antennas.** The PO integration methodology is incorporated in a sequential fashion for the analysis of the dual-reflector antenna system. Initially, the feed illuminates the subreflector, and the currents on the subreflector surface are determined. Subsequently, the near fields scattered from the subreflector are used to illuminate the main reflector, and its induced currents are determined. The main reflector scattered fields are then determined by integrating these currents.

Many coordinate systems are required to allow flexibility in locating and orienting the feed, subreflector, main reflector, and output-pattern generation. The relation among the various coordinate systems is depicted in Fig. 1-4.

### 1.2.1.3  Useful Coordinate Transformations.

In the discussion of the preceding sections (1.2.1 and 1.2.1.2), the analysis is performed using two distinct coordinate systems: reflector and feed coordinates. In addition, it is sometimes convenient to display the computed patterns in yet another coordinate system. Consequently, one must know the transformation equations that permit coordinates and vectors described in one coordinate system to be expressed in terms of some other coordinate system. The transformation may require both translation and rotation. The required transformations are described below. They are the Cartesian-to-spherical transformation and coordinate rotations using Eulerian angles.

The Cartesian-to-spherical transformation is conveniently summarized in matrix form. With the Cartesian components of a vector, **H**, denoted $(H_x, H_y, H_z)$ and the spherical components $(H_r, H_\theta, H_\phi)$, one finds that the transformation is

$$
\begin{bmatrix} H_r \\ H_\theta \\ H_\phi \end{bmatrix} = \begin{bmatrix} \sin\theta\cos\phi & \sin\theta\sin\phi & \cos\theta \\ \cos\theta\cos\phi & \cos\theta\sin\phi & -\sin\theta \\ -\sin\phi & \cos\phi & 0 \end{bmatrix} \begin{bmatrix} H_x \\ H_y \\ H_z \end{bmatrix} \tag{1.2-14}
$$

The inverse transformation is just the transpose of the above matrix.

Rotations are handled by the use of the Eulerian angles $(\alpha, \beta, \gamma)$. These angles describe three successive rotations that bring one Cartesian system

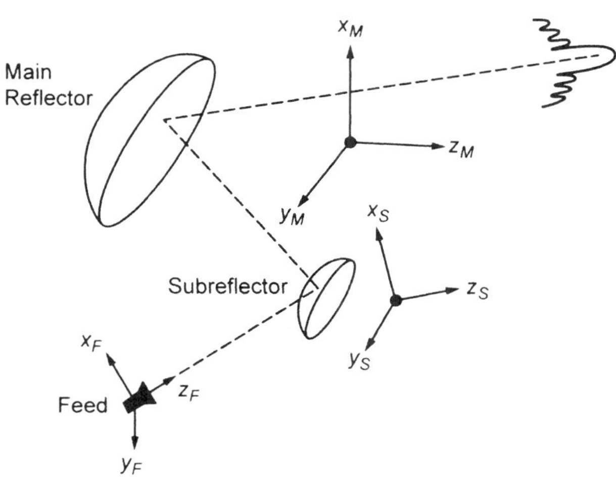


**Fig. 1-4.  Dual-reflector coordinate systems.**

into alignment with another. Let the two systems be denoted $(x_1, y_1, z_1)$ and $(x_2, y_2, z_2)$. As illustrated in Fig. 1-5, the angles are defined as follows:

$\alpha$   describes a positive rotation about the $z_1$-axis, which brings the $x_1$-axis into the x'-axis aligned with the *line of nodes* (the line of intersection between the $(x_2, y_2)$ and $(x_1, y_1)$ planes)

$\beta$   describes a positive rotation about the line of nodes (the x'-axis) that brings the $z_1$-axis to $z_2$

$\gamma$   describes a positive rotation about the $z_2$-axis, which brings the x'-axis to the $x_2$-axis.

The phrase "positive rotation" means the direction of increasing angular measure as defined by the right-hand rule with respect to the axis about which the rotation occurs. Each of the rotations just described is performed using the standard rotation of coordinate formulas of plane analytic geometry.

When these expressions are written in matrix form and applied successively as described above, one obtains the following matrix equation that represents a general 3-D rotation of coordinates:

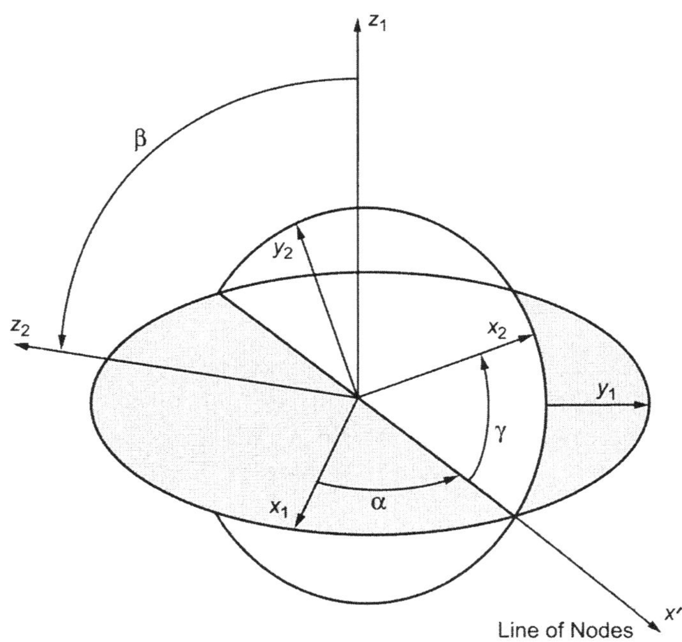


**Fig. 1-5. Euler-angle definitions.**

$$\begin{bmatrix} x_2 \\ y_2 \\ z_2 \end{bmatrix} = \begin{bmatrix} A_{11} & A_{12} & A_{13} \\ A_{21} & A_{22} & A_{23} \\ A_{31} & A_{32} & A_{33} \end{bmatrix} \begin{bmatrix} x_1 \\ y_1 \\ z_1 \end{bmatrix} \qquad (1.2\text{-}15)$$

where the individual matrix elements are

$A_{11} = \cos\gamma\cos\alpha - \sin\gamma\cos\beta\sin\alpha$

$A_{12} = \cos\gamma\cos\alpha + \sin\gamma\cos\beta\cos\alpha$

$A_{13} = \sin\gamma\sin\beta$

$A_{21} = -\sin\gamma\cos\alpha - \cos\gamma\cos\beta\sin\alpha$

$A_{22} = -\sin\gamma\cos\alpha + \cos\gamma\cos\beta\cos\alpha$

$A_{23} = \cos\gamma\sin\beta$

$A_{31} = \sin\beta\cos\gamma$

$A_{32} = -\sin\beta\cos\alpha$

$A_{33} = \cos\beta.$

The inverse transformation is the transpose of the matrix given above.

Although the formulas are presented in terms of coordinate transformations, the transformation matrix is equally valid for the Cartesian components of a vector. Thus, the components of a vector, **H**, transform as

$$\begin{bmatrix} H_{x,2} \\ H_{y,2} \\ H_{z,2} \end{bmatrix} = \begin{bmatrix} A_{11} & A_{12} & A_{13} \\ A_{21} & A_{22} & A_{23} \\ A_{31} & A_{32} & A_{33} \end{bmatrix} \begin{bmatrix} H_{x,1} \\ H_{y,1} \\ H_{z,1} \end{bmatrix} \qquad (1.2\text{-}16)$$

Further information can be found in [7].

### 1.2.1.4 A Numerical Example of Radiation-Pattern Analysis.

In the 1980s, a FORTRAN program was written to perform the linear phase calculations indicated above. The program was extensively verified by comparing with measured data, for example, [8], and many other computer codes. A simple but interesting example is that of an ellipse, shown in Fig. 1-6.

The projected aperture of the ellipse is about 3 m. In the offset plane, centered on the $x_p$ axis, the illumination function is a $\cos^{42}\theta$ pattern function (22.3-dB gain), and the frequency is 31.4 GHz. The ellipse is about $350\lambda$ along the major axis. Figure 1-7 compares the constant-phase approximation for three different grid densities: approximately 4000 and 10,000 and 23,000 triangles.

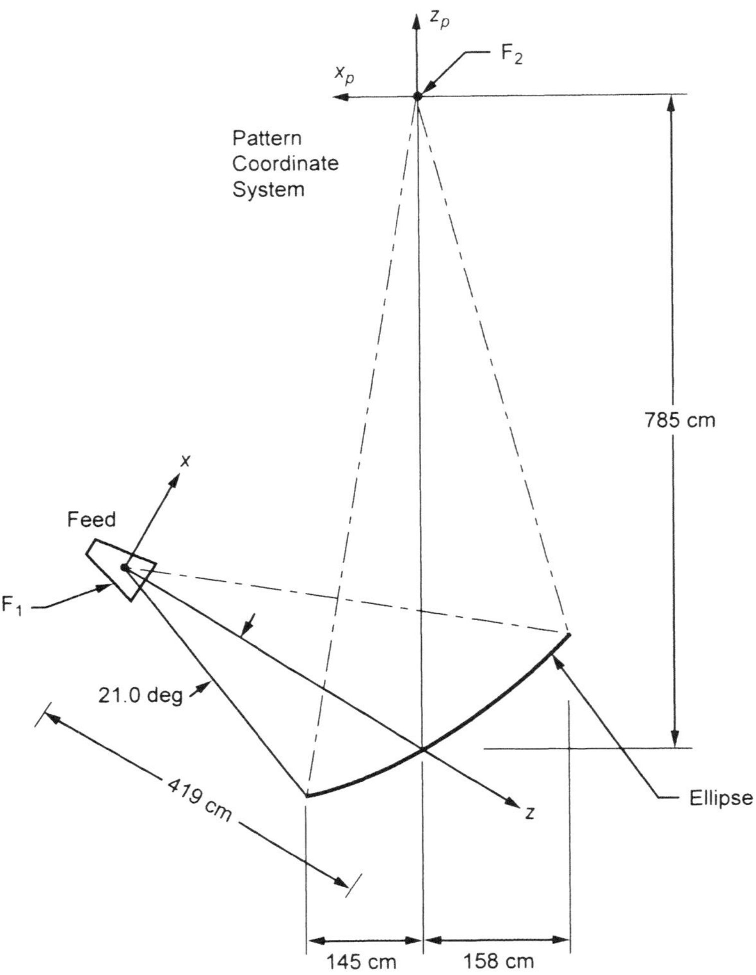



**Fig. 1-6. Ellipse geometry.**

This illustrates a general trend of the method; that is, depending on the size of the triangles, there is an angular limit over which the solution is valid. Figure 1-8 compares the linear-phase approximation with the constant-phase approximation for the 4000-triangle case and demonstrates that the angular range is larger with the linear-phase approximation.

## 1.2.2 Feed-Horn Analysis

An equally critical aspect of the analysis of reflector systems is the ability to accurately compute the radiation pattern of the feed. More details on the design of the feeds will be given later, but the analysis technique for computing the radiation patterns of the feed is summarized below.

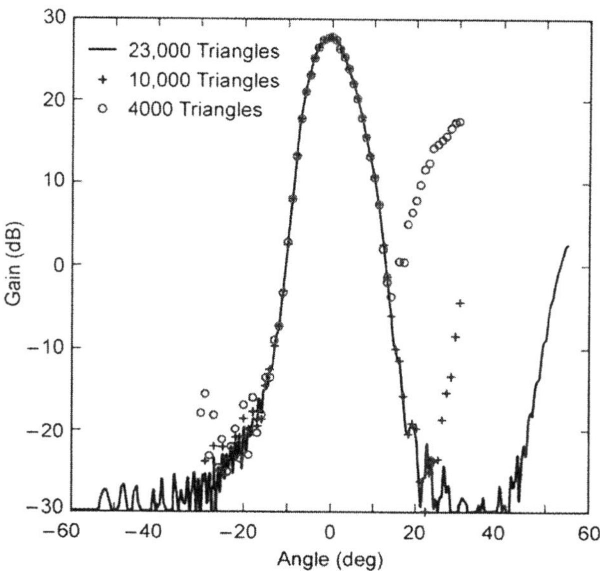

Fig. 1-7.  Ellipse example: constant-phase
approximation for offset plane.

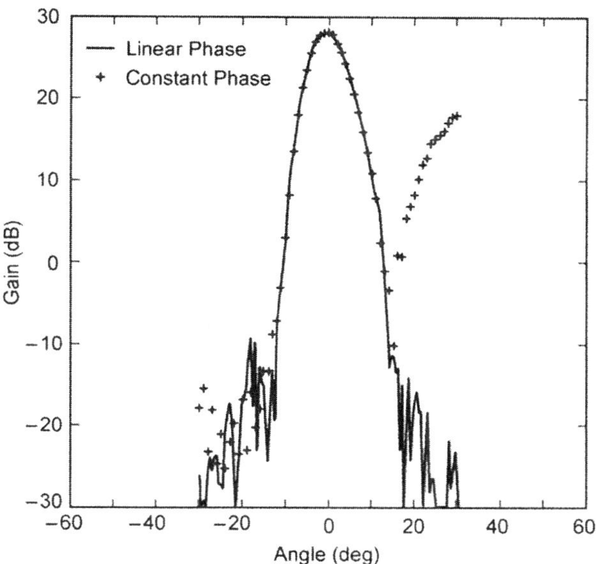

Fig. 1-8.  Ellipse example: constant versus
linear phase for offset plane.

Two types of feed horn possessing equal E- and H-plane patterns are commonly used. The first is the dual-mode feed horn [9], and the second is the corrugated feed horn [10]. In the dual-mode horn, a dominant-mode circular waveguide is connected to another guide of slightly larger diameter, where modes up to (transverse magnetic) $TM_{11}$ may propagate, via a step transition. The step size is chosen to generate the precise amount of $TM_{11}$ mode from the (transverse electric) $TE_{11}$ mode so that when the two modes travel through the flared horn section that follows, the E- and H-plane patterns are equalized. The bandwidth of this feed horn is limited since the two modes must arrive at the horn aperture in phase, and the two modes have phase velocities that vary differently with frequency.

In the corrugated feed horn, the single-mode smooth-wall waveguide is connected to a corrugated waveguide that supports only the (hybrid) $HE_{11}$ mode. Some matching between the waveguides is provided by gradually changing from $\lambda/2$ slot depth to $\lambda/4$ slot depth in a short transition region. Throughout the transition region, only the $HE_{11}$ corrugated waveguide mode may propagate, and the E- and H-plane radiation patterns of this mode become nearly equal when the balanced condition is reached (slot depth = $\sim\lambda/4$). The bandwidth of this horn is larger than that of the dual-mode horn because the transverse electric field patterns and, hence, the radiation pattern of the $HE_{11}$ mode are relatively insensitive to small changes in slot depth around the balanced condition (slot depth = $\sim\lambda/4$). After the $HE_{11}$ mode is established in the single-mode corrugated waveguide, the guide is gradually flared, without changing the slot depth, to the required aperture size.

The corrugated section is analyzed using a computer code developed by Hoppe [11–13]. The analysis follows the method of James [14], expanding the fields inside each fin and slot in terms of circular waveguide modes, and matching the fields at each slot–fin boundary. All of the possible propagating modes as well as a sufficient number of evanescent modes are matched at each boundary, with results for successive edges and waveguide lengths cascading as the analysis moves through the device. In this way, the interaction between the fields of nonadjacent as well as adjacent slots are taken into account. The result of the calculation is a matrix equation relating the reflected and aperture modes to the input modes.

If $\mathbf{a}_1$ is a vector containing the power-normalized amplitudes of the input modes, then we may calculate the reflected modes, $\mathbf{b}_1$, and the aperture modes, $\mathbf{b}_2$, using

$$\mathbf{b}_2 = [S_{21}]\mathbf{a}_1 \tag{1.2-17}$$

$$\mathbf{b}_1 = [S_{11}]\mathbf{a}_1 \tag{1.2-18}$$

Here, $[S_{21}]$ and $[S_{22}]$ are the scattering matrices resulting from the computer run. (See the appendix of [14].) They depend only on frequency and device dimensions, not input modes. We may, therefore, specify any input vector $\mathbf{a}_1$ and calculate the reflected and aperture fields. Using the normalized amplitudes calculated above and the normalized vector functions giving the field distributions for each mode, we find the aperture field $\mathbf{E}_B$. The far field is then calculated by the method described by Silver and Ludwig [15,16].

$$\mathbf{E}_c = \frac{-1}{4\pi} \iint_S \left(- j\mu\omega \left(\hat{n} \times \mathbf{H}_B\right) \phi + \left(\hat{n} \times \mathbf{E}_B\right) \times \nabla\phi\right) ds \qquad (1.2\text{-}19)$$

where

| | | |
|---|---|---|
| $\mu$ | = | free-space permeability |
| $\omega$ | = | angular frequency $= 2\pi f$ |
| $\hat{n}$ | = | unit vector normal to aperture surface |
| $\mathbf{H}_B$ | = | aperture magnetic field |
| $\phi$ | = | $e^{-jkr}/r$ |
| $\mathbf{E}_B$ | = | aperture electric field |
| $\nabla$ | = | gradient operator |
| $f$ | = | frequency |
| $ds$ | = | incremented area on aperture surface |
| $k$ | = | $2\pi/\lambda$ |
| $r$ | = | distance from the origin to the far-field point. |

When $\mathbf{E}_B$ and $\mathbf{H}_B$ are represented in terms of circular waveguide modes, the resulting integrals have already been evaluated by Silver [16]. Therefore, given an input vector and the scattering matrix, we determine the aperture modes and composite far-field patterns. A spherical-wave analysis is then used to get the feed-horn near-field pattern for use in the PO software. Throughout the analysis, care must be taken to ensure proper normalization of the field amplitudes in terms of power. The smooth wall conical feed horn is modeled with the same software, only approximating the horn taper with small steps and zero-depth corrugated slots.

The mode-matching technique for analyzing corrugated horns yields excellent agreement with the measured patterns—so much so, in fact, that if the computed and measured patterns do not match, it is most likely due to measurement and/or manufacturing errors. A recent example of a fairly complicated horn is the X-/X-/Ka-band horn described in [17] and Chapter 9 of this monograph. A comparison of the computed and measured patterns is shown in Figs. 1-9 and 1-10. This type of agreement is typical for a well-manufactured corrugated feed horn.

### 1.2.3   Spherical-Wave Analysis

Spherical-wave-expansion coefficients are normally used as one part of a sequence to analyze an antenna system. The basic role of the technique is to transform far-field patterns to the near-field H-field so that PO may be used for reflectors in the near field of their illumination source.

The theory of spherical waves is described in [18] and will only be briefly outlined here. Any electromagnetic field in a source-free region may be represented by a spherical-wave expansion. In general, the expansion must include both incoming and outgoing waves. If the field satisfies the radiation condition, only outgoing waves will be present, and the expansion will be valid outside the smallest sphere enclosing all sources (the sphere center must be at the coordinate origin used for the expansion). The radial dependence of the spherical waves is then given by the spherical Hankel function $h_n^2(kR)$. The second common case is an expansion valid inside the largest sphere enclosing no sources. In this case, the incoming and outgoing waves are present in equal amounts, producing a radial dependence given by the spherical Bessel function $j_n(kR)$.

Although the spherical-wave expansion can be applied to either of these two most common cases, the version used for the DSN assumes outgoing waves.

In either case, the input data that is used to specify the field is the tangential E-field on the surface of a sphere. For the first case, the data-sphere radius must be greater than or equal to the radius of the sphere enclosing the sources. For far-field data, the data-sphere radius is considered to be infinite. For the second case, the data-sphere radius must be less than or equal to the largest sphere enclosing no sources, and must be greater than zero.

The maximum value of the Hankel function index that is needed to closely approximate the field is roughly equal to $ka$ ($ka + 10$ is typical, but in some cases a lower limit will work), where $a$ is the radius of the sphere enclosing all (or no) sources for the first and second case, respectively.

Input data is specified on a grid of points defined by the intersection of constant contours of $\theta$ and $\phi$. The amplitude and phase of $E_\theta$ and $E_\phi$ are given at each point. The minimum number of $\theta$ values is roughly 1.2 times the maximum value of $n$.

The azimuthal dependence of spherical waves is given by sin $(m\phi)$ and cos $(m\phi)$. In general, $m$ runs from 0 to the maximum value of $n$. Frequently, symmetry can be used to reduce the number of azimuthal terms. A conical feed radiates only $m = 1$ modes, and reflection from a body of revolution will maintain this behavior. Secondly, there is an even and odd $\phi$ dependence, and fre-

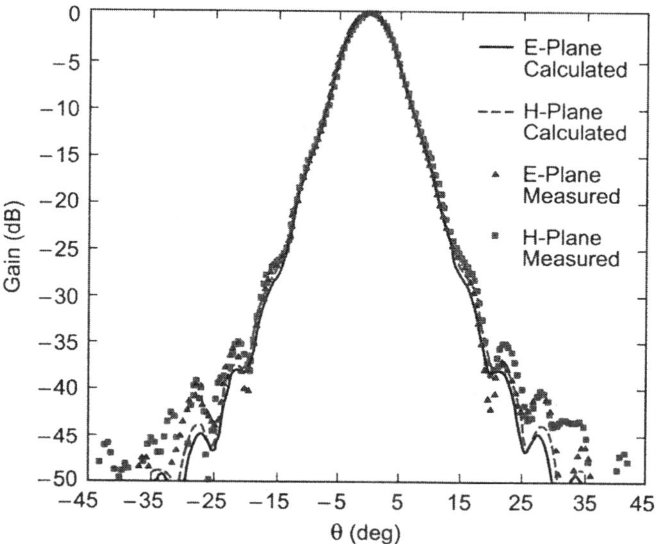

Fig. 1-9. Measured and computed patterns at 8.45 GHz (X-band receive) for the complete feed horn.

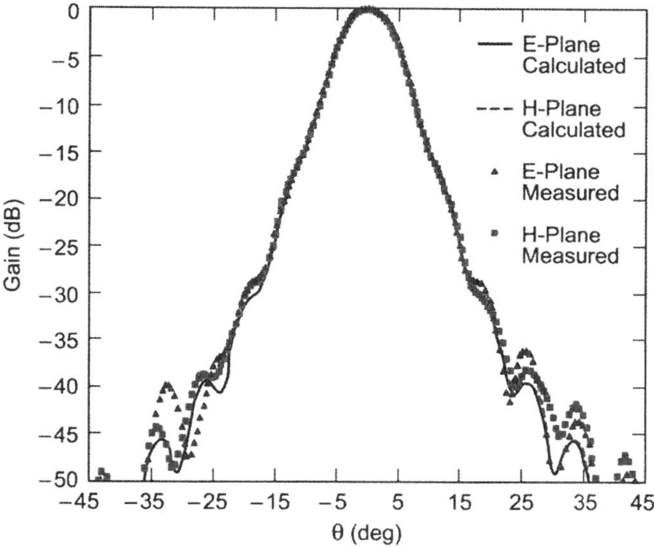

Fig. 1-10. Measured and computed patterns at 7.19 GHz (X-band transmit) for the complete feed horn.

quently, only one will be present. For the even case, $E_\theta$ can be expanded in only $\sin(m\phi)$ terms and $E_\phi$ in $\cos(m\phi)$ terms. For the odd case, this is reversed. The minimum number of $\phi$ values for the data sphere is, in general, $2M + 1$, where $M$ is the maximum value of $m$. Symmetry can be used to reduce this value appropriately. The output of the computer program is the set of spherical-wave-expansion coefficients. These coefficients may then be used to compute the field anywhere within the region of validity. Therefore, the essential utility of the program is to take data consisting of the tangential E-field on a sphere (whose radius may be infinite) and provide the means of computing the field— all three components of E and H—at any other point in the region of validity.

The computer program used is patterned after that in [19].

### 1.2.4    Dual-Reflector Shaping

As initially used in the DSN, the Cassegrain reflector system had a parabolic main reflector and hyperbolic subreflector. The efficiency of these reflectors is primarily determined by (a) the ability of the feed system to illuminate only the reflectors while minimizing the energy that radiates elsewhere and (b) the ability of the feed to uniformly illuminate the parabola. Item (a), above, is termed "spillover efficiency" and (b) "illumination efficiency." The illumination efficiency is 100 percent when the energy density on the entire main reflector aperture is a constant.

Feed-horn patterns always taper gradually from their central maxima to nulls. If all this energy is intercepted by the reflector (for maximum spillover efficiency), the illumination is far from uniform, and the illumination efficiency is very poor. Consequently, any attempt to obtain nearly uniform illumination will result in a great loss of energy in spillover. Therefore, a compromise must be made. A common choice for both a prime focus system and the Cassegrain system is a 10-dB taper of the illumination pattern at the parabolic edge. This selection results in a combination of spillover and illumination efficiency of from about 75 to 80 percent.

It is possible, however, to change the shape of the two reflectors to alter the illumination function and improve efficiency. This methodology is termed dual-reflector shaping and was first introduced by Galindo [20], who demonstrated that one could design a dual-reflector antenna system to provide an arbitrary phase and amplitude distribution in the aperture of the main reflector. Thus, if one chose a uniform amplitude and constant phase, 100 percent illumination efficiency could be achieved. With the feed pattern given, the subreflector size would be chosen to give minimal spillover.

**1.2.4.1  Theoretical Solution for the Symmetric Case.** The complete solution can be found in [20] and [21] and only the uniform aperture case will be summarized below.

The geometry of the symmetric dual-reflector system is shown in Fig. 1-11. Due to circular symmetry, the synthesis reduces to the determination of the intersection curve (of the surface) with the plane through the axis of symmetry, that is, the $x, y$ plane.

The synthesis method uses the analytical expressions of GO principles together with a geometry for the reflectors to develop a pair of first-order, nonlinear ordinary differential equations of the form

$$\frac{dy}{dx} = f(x, y) \tag{1.2-20}$$

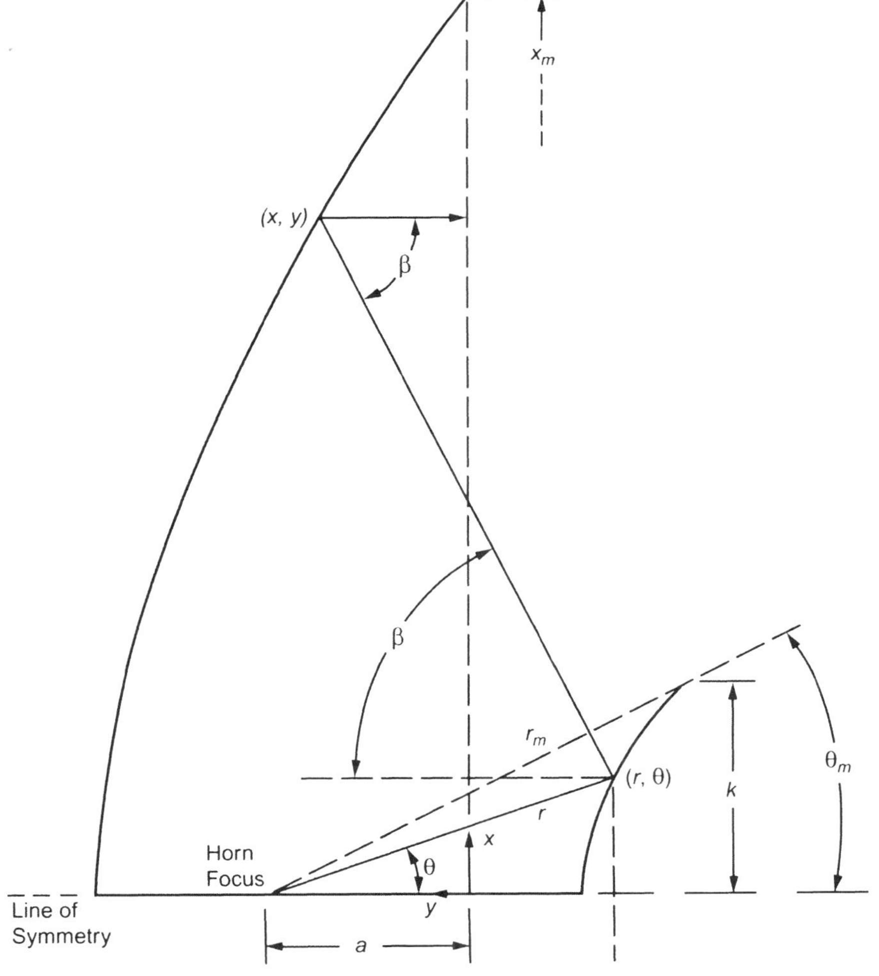


**Fig. 1-11. Coordinate system for shaping.**

which leads to cross sections of each reflector when subject to boundary conditions such as

$$y(x = x_{max}) = 0 \qquad (1.2\text{-}21)$$

which are then solved by a high-speed digital computer.

The optical principles that are used to develop the required equations are that (a) the angle of incidence is equal to the angle of reflection (Snell's Law), (b) energy flow is conserved along the ray trajectories, and (c) surfaces of constant phase form normal surfaces to ray trajectories.

The incident field is assumed to have a spherical-phase function, that is, a phase center and a power-radiation pattern $F(\theta)$. For uniform phase in the aperture, the path length, $r + r' + r''$, must remain constant for all $\theta$. Also, the amplitude function in the aperture $I(x)$ must also be equal to a prescribed distribution or a constant for maximum peak gain.

The equation for equal path lengths in the phase front is obtained from trigonometry:

$$r + y + \frac{x - r \sin \theta}{\sin \beta} = C \text{ (constant)} \qquad (1.2\text{-}22)$$

where $(x,y)$ and $(r,\theta)$ are the coordinates of points on the main reflector and subreflector, respectively.

The application of Snell's law to the two surfaces defines a relationship between the angles shown and the first derivates (slopes) of the surfaces. These are

$$\frac{1}{r}\frac{dr}{d\theta} = \tan\frac{\theta + \beta}{2} \qquad (1.2\text{-}23)$$

$$\frac{-dy}{dx} = \tan\frac{\beta}{2} \qquad (1.2\text{-}24)$$

Since the dual-reflector system is symmetrical about the y-axis, the total power within the increment $d\theta$ of the pattern $F(\theta)$ will be $F(\theta)2\pi \sin \theta d\theta$. Similarly, the total power within the increment $dx$ of the main antenna aperture is $I(x)2\pi dx$, where $I(x)$ is the illumination function of the antenna aperture. Making $I(x)$ constant and equating the total power from $\theta = 0$ to angle $\theta$ to that within $x$, and normalizing by the total power, one obtains

$$x^2 = x_{max}^2 \frac{\int_0^{\theta} F(\theta) \sin \theta d\theta}{\int_0^{\theta_{max}} F(\theta) \sin \theta d\theta} \qquad (1.2\text{-}25)$$

These four equations now have five dependent variables—$x$, $y$, $r$, $\theta$, and $\beta$—and can be solved to provide equations for the surfaces.

Although the above procedure yields an optimum-gain antenna, the FM of a DSN antenna, as stated previously, is $G/T$. Dual-reflector shaping is based upon GO, but due to diffraction, the feed-horn energy scattered from the subreflector does not fall abruptly to zero at the edge of the main reflector. By choosing design parameters such that the geometric ray from the subreflector edge does not go all the way to the outer edge of the main reflector, it is possible to have the feed-horn energy at a very low level relative to the central region of the main reflector, which, in turn, results in the rear spillover noise contribution's becoming acceptably small. This procedure results in slightly lower illumination efficiency, but the significant reduction in noise from spillover also results in an optimum $G/T$ ratio. The first antenna in the DSN to use dual-reflector shaping was the HEF antenna, the design of which is discussed in Chapter 6 of this monograph.

**1.2.4.2  Offset-Shaped Reflector Antennas.** The formulation shown in Section 1.2.4.1 (above) is for circularly symmetric reflector geometries. The exact solution has also been developed for offset geometries [22,23]. Offset geometry will have higher efficiency than symmetric geometry because the central blockage due to the subreflector can be eliminated. In the early 1980s, a 1.5-m antenna with an offset geometry was designed and built; it had an efficiency of 84.5 percent—the highest ever recorded [24]. However, due to the significantly greater mechanical complexity for an offset design, an offset-geometry antenna has never been incorporated into the DSN.

### 1.2.5  Quasioptical Techniques

Multiple reflector systems are typically analyzed by using PO, Gaussian beams, or ray-tracing techniques [25]. While PO offers high accuracy, it does so at the expense of computation time. This trade-off becomes particularly apparent in the analysis of multiple reflector antennas, such as BWG antennas, where PO is used to compute the currents on each reflector, based on the currents of the previous reflector. At the other end of the spectrum are ray tracing approaches that ignore diffraction effects entirely. These methods are fast but sacrifice the ability to accurately predict some effects such as cross-polarization.

More accurate than ray optics but less accurate than PO is Gaussian-beam analysis, which uses an appropriate set of expansion functions to model the field between the reflectors [26]. If the set is chosen wisely, only a few coefficients need to be determined from each reflector current. The field is then computed at the next reflector by using the expansion functions and their coefficients rather than by using the previous reflector current. For a BWG system with no enclosing tubes, an excellent set of expansion functions is the Gaussian-beam-mode set. In many cases, a preliminary design, which includes diffraction effects, may be obtained by considering only the fundamental mode and a thin-lens model for the reflectors. Higher-order modes are included to model the effects of the curved reflector; these effects include asymmetric distortion of the beam, cross-polarization, and beam truncation.

**1.2.5.1  PO Technique.** The PO approach has been described in detail in Section 1.2.1 (above).

**1.2.5.2  Gaussian-Beam Algorithm.** In this section, the use of higher-order Gaussian beams to determine scattering by an arbitrary set of relectors is given. The geometry is depicted in Fig. 1-12. The steps for computing scattering are

1) Using PO, compute the current on the first reflector; the incident magnetic field is provided either by a feed model or by an incident set of Gaussian-beam modes

2) Using ray tracing, compute the direction of propagation for the reflected Gaussian-beam set; using a gut ray in the input direction specified by the feed coordinate system or by the input Gaussian-beam-set propagation

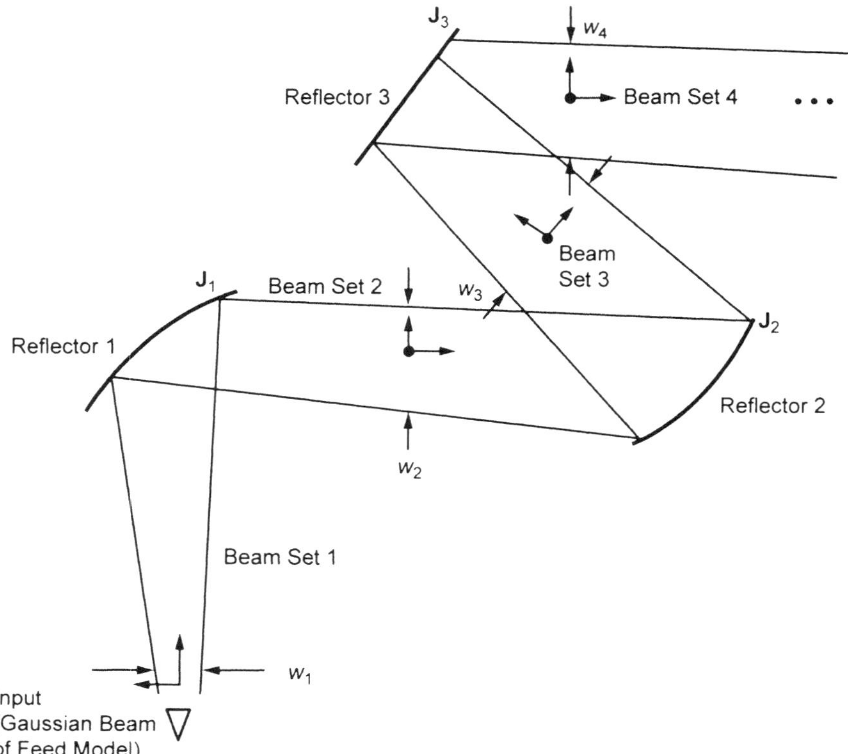


**Fig. 1-12. Gaussian-beam geometry.**

direction and the reflector surface description, compute the gut-ray direction for the output Gaussian-beam set

3) Determine the waist size and location for the output beam set

4) Determine the amplitudes of the individual modes in the output mode set

5) Repeat steps 1 to 4 for each additional reflector in the chain; in each of these cases, the previous Gaussian-beam set provides the input field for the current calculation

6) Using PO, compute the far-field pattern radiated by the final mirror.

One of the key steps in computing scattering of Gaussian-beam-mode sets by arbitrary reflector is the determination of the best choice for the output beam set's waist and its location. The problem is depicted in Fig. 1-13, where $w_{in}$ and $l_{in}$ are the input waist and location, and $w_{out}$ and $l_{out}$ are the same parameters for the output beam set. Once these parameters are determined, computation of the mode amplitudes from the current on the reflector follows easily. Proceed as follows:

1) Compute a waist at the reflector, denoted as $w_{match}$; when the input to the reflector is a Gaussian-beam-mode set, compute the input waist at the point of impact on the reflector; when the input to the reflector is a feed horn, determine an estimate of the waist from the reflector current (the output beam set is required to produce a waist equal to $w_{match}$ at the point of impact; completing this step results in one of the two equations needed to compute the output beam-set parameters)

2) Next, derive a suitable set of points on the reflector to use for field matching or path-length matching (these points are generally chosen from within the waist at the point of impact)

3) Compute a phase center location for the input fields (for a Gaussian-beam-mode set, this is determined from the input radius of curvature at the point of impact; for a feed input, assume the phase center to be at the origin of the feed coordinate system)

4) Using the set of points on the reflector, compute the set of path lengths from the input phase center at each point, as shown in Fig. 1-13

5) Search for the output phase center (output radius of curvature), sweeping along the direction of the output ray (a minimum in the spread of the total path lengths from input phase center to reflector and then to output phase center is sought; generally, a single minimum is found either in front of or behind the reflector)

6) Using the output radius of curvature and the determined waist at the reflector, compute the waist size and location of the output beam set.

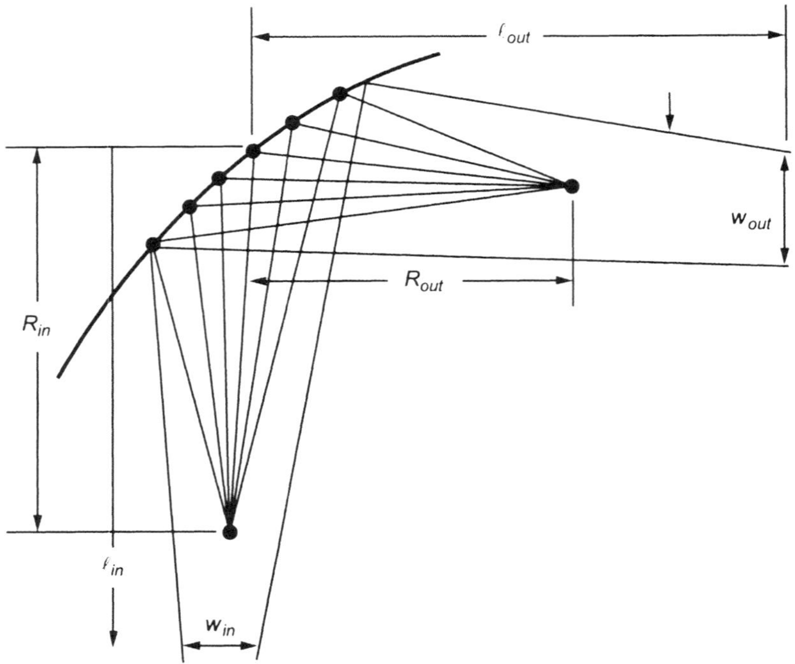



**Fig. 1-13. Ray-optics approach for determining output-beam parameters.**

The approach for determining the amplitudes of Gaussian-beam modes directly from reflector current will now be summarized. Note that two approximations are necessary. First, the reflector current is approximated to be the PO current—a good approximation for large reflectors with low edge illumination. Second, the Gaussian-beam modes are used as solutions to Maxwell's equations in free space—never true but a good approximation if the mode fields are required only in the paraxial region. The Gaussian-beam modes used are given in terms of Laguerre polynomials, as described by Goldsmith [27]. Each mode has a polarization, either $x$ or $y$; a radial index, $p$; and an azimuthal index, $m$.

For two arbitrary fields and their associated sources, denoted by $A$ and $B$, the reciprocity theorem when applied to an arbitrary volume, $V$, and its enclosing surface, $S$, may be stated as follows:

$$\oiint_S \left( \overline{E}_A \times \overline{H}_B - \overline{E}_B \times \overline{H}_A \right) ds =$$

$$\iiint_V \left( \overline{E}_B \bullet \overline{J}_A - \overline{H}_B \bullet M_A - \overline{E}_A \bullet \overline{J}_B + \overline{H}_A \bullet \overline{M} \right) dv$$

(1.2-26)

A half-space completely enclosing the reflector is chosen as $V$, with the surface, $S$, perpendicular to the direction of propagation for the output beam set. For this particular application, we choose the output Gaussian-mode set, with unknown mode amplitudes, as the $A$ field, with the reflector current inside the volume being its source. For the $B$ field, we choose a test field, the conjugate of the $i^{th}$ Gaussian-beam mode now propagating toward the reflector. The source for this field is chosen to be outside the volume, $V$. We then have for the fields,

$$\overline{E}_A = \sum_j a_j \overline{e}_j$$

$$\overline{H}_A = \sum_j a_j \frac{\hat{z}_{out} \times \overline{e}_j}{\eta_0}$$

$$\overline{E}_B = \overline{e}_i^*$$

$$\overline{H}_B = -\frac{\hat{z}_{out} \times \overline{e}_i^*}{\eta_0}$$

(1.2-27)

Using the reciprocity theorem and the orthogonality condition for the Gaussian-beam modes on the infinite surface, $S$, we can obtain the desired equation for the unknown coefficients:

$$a_i = -\frac{1}{2} \iint_{S_{\text{reflector}}} \overline{e}_i^* \bullet \overline{J} ds$$

(1.2-28)

**1.2.5.3  Ray Analysis Algorithm.** A bundle of rays is launched from the feed point to the first reflector. The ray density distribution in angular space is taken to be proportional to the power pattern of the input feed. The rays are then traced through the multiple reflector system. A flat plate is placed a large distance from the final mirror, perpendicular to the central ray. The rays from the final mirror are traced to the flat plate. The density distribution of the rays on the flat plate can be processed to yield the far-field pattern scattered from the last mirror.

**1.2.5.4  An Example.** Consider the three-curved-mirror BWG system used in the NASA/Jet Propulsion Laboratory (JPL) 34-m R&D BWG antenna (see Chapter 7 for a more detailed discussion of this antenna). The physical configuration of the BWG consists of a beam magnifier ellipse followed by an imaging pair of parabolas. The system is designed to operate from 2 to 35 GHz. Figures 1-14 and 1-15 compare the three methods at 2.3 and 32 GHz. Recall that the ray optics technique is frequency independent.

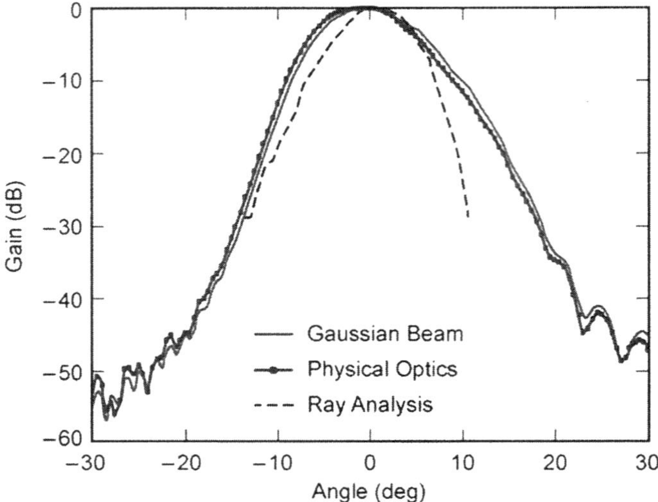

Fig. 1-14.  Comparison of three methods at
2.3 GHz (S-band receive).

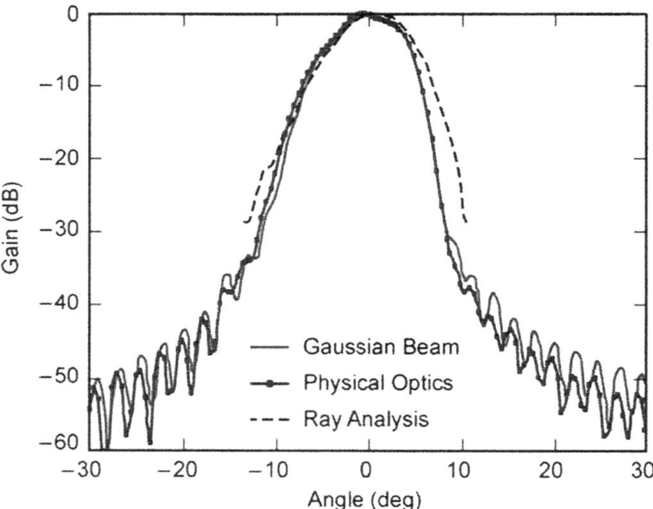

Fig. 1-15.  Comparison of three methods at
32 GHz (Ka-band receive).

Of the three techniques discussed above, the PO technique is the most accurate, albeit most time consuming. For the BWG system, higher-order Gaussian modes provide a very good estimate of performance, with minimal computer time. Ray optics is generally not adequate for diffraction pattern calculations.

### 1.2.6    Dichroic Analysis

The ability to transmit and receive simultaneously at multiple frequency bands is an important requirement for the DSN. It is usually accomplished by using either a dual-band feed horn or separate feed horns and a frequency-selective surface (FSS), typically referred to as a dichroic plate. As dichroic plates are an important aspect of the DSN antenna design, this section presents a technique for analyzing them. For a more complete discussion of DSN use of dichroic plates, see Chapter 5 of this monograph.

The first use of a dichroic plate was as part of the reflex–dichroic feed system used on the DSN's first 64-m antenna. That plate consisted of circular holes in a half-wavelength-thick metallic plate. All subsequent dichroic plates used in the DSN were also thick plates with holes. The discussion that follows treats the most general case of the analysis of a thick dichroic plate with arbitrarily shaped holes at arbitrary angles of incidence. However, most of the plates currently in use in the DSN were designed with programs that only analyzed the simple geometries such as circular [28] or rectangular holes [29]. The technique parallels the development for the rectangular hole case except that the waveguide modes are generated using the finite-element method (FEM).

The analysis of a thick dichroic plate with arbitrarily shaped holes (Fig. 1-16) is carried out in a series of steps:

1) Construct a model of a half-space infinite array

2) Develop a complete set of basis functions with unknown coefficients for the waveguide region (waveguide modes) and for the free-space region (Floquet modes) in order to represent the electromagnetic fields

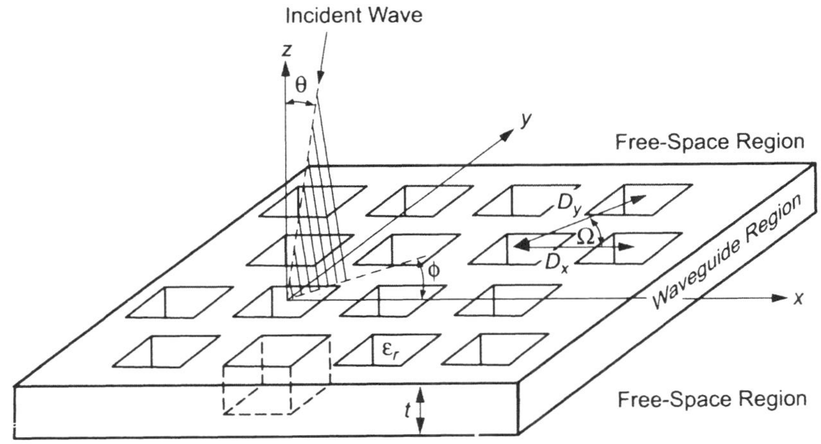


**Fig. 1-16. Geometry of a thick dichroic plate with holes.**

3)  Apply the boundary conditions at the interface between these two regions; use the method of moments to compute the unknown mode coefficients

4)  Calculate the scattering matrix of the half-space infinite array

5)  Move the reference plane of the scattering matrix half-a-plate thickness in the negative $z$ direction

6)  Synthesize a dichroic plate of finite thickness by positioning two plates of half-thickness back to back

7)  Obtain the total scattering matrix by cascading the scattering matrices of the two half-space infinite arrays.

The full details of the analysis are given in [30]. Since the use of Floquet modes and the method of moments to analyze dichroic plates are well documented in many references [28–32], they will not be repeated here. However, a short discussion of the use of FEM to solve the arbitrarily shaped waveguide problem will be presented. Also, an example of the most recent design of a dichroic plate will be shown.

**1.2.6.1  Finite-Element Formulation of the Waveguide Problem.** As is well known, the eigenmodes of a hollow, uniform waveguide may be determined by solving the 2-D scalar Helmholtz equation,

$$\left(\nabla^2 + k^2\right)\Psi = 0 \tag{1.2-29}$$

Here, $\Psi$ represents the magnitude of an axially directed electric or magnetic Hertz vector.

Rather than attempting to solve this eigenvalue problem directly, FEM reframes it in variational terms. As is shown in [33], the solution of the Helmholtz equation with homogeneous boundary conditions is equivalent to minimizing the functional $\mathcal{F}$ defined by

$$\mathcal{F} = \iint_R \left(-\left|\text{grad } \varphi\right|^2 + k^2\varphi^2\right) dxdy \tag{1.2-30}$$

The region of integration is, of course, the waveguide cross section. If a trial solution, $\varphi(x,y)$, is represented geometrically as a function spanning the region, $R$, over the $x,y$ plane, the correct function, $\Psi(x,y)$, is that which yields the smallest possible value of $\mathcal{F}$.

Using a surface function, $\varphi(x,y)$, made up of finite surface elements of a simple kind, FEM approximates the correct solution. For simplicity, all the elementary regions are taken to be triangles. Corresponding to each elementary region, the surface element will be defined by requiring that $\varphi(x,y)$ be a linear function of its values at the vertices; thus, $\varphi$ is defined by

$$\varphi(x, y) = \sum_{i=1}^{N} \alpha_i(x, y)\, \varphi_i \qquad (1.2\text{-}31)$$

$\mathcal{F}$ is now an ordinary function of the parameters that define $\varphi$, in this case the vertex values $\varphi_i$:

$$\mathcal{F}(\varphi) = F(\varphi_1, \varphi_2, \cdots, \varphi_N) \qquad (1.2\text{-}32)$$

The solution is found by minimizing $F$ with respect to all the parameters. That is, it will be required that

$$\frac{\partial \mathcal{F}}{\partial \varphi_m} = 0 \qquad (1.2\text{-}33)$$

for all $m$. If the region, $R$, contains altogether $N$ vertices, this minimization requirement changes Eq. (1.2-30) into a matrix eigenvalue problem of order $N$.

In view of Eq. (1.2-32), the minimization requirement, Eq. (1.2-33), is equivalent to

$$\iint_R \frac{\partial}{\partial \varphi_m} |\text{grad } \varphi|^2 \, dxdy = k^2 \iint_R \frac{\partial}{\partial \varphi_m} \varphi^2 \, dxdy \qquad (1.2\text{-}34)$$

As is shown in [30], if we define the purely geometrical quantities

$$S_{mk} = \iint_R \left( \frac{\partial \alpha_m}{\partial x} \frac{\partial \alpha_k}{\partial x} + \frac{\partial \alpha_m}{\partial y} \frac{\partial \alpha_k}{\partial y} \right) dxdy \qquad (1.2\text{-}35)$$

and

$$T_{mk} = \iint_R \alpha_m \alpha_k \, dxdy \qquad (1.2\text{-}36)$$

which expresses the nature of the region $R$ and the manner of its subdivision into elementary regions, then, in terms of these integral quantities, Eq. (1.2-34) reads

$$S\Phi = k^2 T\Phi \qquad (1.2\text{-}37)$$

where $\Phi$ is the column matrix of vertex values $\varphi_i$, and $S$ and $T$ are square matrices of order $N$, whose elements are $S_{mk}$ and $T_{mk}$, as defined above. Minimization of $\mathcal{F}$ is thus equivalent to the eigenvalue problem, Eq. (1.2-37), and solution of the latter will provide approximate eigenvalues and eigenfunctions for the boundary-value problem. Note that $S$ and $T$ are always symmetric, and the eigenvalues, therefore, always real.

**1.2.6.2  Low-Cost Dichroic Design.** As stated earlier, most of the plates currently in use were designed with programs that analyzed only the simple geometries, such as circular or rectangular holes. Since it is too expensive to experimentally determine FSS parameters, only designs that could be accurately analyzed were chosen, and since it is expensive to empirically design the dichroic plates, the earliest FSS designs used rectangular holes. To achieve the sharp corners of the rectangular holes, it was necessary to use an electrical discharge machining (EDM) manufacturing technique. This manufacturing technique is expensive, and an important use of the arbitrarily shaped analysis is to enable designs that use rounded corners, which can be manufactured less expensively. An example of a manufacturing test sample for an X-/Ka-band dichroic plate is shown in Fig. 1-17. The plate is designed to pass the Ka-band receive frequencies (31.7–32.2 GHz) and reflect both the X-band transmit (7.145–7.190 GHz) and receive (8.4–8.5 GHz), and it is a direct replacement for the X-/Ka-band plate [34] currently in use in the BWG antennas. It was manufactured using an end-milling technique and cost one-fourth the price of an equivalent-sized EDM-manufactured plate. An example of FEM gridding is shown in Fig. 1-18.

Since the analysis of arbitrarily shaped holes is quite new, only designs using rectangular holes are currently installed in DSN BWG antennas. The calculated and measured reflection coefficient of the X-/Ka-band dichroic plate used in the 34-m BWG antennas is shown in Fig. 1-19 [34].

**Fig. 1-17. Manufacture test sample for the X-/Ka-band dichroic plate with rounded corners.**

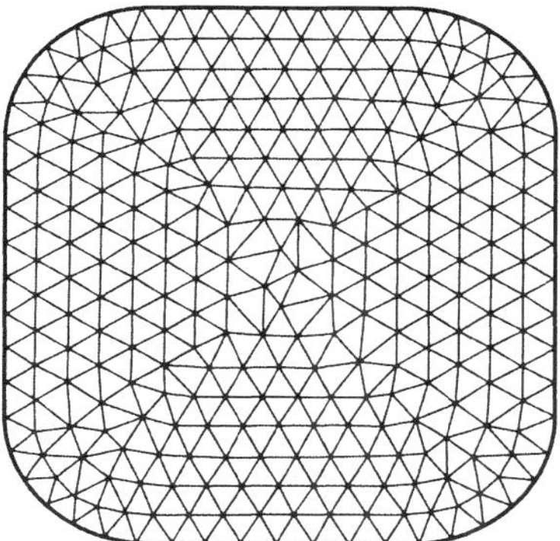

**Fig. 1-18. Finite-element gridding of the rectangular hole with rounded corners.**

### 1.2.7    Antenna Noise-Temperature Determination

The system-noise temperature of a receiving system is the sum of the receiver-noise temperature and the antenna temperature. The individual contributions to the antenna temperature include cosmic radio noise, atmospheric absorption, thermal radiation from the ground, and transmission line loss.

The noise temperature of an antenna is defined [2] as

$$T_a = \frac{P_A}{kB} \tag{1.2-38}$$

where $P_A$ is the noise power delivered by the antenna over a bandwidth, $B$, into a matched termination, and $k$ is Boltzman's constant, equal to $1.380 \times 10^{-23}$ $J/K$.

Also, the effective antenna noise temperature can be expressed as the radiation pattern-weighted environmental black body temperature $T(\Omega)$ averaged over all directions; thus,

$$T_a = \frac{1}{4\pi} \int_{4\pi} T(\Omega)G(\Omega)d\Omega = \frac{\displaystyle\int_{4\pi} T(\Omega)P(\Omega)d\Omega}{\displaystyle\int_{4\pi} P(\Omega)d\Omega} \tag{1.2-39}$$

where $G(\Omega)$ is the antenna gain in the direction $\Omega$, and $P(\Omega)$ is the normalized antenna pattern gain in direction $\Omega$, or

$$P(\Omega) = \frac{G(\Omega)}{G_m} \tag{1.2-40}$$

with $G_m$ the antenna maximum gain.

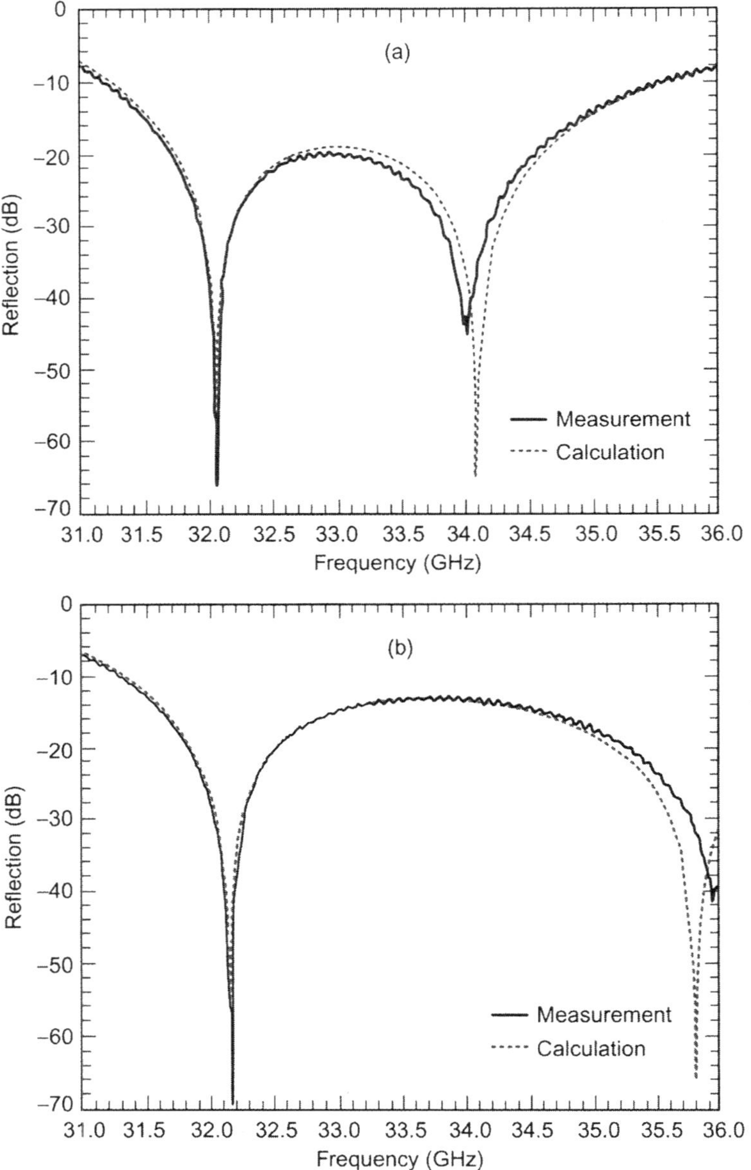

**Fig. 1-19.** Calculated and measured reflection coefficient versus frequency for the X-/Ka-band/KABLE-band dichroic plate: (a) TE polarization and (b) TM polarization.

Evaluation of Eq. (1.2-39) may range from rough approximations or rules of thumb to a very painstaking evaluation by numerical integration of the functions involved.

### 1.2.7.1 Noise Temperature in BWG Systems.

At the start of the BWG antenna project (discussed in Chapter 7), noise-temperature contributions due to BWG mirror spillover were not clearly understood. It was at first assumed that spillover power was eventually absorbed by cold sky or by a cold front-end receiver. However, this assumption was incorrect, and an empirical relationship between the calculated spillover energy and noise temperature was developed [35]. When attempting to use the empirical data from one BWG design to predict performance of another design also proved to be erroneous, it became apparent that an accurate analytic procedure that included the effects of the BWG shroud was required [36–39].

The major contributors to noise temperature in a BWG are the spillover past the mirrors and the conductivity loss in both the BWG shroud walls and reflecting mirrors.

The analytical method extends the approach of [36], which computes the waveguide modes that are propagating in the oversized waveguides. Reference [36] describes a PO integration procedure of the currents of the BWG mirrors, using a Green's function appropriate to the circular-waveguide geometry. Once all the modes in the waveguide are known, it is a simple matter to use standard approximations to determine the attenuation constant and, thus, the conductivity loss if the conductivity of the wall material is known. Also, all energy that propagates toward, but spills past, a BWG mirror is assumed to be lost in the walls of the BWG as well. The noise temperature is computed assuming both loss components are exposed to ambient temperature. The conductivity loss in the reflecting mirrors is a straightforward calculation, and the corresponding noise temperature is linearly added to the other noise contributors.

The BWG tube analysis is conceptually similar to the PO analysis used in reflector-antenna analysis. Currents induced in the BWG mirror are obtained using a standard PO approximation of $J = 2\hat{n} \times H_{inc}$, where $\hat{n}$ is the surface normal and $H_{inc}$ is the incident field. However, the BWG analysis differs in the methods by which the incident field on a mirror and the scattered field radiated from the mirror are calculated.

One approach to calculating the scattered field is to use a dyadic Green's formulation [36], where the field scattered from a BWG mirror is computed using the Green's function appropriate to the cylindrical waveguide geometry.

While it is conceptually convenient to use Green's functions to discuss the comparison with PO, the actual computation using this approach is rather cumbersome. Instead, a simpler method is used to calculate waveguide fields, based upon the reciprocity theorem. The basic problem is to find the fields radiated by an arbitrary current (the PO currents on the reflector) in a cylindrical waveguide. The problem is easily solved by expanding the radiated field in terms of a suitable set of normal modes, with amplitude coefficients determined by an application of the Lorentz reciprocity theorem [38].

An arbitrary field in a waveguide can be represented as an infinite sum of the normal modes for the guide. Let the normal modes be represented by (see Fig. 1-20)

$$\overline{E}_n^{(\pm)} = \left(\overline{e}_n \pm \overline{e}_{zn}\right) e^{\mp \Gamma_n z}$$
$$\overline{H}_n^{(\pm)} = \left(\pm \overline{h}_n + \overline{h}_{zn}\right) e^{\mp \Gamma_n z}$$

(1.2-41)

where $\overline{E}_n^{(+)}$ represents a mode traveling in the $+z$ direction, and $\overline{E}_n^{(-)}$ is a mode traveling in the $-z$ direction. For a basic normal mode description, see [40].

Let the field radiated in the $+z$ direction by the current be represented by

$$\overline{E}^{(+)} = \sum_n a_n \overline{E}_n^{(+)}$$
$$\overline{H}^{(+)} = \sum_n \frac{a_n}{Z_n} \overline{H}_n^{(+)}$$

(1.2-42)

and the field radiated in the $-z$ direction by

$$\overline{E}^{(-)} = \sum_n b_n \overline{E}_n^{(-)}$$
$$\overline{H}^{(-)} = \sum_n \frac{b_n}{Z_n} \overline{H}_n^{(-)}$$

(1.2-43)

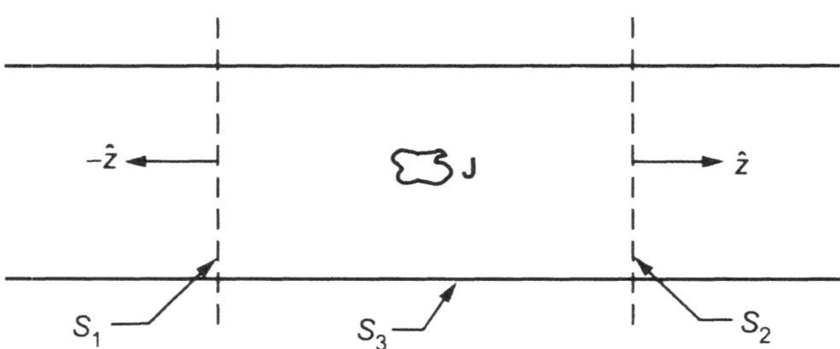


**Fig. 1-20. Geometry for computing waveguide modes.**

An application of the Lorentz reciprocity theorem [38] can show that

$$a_n = -\frac{1}{2}Z_n \int_v \left[\overline{E}_n^{(-)} \bullet \overline{J}\right] dV$$

$$b_n = -\frac{1}{2}Z_n \int_v \left[\overline{E}_n^{(+)} \bullet \overline{J}\right] dV \tag{1.2-44}$$

where $\overline{J}$ is the volume current in the waveguide.

Since we have only surface currents, the integral for the PO currents is over the surface of the reflector. If we let

$$c_n = -\frac{1}{2} \int_s J_s \bullet \overline{E}_n^{(-)} ds \tag{1.2-45}$$

where $J_s$ is the PO currents on the mirror, then

$$\overline{H}^+ = \sum_n c_n \overline{H}_n^{(-)}$$

$$\overline{E}^+ = \sum_n Z_n c_n \overline{E}_n^{(-)} \tag{1.2-46}$$

and the total power contained in the fields is

$$P = \sum_n Z_n |c_n|^2 \tag{1.2-47}$$

If the spillover past the mirror is small (that is, >25-dB edge taper), the PO currents induced on the first mirror are computed in the standard way, by utilizing the free-space near-field radiating H-field of the horn and $J_s = 2\hat{n} \times H_{inc}$. Or they may be computed by utilizing a technique similar to the one just described to compute the propagating modes from the horn, radiating in the oversized waveguide and using the appropriate H-field derived from these modes as the incident field. On subsequent mirrors, PO currents are computed from the H-field derived from the propagating waveguide modes. The technique is summarized in Fig. 1-21, where it should be noted that

$$H = \int_s \overline{J}_1 \bullet \overline{\overline{G}}_{wg} ds = \sum_n c_n \overline{H}_n^{(-)} \tag{1.2-48}$$

Power loss in the conductor is obtained using the standard technique to compute the power dissipated in the conductor per unit length [39] as

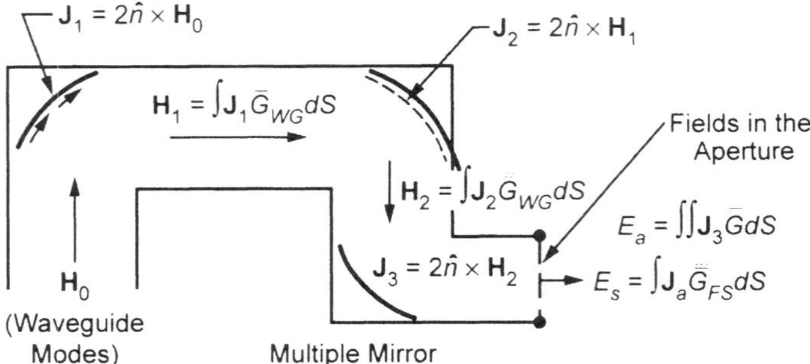



**Fig. 1-21. Computation of fields in the BWG system.**

$$P_d(z) = R \int_0^{2\pi} \left|\overline{H}_t\right|^2 a\,d\phi \qquad (1.2\text{-}49)$$

where

$$R = \sqrt{\frac{\omega\mu}{2\sigma}} \qquad (1.2\text{-}50)$$

and $\sigma$ is the wall conductivity, $a$ the radius, and $\left|\overline{H}_t\right|^2$ the tangential **H** field. It should be noted that $P_d$ is a function of $z$ since $\left|\overline{H}_t\right|^2$ is a function of $z$ (because it is composed of more than one waveguide mode).

Power loss is computed from

$$P = P_o e^{-2\alpha d} \qquad (1.2\text{-}51)$$

where $d$ is the distance from $z_1$ to $z_2$, and the attenuation constant is computed as

$$\alpha d = \frac{\displaystyle\int_{z_1}^{z_2} P_d(z)\,dz}{2P_f} \qquad (1.2\text{-}52)$$

where $P_f$ is the power flow in the waveguide.

To compute noise temperature, it is convenient to separate the total RF power originating from the horn aperture (viewed in transmission, for convenience) and propagating into two parts

$$P_{BWG} = P_m + P_{spill} \qquad (1.2\text{-}53)$$

where $P_{spill}$ is the portion that spills past the mirrors (since the mirrors do not fill the waveguide). The value of $P_{spill}$ can be computed for each mirror by integrating the total power radiated from the induced mirror currents and comparing it to the incident power. Note that the computation uses the induced currents derived from the waveguide modes. It is then assumed that this spillover power is exposed to ambient temperature since it would be lost in the tube due to multiple bounces in a lossy material.

Noise temperature, then, is composed of two parts: the noise due to the spillover power added to the noise from the attenuation of $P_m$ due to the conductivity loss.

Total noise temperature also includes the contribution from the conductivity loss in the mirrors. Noise temperature due to conductivity loss is given as

$$T_m = \left(1 - \frac{2R^2 - 2R\eta + \eta^2}{2R^2 + 2R\eta + \eta^2}\right) T_{amb} \qquad (1.2\text{-}54)$$

where $R$ is computed from the mirror conductivity, $\eta$ is the free-space impedance and $T_{amb}$ is the ambient temperature.

An example of computing noise temperature is given in Table 1-2 for the Deep Space Station (DSS-13) R&D antenna (whose design and performance is fully described in Chapter 7). The table shows the calculated [37] and measured [41] noise temperature at 8.4 GHz (X-band receive) for two different feed horns in the DSS-13 pedestal room. The spillover energy for both feed horns is shown in Table 1-3. Assuming a 290-K ambient temperature, the noise-temperature contribution is shown in Table 1-2. The shroud is steel with conductivity of $0.003 \times 10^7$ mhos/m, and the noise-temperature contribution is also shown in Table 1-2. The mirrors for both feed horns are machined aluminum with conductivity of $3.8 \times 10^{-7}$ mhos/m. The loss for one mirror is 0.001 dB, with a corresponding noise temperature of 0.094 K. The noise-temperature contribution for all six mirrors is 0.564 K.

Observe that the major contributor to noise temperature in a BWG system is the spillover energy past the mirrors. On the other hand, contribution from

**Table 1-2. DSS-13 noise temperature at 8.4 GHz (X-band receive mode): measured versus calculated.**

| Feed-Horn Gain | Noise Temperature (K) | | | | |
| --- | --- | --- | --- | --- | --- |
| | Spillover Energy | Shroud Conductivity | Mirror Conductivity | Total Calculated | Total Measured |
| 22.5-dB | 7.95 | 0.24 | 0.564 | 8.75 | 8.9 |
| 24.2-dB | 0.93 | 0.05 | 0.564 | 1.54 | 2.0 |

**Table 1-3. Spillover energy in the DSS-13 BWG system (X-band transmit mode).**

| | Spillover (percent) | | | |
| --- | --- | --- | --- | --- |
| | 22.5-dB Feed Horn | | 24.2-dB Feed Horn | |
| Mirror | Each Mirror | Total | Each Mirror | Total |
| M6 | 0.36 | 0.36 | 0.00 | 0.00 |
| M5 | 1.02 | 1.38 | 0.12 | 0.12 |
| M4 | 0.25 | 1.63 | 0.00 | 0.12 |
| M3 | 1.04 | 2.67 | 0.20 | 0.32 |
| M2 | 0.04 | 2.71 | 0.00 | 0.32 |
| M1 | 0.03 | 2.74 | 0.00 | 0.32 |

the conductivity loss in both the mirrors and shroud is small. The analytical technique that includes the effects of the shroud has been shown to be a reasonably accurate method to predict noise temperature.

## 1.3   Measurement Techniques

A number of sophisticated measurement techniques are required to properly characterize the performance of a large reflector antenna. First, there is a need to accurately measure the main reflector surface. This measurement is determined either by theodolite or microwave holography. Second, measuring gain or efficiency is required, and third, noise temperature.

### 1.3.1   Theodolite Measurements

Theodolite measurements are typically used in the initial alignment of the antenna after installation of the reflector panels. The method of determining the reflector surface is based on angular measurements made with a theodolite [42,43]. The theodolite is placed at the main reflector vertex and is used to measure the angles to targets installed at the individual reflector panel corners. After the panels are installed on the antenna, the surface panels are drilled using the holes in a metal strap gauge as a template for placing targets.

Errors in theodolite measurements are due to uncertainty in the angular measurement of the target position and radial distance to the target. At a radius of 13 m, which is the surface area median radius for a 34-m antenna, a 0.001-deg error in the angle to the target corresponds to a 0.2-mm error normal to the surface; a 1.0-mm error in the radial distance corresponds to a 0.26-mm error normal to the surface.

A typical error budget [43] for the surface error budget of a 34-m antenna is 0.3-mm root-sum-square (rss). By far, the greatest source of error is in the radial distance measurement provided by using a drill tape. If, instead, a laser range finder is used to measure radial distance, the error could improve from 0.3-mm rms to 0.2-mm rms. In any case, microwave holography has been used to characterize and reset the main reflector surface after initial operation, using the theodolite-set panels.

During 2000 and 2001, a Leica TDM-5000 "total station" theodolite was used to measure all three of the DSN 70-m antennas [44]. The instrument measures vertical and horizontal angles and distances, downloads them to a desktop computer, which converts the coordinates from the spherical to a Cartesian system and can be used to command the instrument to motor to a desired look angle.

Theodolites are used to measure DSN antennas in the following manner: The theodolite is bolted securely to the bracket close to the center of the main reflector. The theodolite's gravity compensator is turned off so that it will rotate with the reflector.

Targets are placed in 12 concentric rings, wherever cables, supports, or other equipment in the feed support structure do not block the view. The targets are made up of 20-mm, square pieces of tape, 0.33-mm thick, attached to the front face of black plastic mounts such that the visible target crosshair is 11.15 mm above and normal to the local reflector surface. The estimated target height variation is on the order of ± 0.38 mm. See [44] for photographs of the targets and visible target placements.

The total station theodolite described above measures distance and angles from the instrument to the target on the reflector surface. The measurement geometry is shown in Fig. 1-22. The range and angle data is used to compute the error normal to the reflector surface for surface adjustment or to determine the half-path-length error for antenna efficiency calculations.

The major error contributors are the angle reading error, distance measurement error, and targeting error. However, by viewing a target as closely as possible to tangent to the surface, the effect of distance measurement error inherent

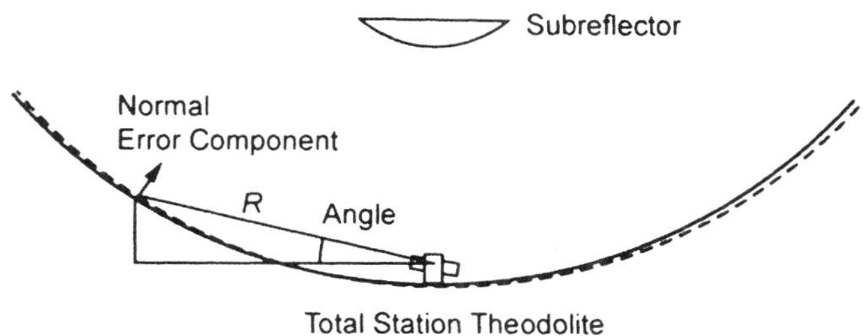


**Fig. 1-22. Reflector measurement geometry.**

in the theodolite is minimized. In the worst case, the farthest target is 36.44 m from the vertex of the main reflector. Using this distance with the estimated operator angle accuracy of ± 3 arcsec gives a peak surface-normal measurement error of ± 0.5 mm.

For a parabolic main reflector with a theodolite at the vertex, the slope of the theodolite line-of-sight to any target is half of the parabola surface slope at the target. For the 70-m shaped main reflector with a nominal focal length of 27.05 m, the edge slope at 35-m radius is 33 deg, so the reflector surface is tilted by as much as 16.5 deg to the theodolite line-of-sight. Based on a distance measurement error of 1.3 mm, this component of the measurement error is 1.3 tan (16.5 deg) = ± 0.38 mm.

Figure 1-23 shows the estimated 1-sigma measurement error normal to the surface as a function of distance from the instrument to the target. The angle, range, and targeting error components are shown with the rss total and the area weighted 1-sigma rms half-path-length error of 0.17 mm.

The reflector coordinate system is based on a reference file created from the nominal target locations. First, the nominal target coordinates are calculated based on the measured reflector surface arc length, circumferential spacing, and reflector radial surface shape. Then, they are combined into a reference file. The theodolite system motors to the nominal location of each target, the target is placed precisely in the theodolite crosshairs, then measured and recorded. When all of the targets are measured the first time at zenith, the data is best-fit to the theoretical coordinates in the reference file to transform the coordinates from a theodolite-centered coordinate system to the reflector system. No reference points, which might indicate the true reflector optics, are used in any of the measurements. The recorded target locations thus

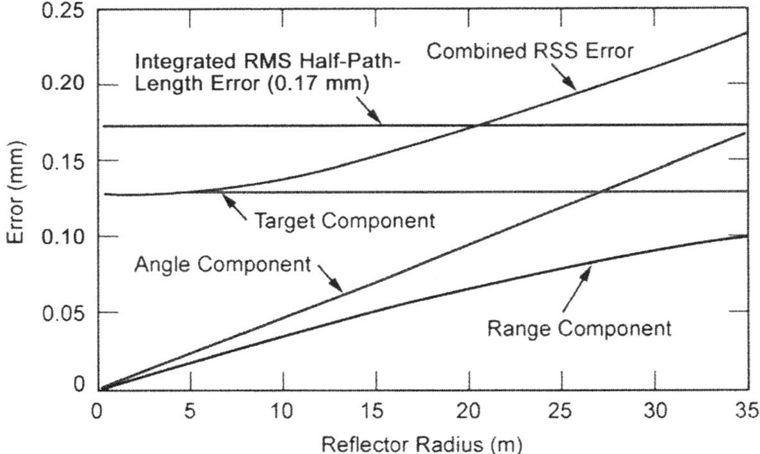


**Fig. 1-23. Measurement error (1 sigma).**

transformed become the new reference target file used in all subsequent measurements.

The measured data is fit to the theoretical surface of the shaped main reflector. Residual surface errors at each point are determined and rms accuracy of the surface computed. Results are presented as half-path-length errors (HPLEs) in millimeters. The HPLE is the usual presentation of surface errors, which relates surface rms to gain loss [45].

Table 1-4 shows the calculation of the rms errors after best-fitting for all three 70-m antennas. The best-fitting applied performs a least-squares fit using six degrees of freedom: translation in x, y and z; rotation about x and y; and focal point adjustment, since there is a motorized subreflector. The first data row is for all measured points. The second row removes the inner panel, which is blocked by the subreflector, and the outer panel, which is not included in the alignment derived from the holographic measurement.

With minor adjustments to the theodolite and a comfortable working platform for the operator, primary surface and subreflector position measurements as described above can be performed at any elevation angle for a tracking antenna. For antenna structures that behave in a linear elastic fashion, measurements at three separate elevation angles provide sufficient information to determine the relative positions of all targets at any other angle. (A common example of nonlinear structural deformation is the bending of bolted flanges.) The process requires the intermediate calculation of the face-up (zenith-pointing) and face-side (horizon-pointing) gravity deformations of the reflector, and their subsequent vector superposition. The best results will be achieved if the elevation angles at which measurements are made are well separated, such as 0, 45, and 90 deg. For the 70-m antennas, the analysis was based on measurements made at 13, 47, and 89 deg.

The gravity deformation of a linear elastic structure at any elevation angle can be derived by vector superposition for the face-up and face-side gravity vectors as

$$\delta = \delta_u \sin \theta + \delta_s \cos \theta \qquad (1.3\text{-}1)$$

where

$\delta_s$ = face-side gravity deformation

$\delta_u$ = face-up gravity deformation.

The face-up and face-side deformations described here are of the type derived from a computer structural analysis in which gravity is "turned on" from a particular direction. As such, the deformations cannot be directly measured in the real world.

**Table 1-4. Measured rms errors for all antennas.**

| RMS | 13-deg Elevation Angle | | | 30-deg Elevation Angle | | | 47-deg Elevation Angle | | | 68-deg Elevation Angle | | | 89-deg Elevation Angle | | |
|---|---|---|---|---|---|---|---|---|---|---|---|---|---|---|---|
| | DSS | | | DSS | | | DSS | | | DSS | | | DSS | | |
| | 14 | 43 | 63 | 14 | 43 | 63 | 14 | 43 | 63 | 14 | 43 | 63 | 14 | 43 | 63 |
| All targets (mm) | 1.32 | 1.30 | 1.42 | N/A[a] | 1.24 | 1.50 | 1.19 | 1.32 | 1.24 | 1.52 | 1.45 | 1.42 | 1.62 | 1.62 | 1.70 |
| Inner and outer targets removed (mm) | 1.07 | 0.97 | 1.22 | N/A | 0.79 | 1.14 | 0.84 | 0.84 | 0.91 | 1.07 | 1.04 | 1.09 | 1.27 | 1.24 | 1.50 |

[a]Not available.

For a linear elastic structure that rotates from angle $\theta_1$ to $\theta_2$, the relative deformation, or motion, will be $\delta_2 - \delta_1$, as shown above. For three measurements, two independent linear equations can be written for $\delta_2 - \delta_1$ and $\delta_3 - \delta_1$

$$\begin{bmatrix} \delta_u \\ \delta_s \end{bmatrix} = \begin{bmatrix} \sin\theta_2 - \sin\theta_1 & \cos\theta_2 - \cos\theta_1 \\ \sin\theta_3 - \sin\theta_1 & \cos\theta_3 - \cos\theta_1 \end{bmatrix}^{-1} \begin{bmatrix} \delta_2 - \delta_1 \\ \delta_3 - \delta_1 \end{bmatrix} \tag{1.3-2}$$

Now that $\delta_u$ and $\delta_s$ are known, for any angle $\theta$, the predicted alignment error at a target is the gravity deformation traveling from a measured reference angle ($\theta_2$) to angle $\theta$ superposed onto the actual measured alignment error at the reference angle ($\delta_\theta$). This can be expressed as

$$\begin{aligned} \delta(\theta) = \delta_\theta &+ \delta_u\left(\sin\theta_2 - \sin\theta\right) \\ &+ \delta_s\left(\cos\theta_2 - \cos\theta\right) \end{aligned} \tag{1.3-3}$$

It is possible to use measurements at more than three angles and develop an overdetermined set of equations for $\delta_u$ and $\delta_s$ that can be solved in a least-squares sense. This would help average errors in the individual datasets.

The gravity analysis of DSS-43 (the 70-m antenna near Canberra, Australia) was performed using measurements from 13, 47, and 89 deg of elevation. The predicted HPLE at each of the measured points was calculated at 5-deg elevation intervals from 0 to 90 deg, plus 13, 30, 47, 68, and 89 deg, using the actual measured data. By definition, the predictions at 13, 47, and 89 deg perfectly matched the input data.

The above process was repeated, recording the predicted gravity deformations with respect to the nominal rigging angle of 47 deg.

The rms data is summarized in Fig. 1-24. The curves show that the measured data at 30 and 68 deg closely matches the predicted values.

Using the all-angle main reflector surface shape data in a PO calculation, with the 70-m feed and subreflector configuration, an efficiency value was determined for all three antennas. The value is shown in Fig. 1-25. This predicted efficiency value is compared to the measured efficiency at DSS-14 shown in Fig. 1-26. It is interesting to note that all three 70-m antennas have similar distortions due to gravity.

### 1.3.2 Microwave Holography

Microwave holography [46–48] has proven to be a valuable technique for characterizing the mechanical and RF performance of large reflector antennas. This technique utilizes the Fourier transform relation between the far-field radiation pattern of an antenna and the aperture-field distribution. Resulting aperture phase and amplitude data can be used to characterize various performance

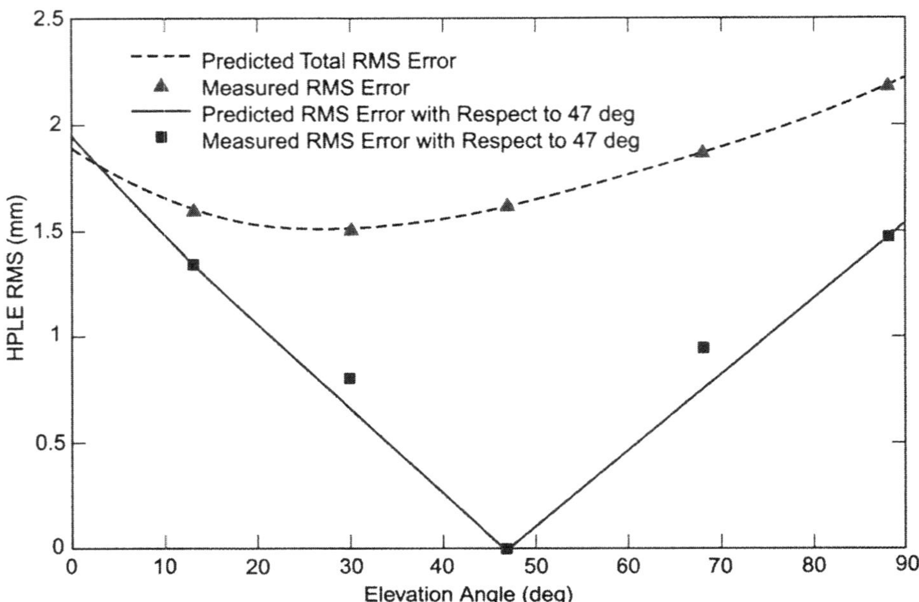

Fig. 1-24.  Summary of predicted versus measured surface rms and gravity deformation for DSS-43.

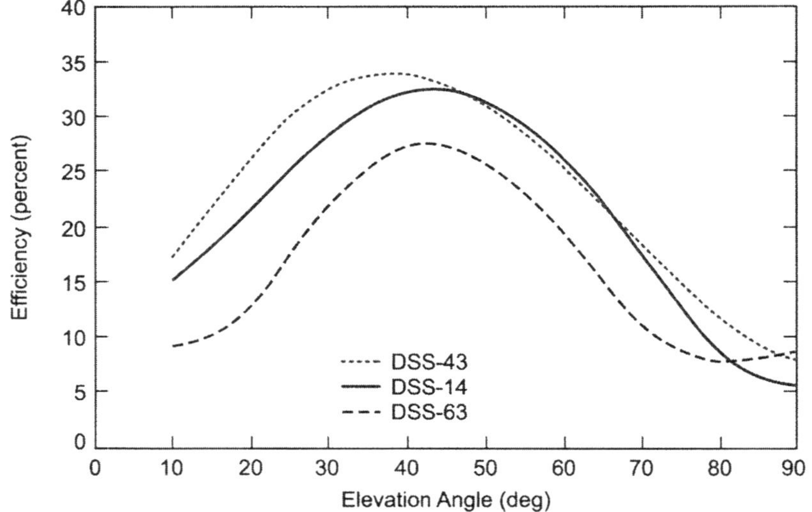

Fig. 1-25.  Computed efficiencies derived from main reflector surface measurements.

parameters, including panel alignment, panel shaping, subreflector position, antenna aperture illumination, and gravity deformation effects.

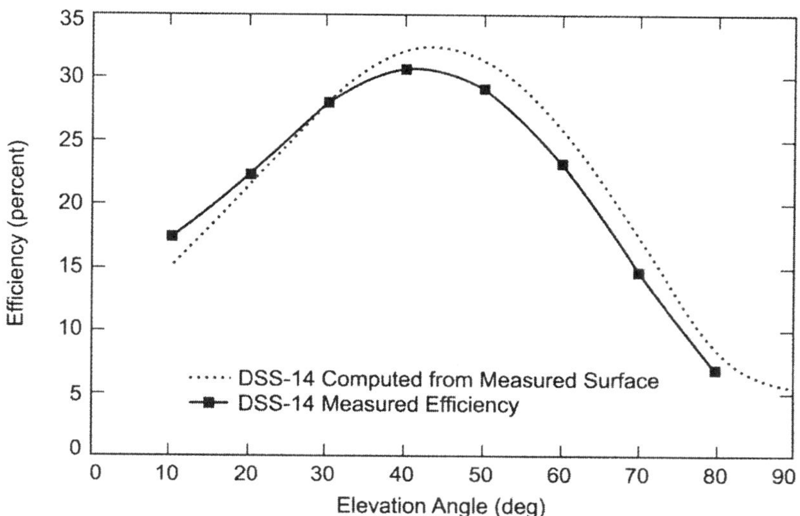


**Fig. 1-26. Computed and measured 70-m antenna efficiencies for DSS-14.**

The raw data (the observable) for this technique is the complex far-field pattern of the antenna under test. For large reflector antennas, geostationary satellites are commonly used as far-field signal sources. Figure 1-27 presents a typical narrowbandwidth receiver system architecture based on a Hewlett-Packard P8510B analyzer and an external phase-locked loop (PLL) (with a variable PLL bandwidth of 1 to 20 Hz) used for receiving the narrowband beacon signal from the satellite. The reference antenna is needed as a phase reference for the measurement as well as to keep the receiver in phase lock with the carrier [49,50].

The mathematical details for microwave holography are given in [47], among others, and are repeated here for completeness.

The mathematical relationship between an antenna far-field radiation pattern ($T$) and the antenna surface-induced current distribution ($J$) is given by the exact radiation integral [51]

$$\vec{T}(u,v) = \iint_{s} \tilde{J}(x',y')\, e^{ikz'} \left[ e^{-jkz'(1-\cos\theta)} \right] e^{j(ux'+vy')} dx'dy' \qquad (1.3\text{-}4)$$

where

$$z'(x',y') \;=\; \text{surface, } S$$
$$(u,v) \qquad\;\; = \;\text{directional cosine space}$$
$$\theta \qquad\qquad\; = \;\text{observation angle.}$$

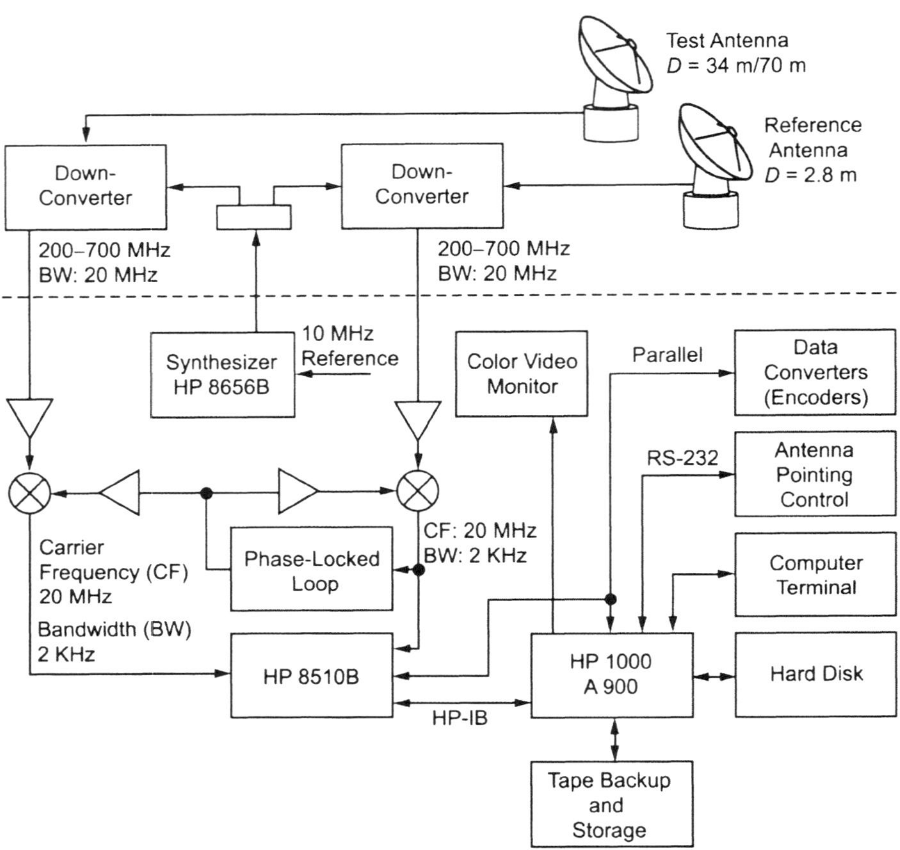



**Fig. 1-27. Holography measurement block diagram.**

For small observation angles of the far-field pattern, this expression becomes a Fourier transform relationship

$$\vec{T}(u,v) = \iint_{s} \tilde{J}\left(x', y'\right) e^{ikz'}\, e^{jk\left(ux'+vy'\right)}\, dx'dy' \qquad (1.3\text{-}5)$$

To derive the residual surface error, GO ray tracing is used to relate the small normal error, $\varepsilon$, directly to an aperture phase error in a main reflector paraboloid geometry (Fig. 1-28). In this small error approximation, it is assumed that the aperture phase error is entirely due to the projected surface errors. In addition, the 3-D structure of the surface error is not recovered; rather, an axial component equal to the average error over the resolution cell size is recovered. The normal error can then be computed using the paraboloid geometry.

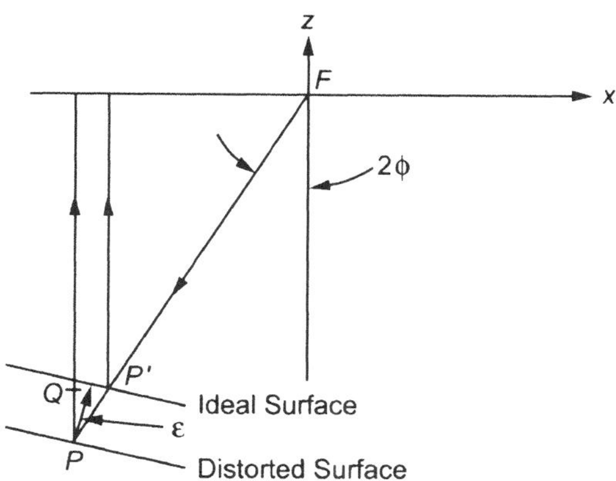


**Fig. 1-28. Surface-distortion geometry.**

The path length is related to the phase as

$$\text{Phase} = \frac{4\pi\varepsilon}{\lambda} \cos\phi \qquad (1.3\text{-}6)$$

with

$$\cos\phi = \frac{1}{\sqrt{1 + \dfrac{x^2 + y^2}{4F^2}}} \qquad (1.3\text{-}7)$$

for a parabola.

Allowing for the removal of a constant phase term and substituting Eq. (1.3-6) into Eq. (1.3-5) yields:

$$T(u,v) = e^{-j2kF} \iint\limits_{s} \left[ \left| \tilde{J}(x',y') \right| e^{j4\pi\varepsilon/\lambda\,\cos\phi} e^{jkz'} \right] e^{jk(ux'+vy')} dx'dy' \qquad (1.3\text{-}8)$$

where $[\ldots] \equiv J_e$ (equivalent current distribution).

For the processing of sampled data, the associated discrete Fourier transform (DFT) is used:

$$T(p\Delta u, q\Delta v) = sxsy \sum_{n=-N1/2}^{N1/2-1} \sum_{m=-N2/2}^{N2/2-1} J(nsx,msy)\, e^{j2\pi\left(\frac{np}{N1} + \frac{mq}{N2}\right)} \qquad (1.3\text{-}9)$$

where

$N1, N2$ = measured data array size

$$sx,\ sy \quad\quad = \quad \text{sampling intervals in the aperture coordinates}$$

$$n,\ m,\ p,\ q \quad = \quad \text{integers indexing the discrete samples}$$

$$\Delta u,\ \Delta v \quad\quad = \quad \text{sampling intervals in } u,\ v \text{ far-field space.}$$

Since the magnitude of the far-field pattern is essentially bounded, the fast Fourier transform (FFT) is generally used for computation and is symbolized here by $F$. Solving for the residual normal surface error and substituting Eq. (1.3-7) yields

$$\varepsilon(x, y) \; = \; \frac{\lambda}{4\pi}\, \sqrt{1 + \frac{x^2 + y^2}{4F^2}}\; \text{Phase}\left\{ e^{j2kF}\, \mathcal{F}^{-1}[T(u,v)] \right\} \qquad (1.3\text{-}10)$$

Accuracy across the holographic map varies with aperture amplitude illumination. Accuracy is better at the center of the dish and gradually worsens toward the edge, where illumination falls off rapidly. For a uniformly illuminated antenna design, accuracy remains relatively constant, becoming quickly worse just at the edge, also where the illumination falls off rapidly. Typically, accuracies of 0.05 mm to 0.10 mm were achieved with a corresponding signal-to-noise ratio (SNR) of 73 dB to 60 dB.

The resulting aperture phase function needs to be corrected for modulo-$2\pi$ phase errors, which occur due to the small measurement wavelength and a large phase error that is partially caused by pointing and subreflector position errors. Next, it is necessary to remove phase errors due to pointing and subreflector position errors, and this is done via the global best-fit paraboloid. Here, the entire dataset is weighted-least-squares-fitted to the paraboloid, permitting six degrees of freedom in the model, three vertex translations, two rotations, and a focal length change. The antenna surface axial rms error is then computed with respect to the position of the fitting paraboloid.

The resultant aperture function at the end of this process is defined here as the effective map, since it includes all phase effects contributing to antenna performance. These frequency-dependent effects include subreflector scattered (frequency-dependent) feed phase function and diffraction effects due to the struts. Removal of the feed phase function and subreflector support structure diffraction effects results in a frequency-independent map, which is defined here as the mechanical map. In general, the panel-setting information is derived from the mechanical map, since the antenna needs to be operated over a range of frequencies. Panel-setting information is derived by sorting all the data points within each panel and performing a rigid-body least-squares fit. The algorithm allows for one translation and two rotations.

### 1.3.3 Aperture Gain and Efficiency Measurements

Determination of aperture efficiency follows a JPL-developed methodology [52,53]. For a radio source of known flux density, the increase in system-noise temperature as determined by the boresight measurements is a measurement of antenna efficiency.

Peak radio source noise temperatures are measured via a boresight algorithm [54]. These measured quantities are then normalized by $T_{100}/C_r$, where $T_{100}$ is the 100 percent antenna efficiency temperature, and $C_r$ is the source size-correction factor [52,55]. Values of the temperature $T_a = T_{100}/C_r$ for radio sources used to calibrate DSN antennas are given in [56]. The effects of atmospheric attenuation must also be considered; for this reason, the final measured values are then corrected for yield vacuum conditions.

The typical method used to measure system-noise temperature in DSN antennas is the seven-point boresight technique. This method moves the antenna sequentially in both the cross-elevation (XEL) and elevation (EL) direction, both on- and off-source. In each direction, the antenna is positioned off-source 10 half-power (one-sided) beamwidths; one half-power beamwidth at the 3-dB point; approximately 0.576 half-power beamwidths at the 1-dB point; on-source; and then similar offsets on the other side. For example, at X-band with the 34-m antenna, the full 3-dB beamwidth is 65 mdeg. The offsets used were 325, 32.5, 18.7, and 0 deg in each direction. At Ka-band, with a full 3-dB beamwidth of 17 mdeg, the offsets were 85, 8.5, 4.9, and 0 mdeg.

For each scan, an off-source baseline is generated from the two off-source points. A Gaussian curve (relative to the baseline) is fitted to the five remaining on-source points, and the peak value of the curve is calculated. Also, the position of the peak is calculated as a measure of the pointing error for that scan. One pair of scans (one XEL and one EL) is considered to be one measurement or data point. The seven pointing offsets for each new scan are corrected for pointing errors found from the previous similar scan, in order to maintain pointing throughout a track.

As defined here, the efficiency is referenced to the input of the low-noise amplifier (LNA) and includes the losses of the feed system. Alternative methods specify efficiency at the aperture of the feed horn or at the antenna aperture itself. The term "aperture efficiency" refers to antenna gain relative to that of a uniformly illuminated circular aperture having the same diameter as the antenna; for example, a 70-percent-efficient antenna has 1.549-dB less peak gain than does the circular aperture.

The radio-source noise-temperature increase measured by the antenna is given by [57]:

$$\Delta T = \frac{\eta S A}{2 k C_r C_p} \tag{1.3-11}$$

where

$\eta$ = antenna aperture efficiency

$S$ = radio source flux (W/m$^2$/Hz)

$A$ = antenna area (m$^2$)

$k$ = Boltzmann constant ($1.3806503 \times 10^{-23}$ J/K)

$C_r$ = source size correction, typically 1.0 for point sources, up to ~1.5 for extended sources, including planets

$C_p$ = pointing correction, assumed = 1.0.

The flux, $S$, is typically given in units of "Janskys," where 1 Jansky (Jy) = $1 \times 10^{-26}$ W/m$^2$/Hz.

Measured $\Delta T$ is compared with the quantity $[(SA)/(2kC_r)]$ to give antenna efficiency. Thus,

$$\eta = \frac{\Delta T}{SA/2kC_r}$$

$$= \frac{\Delta T}{T_{100}/C_r} \tag{1.3-12}$$

The value $T_{100}$ is what would be measured by a perfect antenna looking at a point source emitting the same flux as the observed radio source. The value of $C_r$ is a function of the source structure at a given frequency and the pattern of the antenna at that frequency. For a given antenna, frequency, and radio source, the $T_{100}/C_r$ can thus be specified.

The $\Delta T$ is an on–off source measurement, and Earth's atmosphere attenuates the true source contribution that would be measured under vacuum conditions. The total atmospheric attenuation is estimated from surface weather conditions during all measurements. The surface temperature, pressure, and relative humidity at the site are recorded every half hour. Typical zenith values of attenuation at Goldstone under average clear-sky conditions are:

$$\text{X-band: } A_{zen} = 0.035 \text{ dB}$$

$$\text{Ka-Band: } A_{zen} = 0.115 \text{ dB}$$

The attenuation in decibels at elevation angle, $\theta$, is modeled as:

$$A(\theta) = \frac{A_{zen}}{\sin \theta} \tag{1.3-13}$$

The loss factor at that elevation angle is:

$$L(\theta) = 10^{A(\theta)/10} \tag{1.3-14}$$

The "vacuum $\Delta T$" then becomes:

$$\Delta T = L(\theta)\Delta T \text{ measured} \tag{1.3-15}$$

### 1.3.4 Noise-Temperature Measurements

There are a variety of noise-temperature measurement techniques and instruments used in the DSN. Some of the instruments are a total power radiometer, a noise-adding radiometer, thermal noise standards (cryogenically cooled, ambient, and hot), all well described in [58]. These instruments typically calibrate low-noise systems by injecting a known amount of noise from either a gas tube or solid-state noise diode into the input. The system output power ratio is then used to calculate the operating noise temperature. It can be difficult to determine the equivalent noise temperature of the noise source, as defined at the receive input, because of component and transmission line loss between the noise source and the receiver. The calibration involves either evaluating the noise source and loss separately or calibrating the excess noise temperature directly at the receiver input.

However, a measurement technique that eliminates this problem and that is most often used for the low-noise (<<300 K) receiving systems in the DSN is the ambient termination technique [59]. The operating-noise-temperature calibrations are performed by alternately connecting the LNA input with a waveguide switch between the antenna and an ambient termination. The same effect is accomplished by alternately placing and removing an absorber load over the front of the feed horn.

The operating noise temperature of a receiving system can be expressed as

$$T_{op} = T_a + T_e \tag{1.3-16}$$

where

$T_a$ = antenna temperature (K)

$T_e$ = receiver effective noise temperature (K).

Figure 1-29 shows a simplified block diagram of the instrumentation for operating noise-temperature calibrations. The precision attenuator is adjusted for equal output power when the receiver input is alternately connected to the ambient load and the antenna. The power ratio for an individual measurement is

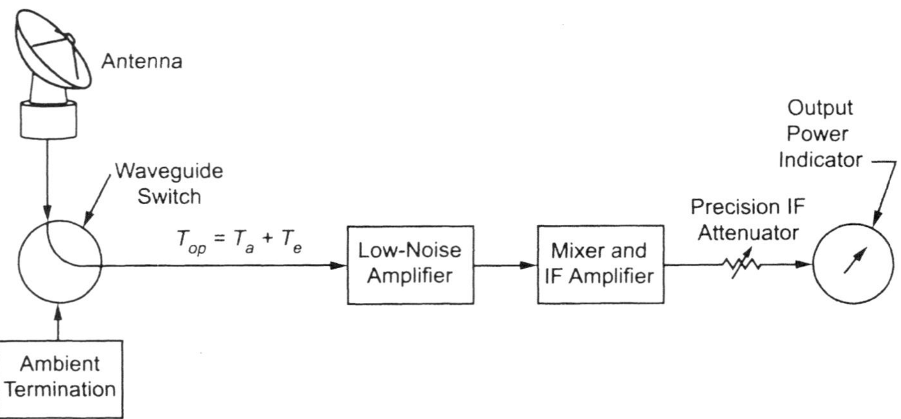



**Fig. 1-29. Simplified block diagram of the operating-noise-temperature calibration instrumentation.**

$$Y_{ap}(i) \; = \; \frac{T_p + T_e}{T_{op}(i)} \tag{1.3-17}$$

where

$T_p$      =    ambient termination physical temperature (K)

$T_e$      =    receiver effective noise temperature (K)

$T_{op}(i)$    =    individual operating noise-temperature measurement (on antenna) (K).

Rearranging Eq. (1.3-17) gives

$$T_{op}(i) \; = \; \frac{T_p + T_e}{Y_{ap}(i)} \tag{1.3-18}$$

A series of measurements is made, and a best estimate in a least-squares sense of $Y_{ap}$ is obtained.

The operating noise temperature is then given by

$$T_{op} \; = \; \frac{T_p + T_e}{Y_{ap}} \tag{1.3-19}$$

The ambient termination temperature is measured, and the receiver temperature is estimated or calibrated. It is not necessary to know $T_e$ accurately if $T_e << T_p$.

In the past, the waveguide-beyond-cutoff attenuator was the most accurate instrument to measure Y-factors. Consequently, the equations were written in terms of Y-factors. Most recently, however, the power meter has been more accurate and convenient; in addition, it mates easily with computers. Measurements in [60,61] do not use Y-factors but deal directly with power meter readings, in watts.

## 1.4 Techniques for Designing Beam-Waveguide Systems

A BWG feed system is composed of one or more feed horns with a series of flat and curved mirrors arranged so that power can be propagated from the feed horn, through the mirrors to the subreflector and main reflector, with minimum losses.

Feeding a large low-noise ground antenna via a BWG system has several advantages over placing the feed directly at the focal point of a dual-reflector antenna. For example, significant simplifications are possible in the design of high-power, water-cooled transmitters and low-noise cryogenic amplifiers. The feed horns do not have to rotate, a normally fed dual reflector does. For this reason, feed horns and other equipment can be located in a large, stable enclosure at an accessible location. Also, since BWG systems can transmit power over considerable distances at very low losses, they are useful in the design of very-high-frequency feed systems.

Since 1991, when the then-new DSS-13 R&D antenna was designed and constructed, the DSN has made extensive use of BWG feed systems. Various systems [62] have been designed using Gaussian-beam analysis, GO, and PO. In addition, more recently, a new technique based on a conjugate-phase-matching focal-plane method has been used. This section will outline each technique, discuss its advantages and disadvantages, and indicate where in the DSN it has been used.

Many existing BWG systems use a quasioptical design, based on Gaussian-wave principles, which optimize performance over an intended operating-frequency range. This type of design can be made to work well with relatively small reflectors (a very few tens of wavelengths) and may be viewed as "bandpass," since performance suffers as the waveglength becomes very short as well as very long. The long-wavelength end is naturally limited by the approaching small $D/\lambda$ of the individual BWG reflectors used; the short-wavelength end does not produce the proper focusing needed to image the feed at the dual-reflector focus. In contrast, a purely GO design has no upper frequency limit, but performance suffers at long wavelengths. Such a design may be viewed as "highpass." The design for DSS-13 (see Chapter 7 of this monograph) was based upon a highpass design while the design for operational BWG antennas (see Chapter 8) was based on a bandpass design.

If there is a need to add frequencies to an existing BWG design outside of the original design criteria, it is sometimes difficult to generate a new solution by a straightforward analysis because of the large number of scattering surfaces required for computation. A unique application of the conjugate phase-matching technique that simplifies the design process is described in Chapter 7, which presents an example of incorporating a low-frequency S-band system into the "highpass" BWG system.

Most BWG systems refocus energy at some point along the path, but this may not be desired in high-power applications. A design technique using PO, described below, does not degrade either the peak or average power handling capability beyond the limitations imposed by either the feed or the dual reflector configuration. The Antenna Research System Task (ARST) antenna (see Chapter 9) was designed for high-power applications.

### 1.4.1 Highpass Design

The design for the highpass BWG system is based on GO criteria introduced by Mizusawa and Kituregawa [63,64], which guarantee a perfect image from a reflector pair. Mizusawa's criteria can be briefly stated as follows: For a circularly symmetric input beam, the conditions on a conic reflector pair necessary to produce an identical output beam are that

- The four loci (two of which may be coincident) associated with the two curved reflectors must be arranged on a straight line
- The eccentricity of the second reflector must be equal to the eccentricity or the reciprocal of the eccentricity of the first reflector.

Figure 1-30 shows some curved reflector pair orientations that satisfy Mizusawa's criteria.

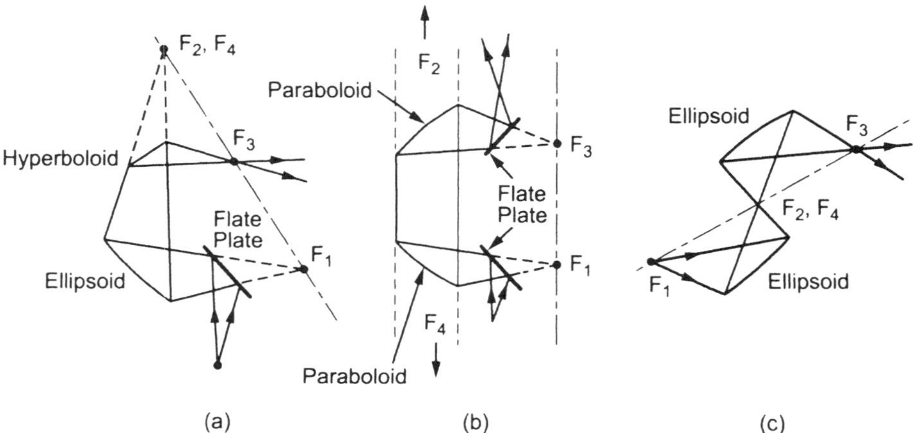


**Fig. 1-30. Examples of two curved-reflector BWG configurations.**

This technique was used for the first DSN BWG antenna (DSS-13), the design of which is shown and described in detail in Chapter 7 of this monograph. The design of DSS-13's center-fed BWG consists of a beam-magnifier ellipse in a pedestal room located below ground level that transforms a 22-dB gain feed horn into a high-gain 29-dB pattern for input to a Mizusawa four-mirror (two flat and two paraboloid-case) BWG system [see Fig. 1-30(b) and 1-31]. The system was initially designed for operation at 8.45 GHz (X-band) and 32 GHz (Ka-band) and loses fewer than 0.2 dB at X-band (determined by

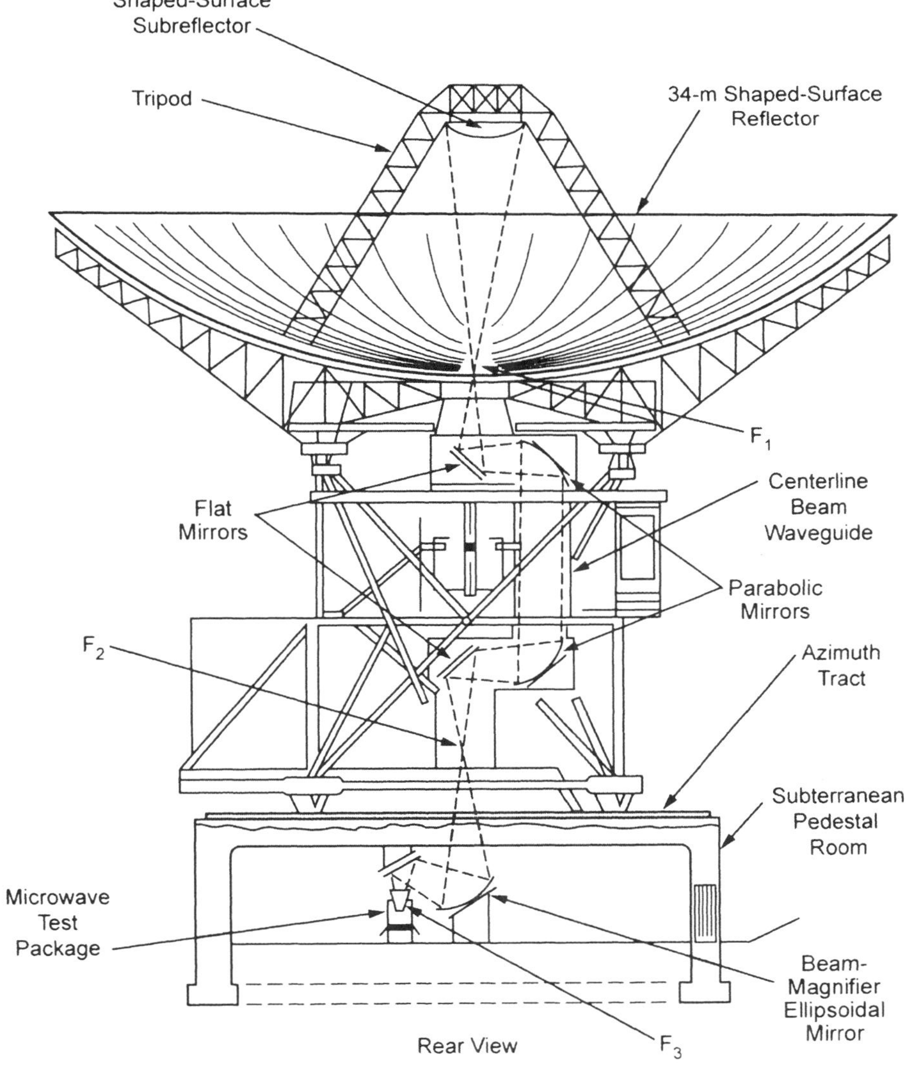


Fig. 1-31. DSS-13 highpass BWG design.

comparing the gain of a 29-dB gain horn feeding the dual-shaped reflector system with that obtained using the BWG system).

Since the upper BWG has an 8-ft (2.44-m)-diameter projected aperture, the mirrors are 70 wavelengths at X-band and 260 wavelengths at Ka-band. These mirrors are large enough to use GO design criteria.

Even though the design was based upon GO, all performance analysis was done using PO computer programs.

## 1.4.2 Focal-Plane Matching

The DSS-13 BWG system was initially designed (see Chapter 7 of this monograph for details of Phase I) for operation at X-band (8.45 GHz) and Ka-band (32 GHz) and utilized a GO design technique. The addition of S-band (2.025–2.120 GHz transmit and 2.2–2.3 GHz receive) was scheduled for a later phase. (At S-band, the mirror diameter is only 19 wavelengths, clearly outside the original GO design criteria.)

If a standard JPL 22-dB S-band feed horn is placed at the input focus of the ellipse, the BWG loss is greater than 1.5 dB. This is primarily because for low frequencies, the diffraction phase centers are far from the GO mirror focus, resulting in a substantial spillover and defocusing loss. The defocusing is especially a problem for the beam-magnifier ellipse, where the S-band phase center at the output of the ellipse is three meters from the GO focus.

A potential solution is to redesign the feed horn to provide an optimum solution for S-band. The question is how to determine the appropriate gain and location for this feed.

A straightforward design by analysis would prove cumbersome because of the large number of scattering surfaces required for the computation. Therefore, a unique application was made of the conjugate phase-matching techniques to obtain the desired solution [65]. Specifically, a plane wave was used to illuminate the main reflector and the fields from the currents induced on the subreflector propagated through the BWG to a plane centered on the input focal point. By taking the complex conjugate of the currents induced on the plane and applying the radiation integral, the far-field pattern was obtained for a theoretical feed horn that maximized antenna gain.

There is no a priori guarantee that the patterns produced by this method will be easily realized by a practical feed horn. However, for DSS-13, the pattern was nearly circularly symmetrical, and the theoretical horn was matched fairly well by a circular corrugated horn.

The corrugated feed-horn performance was only 0.22 dB lower than that of the optimum theoretical horn and about 1.4 dB better than by using the 22-dB feed horn. Subsequently, a system employing the corrugated feed horn was built, tested, and installed in the DSS-13 34-m BWG antenna as part of a simultaneous S-/X-band receiving system [66].

### 1.4.3    Gaussian-Beam Design

While GO is useful for designing systems with electrically large mirrors (>50-wavelength diameter with –20-dB edge taper), some BWGs may be operated at low frequencies, where the mirrors may be as small as 20 wavelengths in diameter. Due to diffraction effects, the characteristics of a field propagated between small BWG mirrors (<30 wavelengths in diameter) will be substantially different from those resulting from the GO solution. For these cases, the Gaussian-beam technique is used.

Solutions to the paraxial wave equation are Gaussian-beam modes. These solutions describe a beam that is unguided but effectively confined near an axis. The zero-order mode is normally used in the design. A major advantage of applying the Gaussian-beam technique is the simplicity of the Gaussian formula, which is easy to implement and requires negligible computation time. Because of the negligible computation time, a Gaussian solution can be incorporated with an optimization routine to provide a convenient method to search the design parameters for a specified frequency range, mirror sizes and locations, and feed-horn parameters.

Goubau [67] gave the first mathematical expression of Gaussian modes derived from the solution of Maxwell's equations described by a continuous spectrum of cylindrical waves. Chu [68] developed the Fresnel zone imaging principle of the Gaussian beam to design a pseudo frequency-independent BWG feed. Betsudan, Katagi, and Urasaki [69] used a similar imaging technique to design large ground-based BWG antennas. McEwan and Goldsmith [70] developed a simple design procedure based on the Gaussian-beam theory for illumination of reflector antennas where the reflector is electrically small or in the near field of a feed.

Although Gaussian-beam analysis is fast and simple, it is less accurate than the PO solution for smaller mirrors (<30 wavelengths in diameter). However, by designing with Gaussian-beam analysis, then checking and adjusting using PO analysis, an accurate and efficient tool can be fashioned. Veruttipong [71] developed such a tool for designing a second 34-m BWG antenna for the DSN. The goal was to provide good performance over the range of 2 to 32 GHz. (As demonstrated by design and measurements, the use of the GO technique in conjunction with the focal-plane-matching technique produced an antenna with virtually the same characteristics as the antenna designed using Gaussian-beam analysis [see Chapters 7 and 8].)

The design of this antenna is similar to that of DSS-13 (see Fig. 1-31) in that it uses three curved mirrors (one in the basement room and two rotating in azimuth) and a 34-m dual-shaped reflector antenna (Fig. 1-32). Multiple-frequency operation is provided by the use of dichroic mirrors. The desire is to have the radius of curvature and –18-dB beam diameter of the Gaussian beam

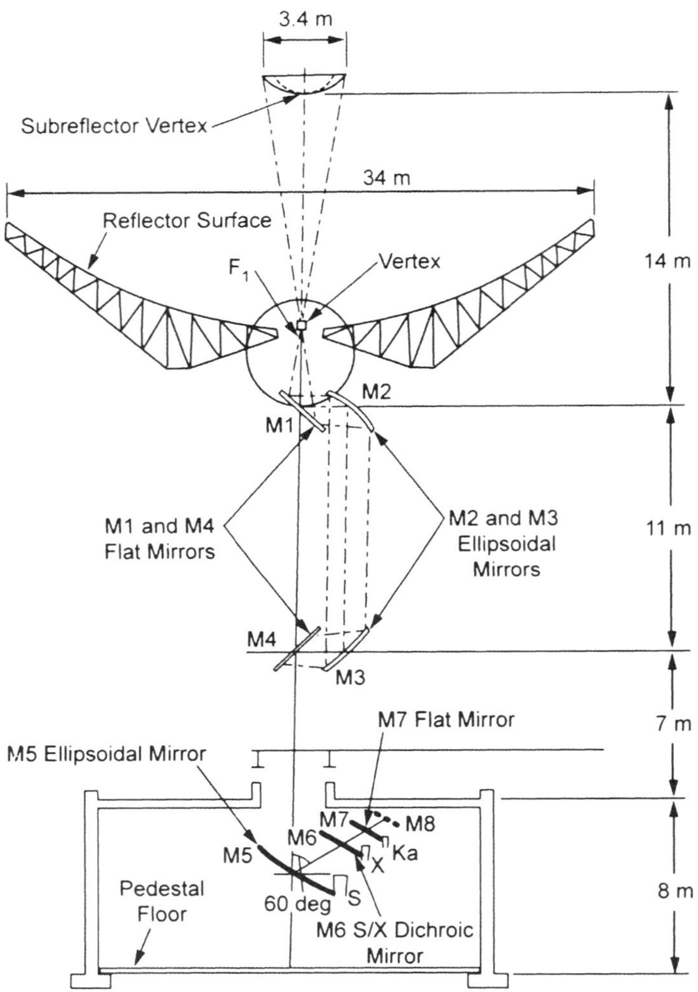


**Fig. 1-32. Operational BWG antenna.**

at the subreflector be the same at all frequencies. The size and locations of the mirrors are relatively fixed because of the basic structure geometry, so the pertinent variables are the feed-horn diameters, horn positions, and mirror curvatures. Approximating the mirrors by a thin-lens formula (Fig. 1-33) and using Gaussian-mode analysis to iterate the various design parameters, a design is achieved that meets the initial design constraint of identical patterns at the subreflector. Using the technique described above, a 34-m BWG antenna was built and measured at S-, X-, and Ka-bands (see Chapter 8 for details).

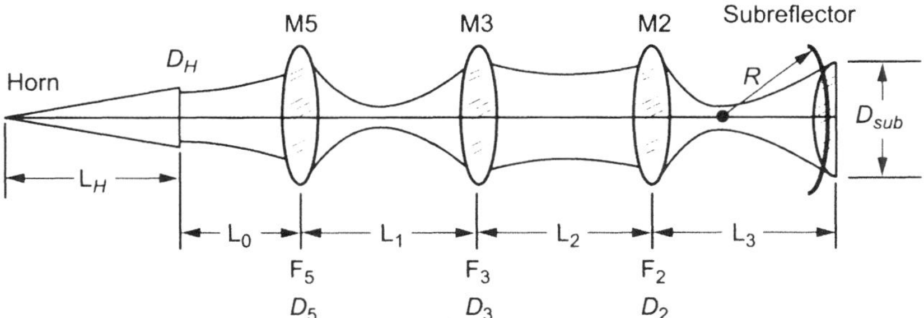



**Fig. 1-33. Parameters for BWG Gaussian design.**

### 1.4.4    High-Power Design

Observe that the GO technique used to design the BWG system images (refocuses) the feed horn at a location near the main reflector (position $F_1$ in Fig. 1-31).

The refocusing of the feed system energy near the main reflector has two distinct disadvantages. For one, although the phase center is inside the feed, the peak field point is generally at or in front of the feed itself. Thus, this peak power point is imaged in front of the main reflector. If the peak field point can be contained within the BWG system, there is the possibility to fill the BWG tube with a gas to enhance the peak power-handling capability of the system. Also, when energy is reflected from a surface such as a BWG mirror, there is a 6-dB enhancement of the power near the reflector, since the incident and reflector field add coherently near the reflection point.

It is thus important to have the energy near the mirrors at least 6 dB below the peak point in order to prevent the mirrors from degrading the power-handling capability of the system. This is difficult to do if the energy is refocused in the BWG system. For these reasons, an unconventional design required for the BWG optics. Figure 1-34 shows the geometry of optics for the high-power design antenna ([72] and Chapter 9).

For this design, only one curved mirror—a paraboloid—is used, along with three flat mirrors. The radiation from the feed horn is allowed to spread to the paraboloid, where it is focused to a point at infinity. That is, after reflection, a collimated beam exists that is directed to the subreflector by the three flat reflectors. The energy is thus spread over the entire 9-ft (2.74-m) diameter of the BWG tube. The beam reflected by the paraboloid does not begin to spread significantly due to diffraction until it exists through the main reflector. Additional spreading occurs in the region between the main reflector and the subreflector. Since a collimated beam exists beyond the first mirror, the design of this antenna is closely related to a near-field Cassegrain design, if the feed sys-

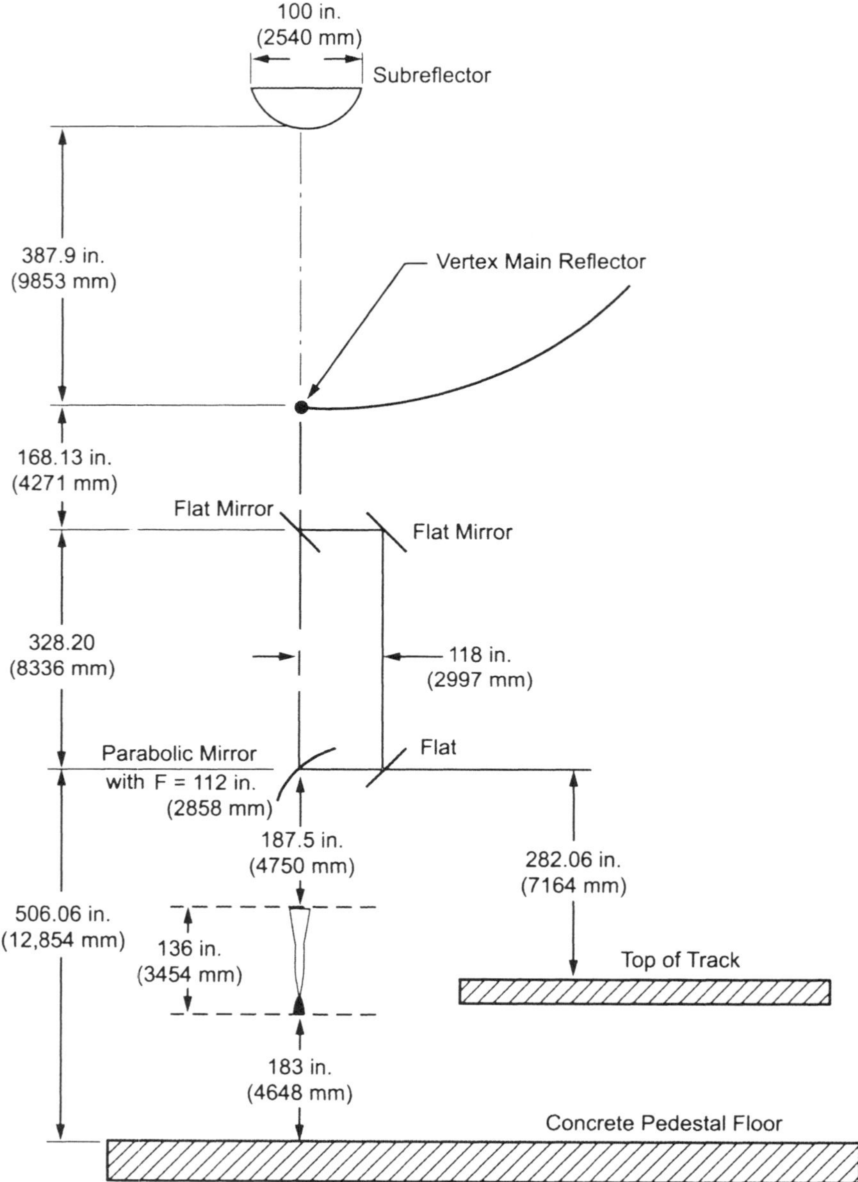



**Fig. 1-34.  Design for a high-power BWG.**

tem is defined to include both the feed horn and a parabolic mirror. The main reflector and subreflector are then determined through a standard symmetric uniform aperture synthesis program using the PO-determined input pattern radiated from the feed and four mirrors. Design and performance are described in detail in Chapter 9.

## 1.5  Summary

Since it is usually impractical to build a scale-model antenna to test new designs, accurate analytical techniques are required. These analytical techniques have evolved to the point where they can reliably and accurately predict the performance of a new design, thus eliminating the need for trial and error. The measurement techniques that have been developed can accurately characterize the performance of the antennas.

The remainder of this monograph describes in detail the RF design and performance of the large antennas of the DSN and illustrates the use of these techniques. Included are comparisons made between the predicted and measured performance of the antennas.

# References

[1]  R. F. Harrington, *Time-Harmonic Electromagnetic Fields*, New York: McGraw-Hill, 1961.

[2]  W. V. T. Rusch and P. D. Potter, *Analysis of Reflector Antennas*, New York: Academic Press, 1970.

[3]  W. A. Imbriale and R. E. Hodges, "Linear-Phase Approximation in the Triangular Facet Near-Field Physical Optics Computer Program," *Telecommunications and Data Acquisition Progress Report 42-102*, vol. April–June 1990, http://tmo.jpl.nasa.gov/progress_report/issues.html  Accessed December 2001.

[4]  W. A. Imbriale and R. E. Hodges, "The Linear Phase Triangular Facet Approximation in Physical Optics Analysis of Reflector Antennas," *Applied Computational Electromagnetic Society*, vol. 6, no. 2, pp. 74–85, Winter 1991.

[5]  A. W. Rudge, K. Milne, A. D. Olver, and P. Knight, *The Handbook of Antenna Design, Volumes 1 and 2*, London: Peter Peregrinus, 1982.

[6]  S. W. Lee and R. Mittra, "Fourier Transform of a Polygonal Shape Function and Its Application in Electromagnetics," *IEEE Transactions on Antennas and Propagation,* vol. 31, no. 1, pp. 99–103, January 1983.

[7]  Y. Rahmat-Samii, "Useful Coordinate Transformations for Antenna Applications," *IEEE Transactions on Antennas and Propagation*, vol. 27, pp. 571–574, July 1979.

[8]   J. R. Withington, W. A. Imbriale, and P. Withington, "The JPL Beam-waveguide Test Facility," *Antennas and Propagation Society Symposium*, London, Ontario, Canada, pp. 1194–1197, June 24–28, 1991.

[9]   P. D. Potter, A new horn antenna with suppressed sidelobes and equal beamwidths, *Microwave Journal*, pp. 71–78, June 1963.

[10]  S. A. Brunstein, "A New Wideband Feed Horn with Equal E- and H-plane Beamwidths and Suppressed Sidelobes," *Space Programs Summary 37-58, Vol. II, The Deep Space Network*, Jet Propulsion Laboratory, Pasadena, California, pp. 61–64, July 1969.

[11]  D. Hoppe, "Scattering Matrix Program for Circular Waveguide Junctions," *Cosmic Software Catalog*, NASA-CR-179669, NTO-17245, National Aeronautics and Space Administration, Washington, D.C., 1987.

[12]  D. Hoppe, "Modal Analysis Applied to Circular, Rectangular and Coaxial Waveguides," *Telecommunications and Data Acquisition Progress Report 42-95*, vol. July–September 1988, http://tmo.jpl.nasa.gov/progress_report/issues.html  Accessed December 2001.

[13]  D. Hoppe, W. Imbriale, and A. Bhanji, "The Effects of Mode Impurity on Ka-band System Performance," *Telecommunications and Data Acquisition Progress Report 42-80*, vol. October–December 1984, http://tmo.jpl.nasa.gov/progress_report/issues.html  Accessed December 2001.

[14]  G. L. James, "Analysis and Design of $TE_{11}$ and $HE_{11}$, Corrugated Cylindrical Waveguide Mode Converters," *IEEE Transactions on Microwave Theory and Techniques*, vol. MTT-29, pp. 1059–1066, October 1981.

[15]  A. C. Ludwig, "Radiation Pattern Synthesis for Circular Aperture Horn Antennas," *IEEE Transactions on Antennas and Propagation*," vol. AP-14, pp. 434–440, July 1966.

[16]  S. Silver, *Microwave Antenna Theory and Design*, Radiation Laboratory Series, vol. 12, New York: McGraw-Hill, pp. 336–338, 1949.

[17]  P. H. Stanton, D. J. Hoppe, H. Reilly, "Development of a 7.2-, 8.4-, and 32-Gigahertz (X-/X-/Ka-Band) Three-Frequency Feed for the Deep Space Network," *Telecommunications and Mission Operations Progress Report 42-145*, vol. January–March 1991, http://tmo.jpl.nasa.gov/progress_report/issues.html  Accessed December 2001.

[18]  A. C. Ludwig, "Spherical Wave Theory," section in *Handbook of Antenna Design* (A. W. Rudge, et al., editors), London: Peter Peregrinus, 1982.

[19]  A. C. Ludwig, *Calculation of Scattered Patterns from Asymmetrical Reflectors*, Ph.D. dissertation, University of Southern California, Los

Angeles, 1969. Also *Technical Report 32-1430,* Jet Propulsion Laboratory, Pasadena, California, February 1970.

[20] V. Galindo, "Design of Dual-Reflector Antennas with Arbitrary Phase and Amplitude Distributions," *IEEE Transactions on Antennas and Propagation*, vol. AP-12, pp. 403–408, July 1964.

[21] W. F. Williams, "High Efficiency Antenna Reflector," *Microwave Journal*, vol. 8, pp. 79–82, July 1966.

[22] V. Galindo-Israel, W. A. Imbriale, and R. Mittra, "On the Theory and Synthesis of Single and Dual Offset Shaped Reflector Antennas," *IEEE Transactions on Antennas and Propagation*, vol. AP-35, no. 8, pp. 887–896, August 1987.

[23] V. Galindo-Israel, W. A. Imbriale, R. Mittra, and K. Shogen, "On the Theory of the Synthesis of Offset Dual-Shaped Reflectors—Case Examples," *IEEE Transactions on Antennas and Propagation*, vol. 39, no. 5, pp. 620–626, May 1991.

[24] A. G. Cha, "The JPL 1.5-meter Clear Aperture Antenna with 84.5 Percent Efficiency," *Telecommunications and Data Acquisition Progress Report 42-73*, vol. January–March 1983, http://tmo.jpl.nasa.gov/progress_report/issues.html Accessed December 2001.

[25] W. A. Imbriale and D. J. Hoppe, "Computational Techniques for Beam Waveguide Systems," *2000 IEEE Antennas and Propagation International Symposium*, Salt Lake City, Utah, pp. 1894–1897, July 16–21, 2000.

[26] W. A. Imbriale and D. J. Hoppe, "Recent Trends in the Analysis of Quasioptical System," AP2000 Millennium Conference on Antennas and Propagation, Davos, Switzerland, April 9–14, 2000.

[27] P. F. Goldsmith, *Quasioptical Systems*, New York: IEEE Press, pp. 26–29, 1998.

[28] C. C. Chen, "Transmission of Microwave Through Perforated Flat Plates of Finite Thickness," *IEEE Transactions on Microwave Theory and Techniques*, vol. MTT-21, no. 1, pp. 1–6, January 1973.

[29] J. C. Chen "Analysis of a Thick Dichroic Plate with Rectangular Holes at Arbitrary Angles of Incidence," *Telecommunications and Data Acquisition Progress Report 42-104,* vol. October–December 1990, http://tmo.jpl.nasa.gov/progress_report/issues.html Accessed December 2001.

[30] W. A. Imbriale, "Analysis of a Thick Dichroic Plate with Arbitrarily Shaped Holes," *InterPlanetary Network Progress Report 42-146*, vol. April–June 2001, http://tmo.jpl.nasa.gov/progress_report/issues.html Accessed December 2001.

[31]  V. D. Agrawal and W. A. Imbriale, "Design of a Dichroic Cassegrain Subreflector," *IEEE Transactions on Antennas and Propagation*, vol. AP-27, no. 4, pp. 466–473, July 1979.

[32]  J. P. Montgomery, "Scattering by an Infinite Periodic Array of the Conductors on a Dielectric Sheet," *IEEE Transactions on Antennas and Propagation*, vol. AP-23, no. 1, pp. 70–75, January 1975.

[33]  R. Silvester, "Finite-element Solution of Homogeneous Waveguide Problems," *Alta Frequenza*, vol. 38, pp. 313–317, 1969.

[34]  J. C. Chen, P. H. Stanton, and H. F. Reilly, "Performance of the X-/Ka-/KABLE-Band Dichroic Plate in the DSS-13 Beam Waveguide Antenna," *Telecommunications and Data Acquisitions Progress Report 42-115*, vol. July–September 1991, http://tmo.jpl.nasa.gov/progress_report/issues.html   Accessed December 2001.

[35]  W. Veruttipong and M. Franco, "A Technique for Computation of Noise Temperature Due to a Beam Waveguide Shroud," *1993 IEEE International Antennas and Propagation Symposium Digest*, The University of Michigan at Ann Arbor, pp. 1659–1662, June 28–July 2, 1993.

[36]  A. G. Cha and W. A. Imbriale, "A New Analysis of Beam Waveguide Antennas Considering the Presence of the Metal Enclosure," *IEEE Transactions on Antennas and Propagation*, vol. 40, no. 9, pp. 1041–1046, September 1992.

[37]  W. A. Imbriale, "On the Calculation of Noise Temperature in Beam Waveguide Systems," *Proceedings of the International Symposium on Antennas and Propagation*, Chiba, Japan, pp. 77–80, September 24–27, 1996.

[38]  W. Imbriale, W. Veruttipong, T. Otoshi, and M. Franco, "Determining Noise Temperature in Beam Waveguide Systems," *Telecommunications and Data Acquisition Progress Report 42-116*, vol. October–December 1993, http://tmo.jpl.nasa.gov/progress_report/issues.html   Accessed December 2001.

[39]  W. A. Imbriale, T. Y. Otoshi, and C. Yeh, "Power Loss for Multimode Waveguide and Its Application to Beam-Waveguide Systems," *IEEE Transaction on Microwave Theory and Techniques*, vol. 46, no. 5, pp. 523–529, May 1998.

[40]  R. F. Harrington, *Time-Harmonic Electromagnetic Fields*, New York: McGraw-Hill, pp. 381–391, 1961.

[41]  D. A. Bathker, W. Veruttipong, T. Y. Otoshi, and P. W. Cramer, Jr., "Beam Waveguide Antenna Performance Predictions with Comparisons

to Experimental Results," *Microwave Theory and Techniques, Special Issue (Microwaves in Space)*, vol. MTT-40, no. 6, pp. 1274–1285, 1992.

[42] R. Levy, "Structural Engineering of Microwave Antennas," New York: IEEE Press, pp. 289–296, 1996.

[43] R. Levy, "DSS-13 Antenna Structure Measurements and Evaluation," JPL D-8947 (internal document), Jet Propulsion Laboratory, Pasadena, California, October 1, 1991.

[44] M. Brenner, M. J. Britcliffe, W.A. Imbriale, "Gravity Deformation Measurements of 70m Reflector Surfaces," Antenna Measurement Techniques Association 2001 Conference, Denver, Colorado, October 21–26, 2001.

[45] J. Ruze, "Antenna Tolerance Theory—A Review," *Proceedings of the IEEE*, vol. 54, pp. 633–640, 1966.

[46] D. J. Rochblatt and B. L. Seidel, "DSN Microwave Antenna Holography," *Telecommunications and Data Acquisition Progress Report 42-76*, vol. October–December 1983, http://tmo.jpl.nasa.gov/progress_report/issues.html Accessed December 2001.

[47] D. J. Rochblatt, "A Microwave Holography Methodology for Diagnostics and Performance Improvement for Large Reflector Antennas," *Telecommunications and Data Acquisition Progress Report 42-108*, October–December 1991, http://tmo.jpl.nasa.gov/progress_report/issues.html Accessed December 2001.

[48] D. J. Rochblatt and B. L. Seidel, "Performance Improvement of DSS-13 34-meter Beam-Waveguide Antenna Using the JPL Microwave Methodology," *Telecommunications and Data Acquisition Progress Report 42-108*, vol. October–December 1991, http://tmo.jpl.nasa.gov/progress_report/issues.html Accessed December 2001.

[49] D. J. Rochblatt, "System Analysis for DSN Microwave Antenna Holography," *Telecommunications and Data Acquisition Progress Report 42-97*, vol. January–March 1989, http://tmo.jpl.nasa.gov/progress_report/issues.html Accessed December 2001.

[50] D. J. Rochblatt and Y. Rahmat-Samii, "Effects of Measurement Errors on Microwave Antenna Holography," *IEEE Transactions on Antennas and Propagation,* vol. 39, no. 7, pp. 933–942, July 1991.

[51] Y. Rahmat-Samii, "Surface Diagnosis of Large Reflector Antennas Using Microwave Metrology—An Iterative Approach," *Radio Science*, vol. 19, no. 5, pp. 1205–1217, September–October 1984.

[52] M. J. Britcliffe, L. S. Alvarez, D. A. Bathker, P. W. Cramer, T. Y. Otoshi, D. J. Rochblatt, B. L. Seidel, S. D. Slobin, S. R. Stewart, W. Veruttipong,

and G. E. Wood, "DSS 13 Beam Waveguide Atnenna Project: Phase 1 Final Report," JPL D-8451 (internal document), Jet Propulsion Laboratory, Pasadena, California, May 15, 1991.

[53]  S. D. Slobin, T. Y. Otoshi, M. J. Britcliffe, L. S. Alvarez, S. R. Stewart, and M. M. Franco, "Efficiency Calibration of the DSS 13 34-meter Beam-Waveguide Antenna at 8.45 and 32 GHz," *Telecommunications and Data Acquisition Progress Report 42-106*, vol. April–June 1991, http://tmo.jpl.nasa.gov/progress_report/issues.html  Accessed December 2001.

[54]  L. S. Alvarez, "Analysis and Applications of a General Boresight Algorithm for the DSS 13 Beam-Waveguide Antenna," *Telecommunications and Data Acquisition Progress Report 42-111*, vol. July–September 1992,  http://tmo.jpl.nasa.gov/progress_report/issues.html   Accessed December 2001.

[55]  P. H. Richter and S. D. Slobin, "DSN 70-meter Antenna X- and S-band Calibration Part I: Gain Measurements," *Telecommunications and Data Acquisition Progress Report 42-97*, vol. January–March 1989, http://tmo.jpl.nasa.gov/progress_report/issues.html   Accessed December 2001.

[56]  P. Richter, "DSN Radio Source List for Antenna Calibration," JPL D-3801, Rev. C (internal document), Jet Propulsion Laboratory, Pasadena, California, August 19, 1993.

[57]  J. A. Turegano and M. J. Klein, "Calibration Radio Sources for Radio Astronomy: Precision Flux Density Measurements at 8420 MHz," *Astronomy and Astrophysics*, vol. 86, pp. 46–49, 1980.

[58]  C. Stelzried, "The Deep Space Network—Noise Temperature Concepts, Measurements, and Performance," JPL Publication 82-33, Jet Propulsion Laboratory, Pasadena, California, September 15, 1982.

[59]  C. T. Stelzried, "Operating Noise-Temperature Calibrations of Low-Noise Receiving Systems," *Microwave Journal*, vol. 14, no. 6, p. 41–48, June 1971.

[60]  C. T. Stelzried and M. J. Klein, "Precision DSN Radiometer Systems: Impact on Microwave Calibrations," *Proceedings of the IEEE*, vol. 82, no. 5, pp. 776–787, May 1994.

[61]  C. T. Stelzried and M. J. Klein, corrections to "Precision DSN Radiometer Systems: Impact on Microwave Calibrations," *Proceedings of the IEEE*, vol. 84, no. 8, p. 1187, August 1996.

[62]  W. A. Imbriale, "Design and Applications of Beam Waveguide Systems," *1997 IEEE Aerospace Conference*, vol. 3, Snowmass, Colorado,    pp. 121–134, February 1–8, 1997.

[63] M. Mizusawa and T. Kitsuregawa, "A Beam-waveguide Feed Having a Symmetric Beam for Cassegrain Antennas," *IEEE Transactions on Antennas and Propagation*, vol. AP-21, no. 6, pp. 844–846, November 1973.

[64] T. Veruttipong, J. R. Withington, V. Galindo-Israel, W. A. Imbriale, and D. Bathker, "Design Considerations for Beam-waveguide in the NASA Deep Space Network," *IEEE Transactions on Antennas and Propagation*, vol. AP-36, no. 12, pp. 1779–1787, December 1988.

[65] W. A. Imbriale and J. S. Esquival, "A Novel Design Technique for Beam-waveguide Antennas," *1996 IEEE Aerospace Applications Conference Proceedings*, vol. 1, Aspen, Colorado, pp. 111–127, February 3–10, 1996.

[66] W. A. Imbriale, M. S. Esquivel, and F. Manshadi, "Novel Solutions to Low-frequency Probelms with Geometrically Designed Beam-waveguide systems," *IEEE Transactions on Antennas and Propagation*, vol. 46, no. 12, pp. 1790–1796, December 1998.

[67] G. Goubau and F. Schwering, "On the Guide Propagation of Electromagnetic Wave Beams," *Institute of Radio Engineers Transactions on Antennas and Propagation*, vol. AP-9, no. 5, pp. 248–256, May 1961.

[68] T. S. Chu, "An Imaging Beam Waveguide Feed," *IEEE Transactions on Antennas Propagation*, vol. AP-31, no. 4, pp. 614–619, July 1983.

[69] S. Betsudan, T. Katagi, and S. Urasaki, "Design Method of Four Reflector-Type Beam Waveguide Feeds," *Japanese Electronics and Communications Society Journal*, vol. J67-B, no. 6, pp. 622–629, June 1984.

[70] N. J. McEwan and P. F. Goldsmith, "Gaussian Beam Techniques for Illuminating Reflector Antenna," *IEEE Transactions on Antennas and Propagation*, vol. 37, no. 3, pp. 297–304, March 1989.

[71] W. Veruttiong, J. C. Chen, and D. A. Bathker, "Gaussian Beam and Physical Optics Iteration Technique for Wideband Beam Waveguide Feed Design," *Telecommunications and Data Acquisition Progress Report 42-105*, vol. January–March 1991, http://tmo.jpl.nasa.gov/progress_report/issues.html Accessed December 2001.

[72] W. A. Imbriale, D. J. Hoppe, M. S. Esquivel, and B. L. Conroy, "A Beamwaveguide Design for High-Power Applications," *Intense Microwave and Particle Beams III*, proceedings of the SPIE meeting, Los Angeles, California, pp. 310–318, January 20–24, 1992.

# Chapter 2[1]
# Deep Space Station 11: Pioneer—
# The First Large Deep Space Network
# Cassegrain Antenna

## 2.1  Introduction to the Cassegrain Concept

The two-reflector system invented by Nicholas Cassegrain has been used extensively in optical telescopes, primarily to achieve a long effective focal length with a convenient physical configuration. During the late 1950s, widespread interest developed in the use of this type of system for microwave frequencies. An excellent tutorial paper by Hannan [1] from 1961 discusses the Cassegrain antenna from a geometric optics standpoint and mentions its possible advantages over a focal-point feed system: superior physical configuration, greater flexibility in feed-system design, and possible longer equivalent focal length for simultaneous lobing applications. The specific advantages of the two-reflector system applications, that is, minimum rear hemisphere radiation and high front-to-back ratio, were pointed out by Foldes and Komlos in 1960 [2]. They discuss detailed experimental measurements relating to the deviation of the feed-system performance from that which would be expected based on geometric optics; their design was, in fact, experimentally optimized, taking into account the empirically determined diffraction effects of the optical subreflector. Under Jet Propulsion Laboratory (JPL) sponsorship, this work was extended by Foldes to large low-noise antenna applications [3]. At the same

---

<sup></sup>[1]Based on "Evolution of the Deep Space Network 34-M Diameter Antennas," by William A. Imbriale, which appeared in *Proceedings of the IEEE Aerospace Conference*, Snowmass, Colorado, March 21–28, 1998. (© 1998 IEEE)

71

time, experimental work was being done at JPL with a low-noise, 26-m Cassegrain system, described by Potter [4]. Potter showed that the major factors in choosing the feed-system configuration were the forward sidelobe distribution, which must be controlled to reduce the effect of solar noise interference, and the backlobe level, which must be controlled to reject black-body radiation from the antenna environment. The 26-m antenna operating at 960 MHz has an aperture efficiency of approximately 50 percent and a measured zenith noise temperature of 9.5 K.

## 2.2  Factors Influencing Cassegrain Geometry

The basic geometry of the Cassegrain system is shown in Fig. 2-1. A hyperboloidal subreflector is interposed between the focal point of the reflector and the reflector surface and provides a constant path length for the rays from the feed to the aperture plane. Based upon geometrical optics, it can be seen that the Cassegrain system has no spillover. However, due to the finite size of the feed, there *is* forward spillover past the subreflector. Since the subreflector is only moderately sized in terms of wavelengths, there is also diffraction spillover past the main reflector. For tracking missions involving the planets Mercury or Venus, the geometry is such that the antenna may be pointed to within a few degrees of the Sun. Since the Sun is an extremely strong noise source, it is important that the antenna sidelobe level be well controlled in the solar region.

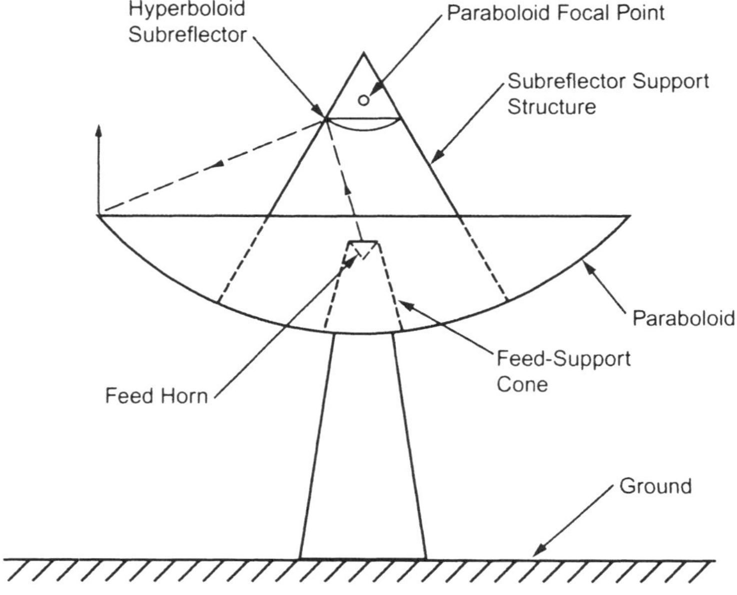


**Fig. 2-1. Basic Cassegrain geometry.**

In addition to the normal aperture-distribution diffraction pattern, the Cassegrain system introduces spurious sidelobe energy from two separate mechanisms: subreflector aperture blockage and forward spillover around the subreflector. The aperture distribution with blockage is thought of as being composed of the linear superimposition of the aperture distribution without blockage, minus the blockage aperture distribution. Potter [4] demonstrates that, at higher frequencies, it is reasonable to use the first sidelobe level of the blocking pattern as the criterion for maximum subreflector size, since the angular distribution of the blocking energy is very narrow.

Generally speaking, in order to illuminate the paraboloid efficiently, the hyperboloid must be illuminated more or less uniformly. On the other hand, in order to avoid undue forward spillover energy loss, the feed horn must be of sufficiently high gain to contain 70 to 90 percent of its energy within the area subtended by the subreflector. A typical compromise between these two considerations results in subreflector edge illumination about 10 dB below the central illumination. The spillover energy-peak gain relative to isotropic gain will thus be approximately given by the feed-horn gain reduced by the subreflector edge-taper ratio. It is clear that the feed-horn gain must be minimized to reduce the forward spillover. This may be accomplished by (a) using the largest subreflector possible and (b) using a geometry that places the feed-horn aperture as close to the subreflector as is structurally practical. The final configuration choice, therefore, involves the following steps:

1) Choose the largest subreflector possible, based on analysis of the blocking sidelobe level

2) Use the smallest horn aperture-to-hyperboloid spacing that is structurally practical and does not further increase the blockage by feed-horn shadow on the main reflector caused by the central rays from the subreflector.

Observe that in Cassegrain geometry, the subreflector subtended angle is small (almost always less than 20 deg), resulting in the need for a modest gain horn (~20 dB) that leads to a fairly high forward spillover. Potter [5] suggested an improvement in the subreflector design, involving the use of a flange that would both reduce forward (less noise from the Sun) and rear (less noise from the ground) spillover while simultaneously increasing aperture efficiency. The geometry of the beam-shaping flange is shown in Fig. 2-2. The forward spillover is reduced because of the larger extended angle, and the rear spillover is reduced because the energy radiated from the flange near the edge of the dish subtracts from the energy radiated from the hyperboloid, effectively steepening the slope at the edge of the dish and consequently reducing the rear spillover. Both effects are actually due to the increased size of the subreflector. Based upon scale model tests, an 18.4-deg flange angle was chosen by Potter as a suitable compromise between aperture efficiency and low-noise performance.

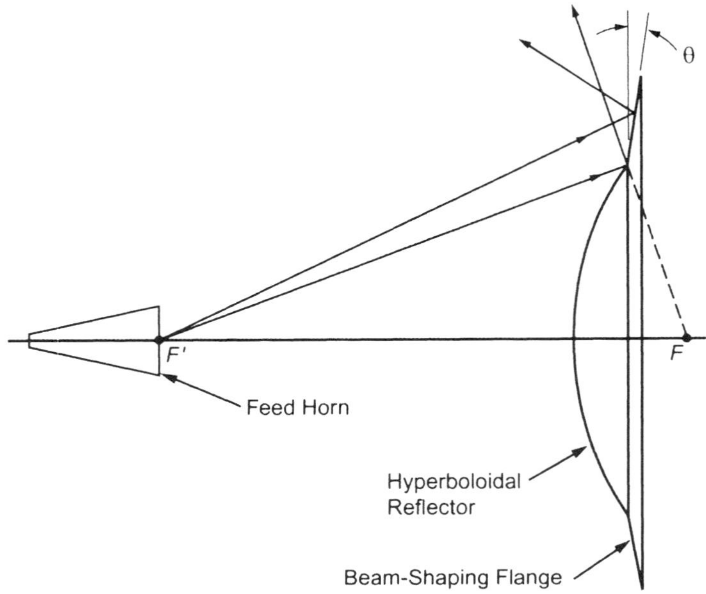


**Fig. 2-2.  Geometry of a beam-shaping flange.**

## 2.3   The DSS-11, 26-Meter Cassegrain System

In order to obtain firsthand knowledge of the performance and operational convenience of a well-designed, low-noise Cassegrain system, JPL installed and tested one on the Goldstone 26-m polar mount antenna known as Deep Space Station 11 (DSS-11). The basic design and mechanical properties of this antenna have been previously described [6]. As was typical of many of the early Deep Space Network (DSN) antennas, it was named after the first mission it supported, the Pioneer 3 and 4 lunar missions.

Although the surface tolerance of the antenna is sufficiently accurate to allow efficient operation in the S-band (2.2–2.3 GHz) region, the feed system operational frequency was chosen to be L-band (960 MHz) to allow its use for tracking the Ranger and Pioneer spacecraft. Figure 2-3 shows the feed system configuration. Figure 2-4 is a photograph of the installation. The feed system is composed of five basic components: subreflector, subreflector mount and alignment system, feed-horn, support cone, and transmission line system.

The subreflector, in this case, consists of an 8-ft (2.438-m)-diameter hyperboloid together with a beam-shaping extension flange [5] and a vertex plate in the central region [2]. Because of the small size of the hyperboloid in this system (approximately 8 wavelengths), it was necessary to use the beam-shaping

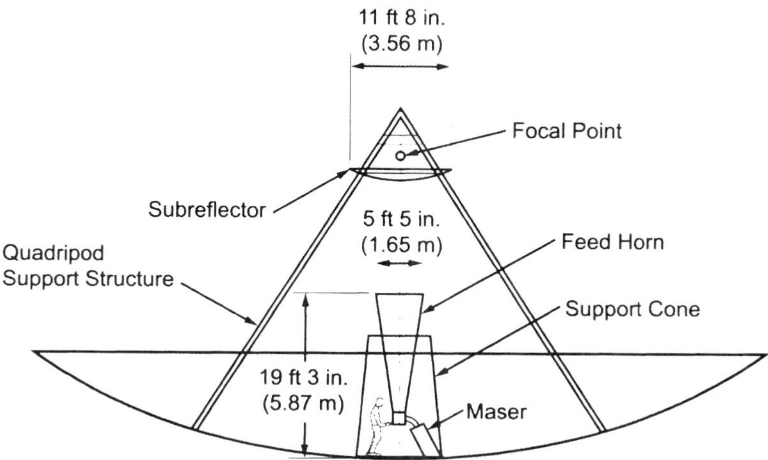


**Fig. 2-3. Feed-system configuration of the DSS-11 twenty-six-meter Cassegrain antenna.**

**Fig. 2-4. The Cassegrain installation at DSS-11.**

flange to effect a low spillover. The vertex matching device was found necessary to prevent reflection of energy by the subreflector back into the feed horn.

The subreflector mount consists of a square trusswork with a three-point jackscrew support. The three jackscrews are chained together and motorized with remote control. The system is thus designed for rapid focus adjustment on far-field radio star sources. Boresight adjustments are made by removing the chain and independently adjusting the three jackscrews.

The feed horn is a pyramidal horn of circular cross-section, with a 5.3-wavelength aperture and a 9.5-deg half-flare angle. Symmetry about the antenna axis is used not only in the feed horn but also in the subreflector, allowing complete polarization flexibility. A maximum horn aperture and minimum flare angle, constrained in this case by structural limitations, was utilized in order to minimize the energy spilled past the subreflector. The latter must be minimized not only to prevent undue aperture-efficiency degradation due to loss of energy, but also to prevent an unduly high forward sidelobe level, which would make the system susceptible to solar noise jamming.

In order to minimize loss, the transmission-line system uses WR975 (9.75-in. [248-mm]) rectangular waveguides wherever practical and rigid 3-1/8-in. (79-mm) coaxial line elsewhere. A turnstile junction [7] is used as a polarizer. This junction, together with two rotary joints, provides for dual polarization with either right- and left-hand circular or two orthogonal rotatable linear modes.

The measured and predicted noise-temperature contributions for the 960-MHz Cassegrain system are given in Table 2-1. The aperture efficiency corresponding to a given feed system illumination function may be established by simple graphical integration techniques [9]. For the 960-MHz Cassegrain system, this method predicted a 58 percent efficiency as compared with a measured value of 50 percent ± 8 percent.

**Table 2-1. Cassegrain noise contributions, 960 MHz (antenna at zenith).**

| Source of Noise | Fractional Loss | Physical Temperature (K) | Predicted Noise Contribution (K) | Measured Value (K) |
|---|---|---|---|---|
| Extra-atmospheric | N/A[a] | N/A | 1.5 (estimate) | 3.5 |
| Atmosphere | N/A | N/A | 2 K [8] | Not available |
| Direct spillover | 0.013 | 240[b] | 3 | Not available |
| Scattering from subreflector support | 0.010 | 240 | 2.5 | Not available |
| Total | | | 9 | 9.5 ± 2 |

[a]Not applicable.
[b]Includes ground reflection coefficient.

DSS-11 was decommissioned in 1981. In 1985, the National Park Service declared the site a national historic landmark.

# References


[1] P. W. Hannan, "Microwave Antennas Derived from the Cassegrain Telescope," *IRE Transactions on Antennas and Propagation*, vol. AP-9, pp. 140–153, March 1961.

[2] P. Foldes and S. G. Komlos, "Theoretical and Experimental Study of Wideband Paraboloid Antenna with Central Reflector Feed," *RCA Review*, vol. XXI, pp. 94–116, March 1960.

[3] P. Foldes, "The Capabilities of Cassegrain Microwave Optics Systems for Low Noise Antennas," *Proceedings of the Fifth AGARD Avionics Panel Conference*, Oslo, Norway, vol. 4, New York: Pergamon Press, pp. 319–352, 1962.

[4] P. D. Potter, "The Application of the Cassegrain Principle to Ground Antennas for Space Communications," *IRE Transactions on Space Electronics and Telemetry*, vol. SET-8, pp. 154–158, June 1962.

[5] P. D. Potter, *A Simple Beamshaping Device for Cassegrainian Antennas*, Technical Report TR 32-214, Jet Propulsion Laboratory, Pasadena, California, January 1962.

[6] R. Stevens, W. D. Merrick, and K. Linnes, "Ground Antenna for Space Communication System," *IRE Transactions on Space Electronics and Telemetry*, vol. SET-6, pp. 45–54, March 1960.

[7] R. S. Potter, *The Analysis and Matching of the Trimode Turnstile Waveguide Junction*, NRL Report 4670, Naval Research. Laboratory, Washington, D.C., December 1955.

[8] D. C. Hogg, "Effective Antenna Temperature Due to Oxygen and Water Vapor in the Atmosphere," *Journal of Applied Physics*, vol. 30, September 1959.

[9] R. S. Potter, *The Aperture Efficiency of Large Paraboloidal Antennas as a Function of their Feed System Radiation Characteristics*, Technical Report TR 32-149, Jet Propulsion Laboratory, Pasadena, California, October 1961.

# Chapter 3[1]
# Deep Space Station 12: Echo

The Echo site, at Goldstone, California, has been in operation since 1960 and is named for its initial operational support of Project Echo, an experiment that transmitted voice communications coast to coast by bouncing signals off the surface of a passive balloon-type satellite. The original 26-m antenna erected for the Echo experiment was moved in 1962 to the nearby Venus site.

The present Echo antenna, originally 26 m in diameter, was erected in late 1962 and was extended to 34 m in 1979. The antenna was patterned after radio-astronomy antennas then in use at the Carnegie Institution of Washington (D.C.) and the University of Michigan. The main features borrowed from the radio-astronomy antenna design are the mount and the celestial-coordinate pointing system. The axis of the polar, or hour-angle gear wheel, which moves the antenna east and west, is parallel to the polar axis of Earth and points precisely at Polaris, the North Star. The declination gear wheel, which moves the antenna north and south, is mounted on an axis that parallels Earth's equator. A spacecraft in deep space appears in the sky much like any celestial object, rising in the east and setting in the west, generally 7 to 14 hours later. The desirable feature of the polar-mount antenna is that (usually) once the declination angle has been set, only the hour-angle gear wheel is driven in order to track the spacecraft. The rate of movement, which counteracts Earth's rotation rate, is approximately 0.004 deg/s. Figure 3-1 is a photograph of the Echo antenna.

The Echo antenna stands approximately 35-m high and weighs approximately 270,000 kg. The operating speed ranges from 0.25 deg/s down to

---

[1] Based on "Evolution of the Deep Space Network 34-M Diameter Antennas," by William A. Imbriale, which appeared in *Proceedings of the IEEE Aerospace Conference*, Snowmass, Colorado, March 21–28, 1998. (© 1998 IEEE)

**Fig. 3-1. Photograph of the Echo antenna.**

0.001 deg/s. The antenna can be operated automatically or manually and has a pointing accuracy of 0.020 deg in winds up to 70 km/h.

As did the entire DSN, DSS-12 began operation at the 960-MHz L-band frequency chosen for the Pioneer 3 and 4 lunar missions. It was equipped with a 890–960-MHz linearly polarized feed horn and a 10-kW transmitter.

In mid-1962, Eberhardt Rechtin recommended the development of an S-band (2.110–2.120 GHz transmit and 2.29–2.30 GHz receive) configuration as a standard for the DSN [1]. S-band was chosen, both because of the availability of the spectrum and the frequency's application to future deep-space communications design. For example, galactic background noise was high at L-band. The higher S-band frequency decreases the noise while providing additional advantages by increasing the gain of spacecraft antennas. Therefore, the Echo site was converted to S-band in 1965.

An important requirement of the conversion was a closed-loop device for automatically pointing the antennas at a space probe. In order to track the space probe automatically, the antenna had to possess an electrical feed capable of using the space probe signal for driving the servo control system. An S-band Cassegrain monopulse (SCM) feed horn provides this capability.

## 3.1 The S-Band Cassegrain Monopulse Feed Horn

An SCM feed horn [2–4], developed for incorporation into the Cassegrain cone assembly, was designed and built to convert the 26-m Echo antenna to S-band. The feed horn contains both a sum channel for telemetry reception and a difference channel to provide error signals for antenna pointing. The multimode monopulse feed horn uses the basic (transverse electric) $TE_{10}$ mode to provide a reference (sum) channel radiation. In addition, the $TE_{12}$, (transverse magnetic) $TM_{12}$, and $TE_{30}$ modes are used for feed sidelobe suppression. The sum pattern gain enables a larger portion of the feed energy to reside in the main beam, thus minimizing the feed forward spillover losses. Monopulse capability is achieved in the same feed horn by combining the $TE_{11}$ and $TM_{11}$ modes to form the E-plane difference pattern, and the $TE_{20}$ mode for the H-plane difference pattern.

To produce the desired patterns, the relative phase and amplitude of the modes must be generated and controlled by the monopulse circuitry. These modes must also be matched into the multimode feed horn. The monopulse bridge that feeds the common-aperture section is a standard four-guide monopulse circuit providing dual polarization. More details of the feed design and performance measurements are given in [2–4].

After development of the feed design, a contract was awarded to the Hughes Aircraft Company to fabricate the SCM cone assemblies for use on the 26-m antennas. The cone assembly consisted of the feed horn and monopulse circuitry, the S-band traveling-wave maser (TWM), and a diplexer to provide for the 10-kW transmitter input. Table 3-1 presents aperture efficiency and antenna gain data for four different antennas on three continents. The Pioneer antenna (DSS-11) was the oldest antenna in the DSN and had the poorest surface, and the Canberra (Australia) antenna (DSS-42) had the first precision surface to be used by the DSN. Thus, the apparent variation in aperture efficiency between antennas has some correlation with known antenna characteristics and does not necessarily reflect a difference in the four SCM cone assemblies involved.

**Table 3-1. Antenna aperture efficiency and gain at 2.295 GHz, SCM feed system.**

| Parameter | Deep Space Station | | | |
|---|---|---|---|---|
| | 11 | 41 | 42 | 51 |
| Aperture efficiency (%) | 51 | 55 | 58 | 52 |
| Aperture efficiency (dB) | –2.92 | –2.59 | –2.37 | –2.84 |
| Antenna gain (dB) | 53.0 | 53.3 | 53.5 | 53.0 |

The efficiencies and gains given in Table 3-1 were measured at the SCM sum channel selector switches, and with a small correction for the total losses between the feed horn and maser (estimated to be 0.18 dB), the actual antenna gain at the feed-horn output is about $53.4 \pm 0.3$ dB, or an aperture efficiency of about $57 \pm 4$ percent. The complete system temperature of the SCM cone assembly with the TWM has been determined to be $42 \pm 3$ K with the antenna at zenith. The antenna temperature, measured at the SCM feed output, has been determined to be approximately 13 K.

## 3.2   The 26-Meter S-/X-Band Conversion Project

The addition of X-band receive (8.4–8.5 GHz) capability to the existing 26-m stations was in response to the support requirements for the outer planet missions. Due to the vastly increased distances from Earth, high frequencies must be employed by the spacecraft and the DSN to determine spacecraft trajectory and to achieve a greater data return. Also, several spacecraft require 64-m station support from the same area of the sky due to the combination of very long flight times and the relative position of the planets, which enables the use of intervening gravity-assist trajectories. The solution to alleviating the 64-m station overload was to increase the X-band capability of the DSN, either by adding stations or by converting existing S-band-only stations to S-/X-band capability. The latter, more economical, approach was selected [5].

The extension of the antenna diameter to 34 m provided a level of X-band performance sufficient to partially alleviate the 64-m subnet overload by (a) supporting S-/X-band navigation data acquisition out to the planet Saturn and (b) simultaneously providing a usable real-time video capability out to the planet Jupiter. This antenna diameter also proved to be the smallest diameter size suitable for arraying with the 64-m antenna.

The radio frequency (RF) modifications to the antenna system primarily included increasing the aperture diameter from 26 to 34 m and adding a simultaneous S-/X-band capability. The new redesign included a reflex–dichroic feed system similar to the one being used on the 64-m subnet [6]. In the 34-m case, a single new Cassegrain feed-horn housing, large enough to enclose both the S-band and X-band feed systems, was designed [7]. In order to minimize aperture blockage, the feed-horn focal point location was substantially moved toward the parabolic vertex and offset from the paraboloid axis. As a result, a new subreflector and quadripod were required.

Figure 3-2 details the S-/X-band feed geometry. Figure 3-3 shows a photo of the feed system. The original 26-m-diameter paraboloid focal length of 432 in. (10.97 m) is maintained and new outer panels according to the required contour are added. The initial 26-m system $F/D$ ratio of 0.4235 is, therefore,

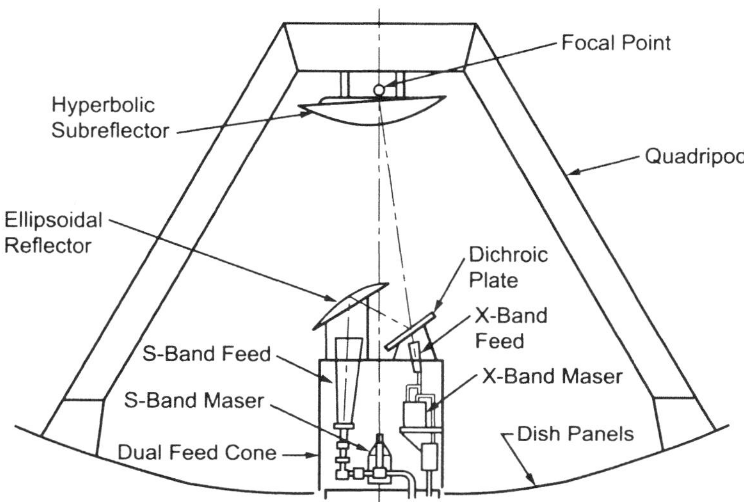


Fig. 3-2. The S-/X-band feed system.

substantially reduced to a new value of 0.3227, yielding a much deeper dish. In order to minimize central blockage caused by the substantial size of the single large feed cone and associated reflex–dichroic reflectors, the original design value for the separation of primary and secondary focal points of 228 in. (5.79 m) was increased to 276 in. (7.01 m). This results in a feed-horn position above the paraboloid vertex 48.0 in. (121.92 cm) lower than in the previous 26-m designs. In contrast to the 64-m system, where the subreflector is rotatable about the paraboloid axis (to obtain switching along the three separate feed horns used in that system), the 34-m design uses a fixed subreflector, permanently focused upon the X-band feed-horn focal point, located below the dichroic plate. Four of the elements—the S- and X-band feed horns, the ellipsoidal reflector, and the planar dichroic reflector—as well as the critical relative positions of the four elements, are borrowed directly from the previous 64-m designs [8,9].

With these critical positions held constant, the final optics design was determined by rotating the subassembly about the primary focal point at the radius value equal to 276 in. (7.01 m) until an approximately equal central blockage condition was achieved. This occurred with an offset angle of 7.6677 deg. This offset angle then determined the design of the subreflector.

The subreflector consists of three parts: the hyperboloidal surface contour, the nonoptic peripheral flange, and the nonoptic vertex plate. The peripheral flange was iteratively designed using computed scattered patterns, taking into account the decrease in power spilled past the rim of the paraboloid at both S- and X-bands. This is a key parameter in determining the noise performance

**Fig. 3-3. The feed system.**

of the finished design. The vertex plate was added to decrease RF illumination into the central zone, containing the feed horn and reflex reflectors.

### 3.2.1    Performance Predictions

A detailed description of the methods used for analysis of the computed performance is given in [10] and will only be briefly summarized along with the presentation of the final results. The basic analysis technique uses the physical optics algorithm to determine the scattering from the reflecting surfaces. For S-band, starting with the E- and H-plane radiation patterns from a standard Jet Propulsion Laboratory (JPL) corrugated conical feed horn, the feed horn patterns are scattered off the ellipsoidal reflector and flat plate. The output is then scattered off the subreflector, which includes a vertex plate and flange. The final output is then used to evaluate the overall illumination efficiency of the main paraboloid. It should be noted that a complex set of antenna patterns is used throughout this process, with azimuthal modes of $m = 0$, $m = 1$, $m = 2$, and $m = 3$ included. This preserves and sums the beam asymmetry accumulation as it develops from each surface.

The azimuthal analysis of feed radiation patterns is well documented in previous literature [10], showing that only $m = 1$ components of the pattern contribute to antenna gain. Energy contained in the $m \neq 1$ components represents a degradation of antenna gain, and in a receive mode, a possible source of system-noise-temperature degradation.

The X-band system evaluation is much simpler because no asymmetric ellipsoid is present; the asymmetric subreflector introduces the need for azimuthal modal decomposition analysis. Using the antenna patterns that were generated as described above, and at each stage preserving the azimuthal modes $m = 0$, 1, 2, and 3, the various antenna efficiency components are computed, as shown in Table 3-2.

The predicted paraboloid and subreflector surface tolerance ranges from a low of 0.58 mm at reasonable elevation angles, with no wind or thermal, to a maximum of 1.0 mm at 60-deg elevation, with wind and thermal distortion included. This maximum gives a gain loss at X-band of 0.51 dB and at S-band of 0.04 dB. The higher root-mean-square (rms) of 1.0 mm was used for the antenna gain in Table 3-2.

**Table 3-2. The 34-m reflex feed-system-gain theoretical calculations for S- and X-bands.**

| Efficiency Component | X-Band (8.415 GHz) | | S-Band (2.295 GHz) | |
|---|---|---|---|---|
| | Ratio | dB | Ratio | dB |
| Forward spillover | 0.96005 | −0.1771 | 0.98231 | −0.0775 |
| Real spillover | 0.99568 | −0.0188 | 0.99637 | −0.0158 |
| Nonuniform phase illumination | 0.84289 | −0.7423 | 0.83582 | −0.7789 |
| Nonuniform phase illumination | 0.91297 | −0.3954 | 0.96398 | −0.1593 |
| Cross polarization | 0.99942 | −0.0025 | 0.99975 | −0.0011 |
| Mode conversion ($m = 1$) | 0.99038 | −0.0420 | 0.97503 | −0.1098 |
| Central blockage | 0.97710 | −0.1006 | 0.96510 | −0.1543 |
| Quadripod blockage (optical) | 0.91700 | −0.3760 | 0.91700 | −0.3760 |
| Paraboloid/subreflector surface tolerance loss (1.0-mm rms) | 0.88209 | −0.5448 | 0.99096 | −0.0394 |
| Ellipsoid reflector tolerance loss (0.51-mm rms) | N/A[a] | N/A | 0.99759 | −0.0105 |
| Transmission loss through planar reflector | 0.9950 | −0.0218 | N/A | N/A |
| Waveguide dissipative loss | 0.97570 | −0.1070 | 0.97720 | −0.1000 |
| Overall efficiency | 0.55870 | −2.5290 | 0.65720 | −1.8230 |
| Gain for 100% efficiency | N/A | 69.537 | N/A | 58.252 |
| Overall gain (dB) | N/A | 67.01 | N/A | 56.43 |

[a]Not applicable.

The gain values are defined at the input to the maser preamplifiers. The S-band gain as indicated in Table 3-2 is for the S-band low-noise system. The expected gain for the S-band diplex mode is 0.09 dB less. The expected tolerance for X-band is +0.3 dB to –0.9 dB, and the S-band gain tolerance is ± 0.6 dB. These tolerances include the major error sources that will be incurred during field measurements using natural radio sources. As stated in the section on azimuthal mode analysis, beam squint at S-band, due to higher-order mode generation as a result of the asymmetric design, will degrade S-band gain by 0.34 dB below the peak value indicated in Table 3-2. Thus, the operational system, assuming X-band beam peak pointing, is predicted to operate with +67.0-dB X-band gain and +56.1-dB S-band gain (low-noise path) or +56.0-dB S-band gain (diplex path).

### 3.2.2  Performance Measurements

It should be noted that when DSS-12 was first completed, a series of star tracks was performed to measure the gain and system temperature, the result of which showed the maximum gain to be 65.7 dB, significantly below the predicted value. To explore the problem, a 34-m X-band gain study team was formed in 1980. The team discovered that the quality of the outer panels of the main reflector was significantly worse than was believed when the panels were first delivered. Removing 14 of the 96 new panels and retesting those removed showed rms values ranging from 1.02 mm to 2.54 mm, with a composite value of 1.4 mm. This could account for 0.75 dB of the loss, so new panels were procured, installed, and aligned. Also, it was discovered that there was an incorrect map of subreflector position as a function of elevation angles. Consequently, each antenna was calibrated to provide a permanent optimum subreflector position map. After the above-mentioned corrections, results shown in Table 3-3 were obtained, where S-band = 2.290–2.300 GHz and X-band = 8.400–8.450 GHz. S-band polarization is selectable right-hand circular polarization (RCP), left-hand circular polarization (LCP), and manually rotatable linear; and X-band polarization is selectable RCP and LCP.

## 3.3  The Goldstone–Apple Valley Radio Telescope

In 1994, the Echo antenna was decommissioned and converted into a radio telescope to be utilized as an educational resource for school children throughout the country. Technicians removed unneeded equipment from the antenna and installed new cables and a radio-astronomy receiving system. At the same time, engineers adapted the DSN's remote-control software for use on the radio telescope and developed training materials on how to operate it.

**Table 3-3. Measured performance of the 34-m S-/X-band system.**

| Predicted Performance | Measured Performance | | |
| --- | --- | --- | --- |
| | DSS-12 | DSS-42 | DSS-61 |
| S-band: | | | |
|    Gain (56.1 +0.6/–0.6 dB) | 56.6 dB | 56.6 dB | 56.3 dB |
|    Temperature (27.5 ± 2.5 K) | 27.4 K | 27.4 K | 27.5 K |
| X-band: | | | |
|    Gain (66.9 +0.3/–0.9 dB) | 66.6 dB | 66.6 dB | 66.3 dB |
|    Temperature (25 ± 2.0 K) | 23.7 K | 22.3 K | 18.5 K |

The Goldstone–Apple Valley Radio Telescope (GAVRT) program is a curriculum-driven educational program operated by a partnership between the Lewis Center for Educational Research (LCER), NASA, JPL, and the Apple Valley (California) Unified School District. The GAVRT program offers the unique opportunity for first through twelfth graders across America to take control of a dedicated 34-m radio telescope to pursue legitimate scientific inquiry. By connecting via the Internet to "mission control" at LCER in Apple Valley, students assume command of the massive instrument located at NASA's deep space communications complex at Goldstone to observe the Sun, stars, and other astronomical objects.

Students conduct actual scientific observations in radio astronomy. The data is then compiled and made available by JPL to scientists, astronomers, teachers, and students around the world. For each student astronomer, personal discovery is likely, and scientific discovery is possible.

# References


[1] N. Renzetti, *A History of the Deep Space Network*, JPL TR 32-1533, Jet Propulsion Laboratory, Pasadena, California, September 1971.

[2] *S-Band Cassegrain Monopulse (SCM) Cone Assembly Final Report*, JPL 64-319 (internal document), Jet Propulsion Laboratory, Pasadena, California, July 10, 1964.

[3] *S-Band Acquisition Antenna System Final Report*, Document FR-64-14-109, Hughes Aircraft Company, Fullerton, California, June 12, 1964.

[4]  "S-Band Cassegrain Monopulse Feed Development," *Space Programs Summary 37-33, Vol. III*, Jet Propulsion Laboratory, Pasadena, California, pp. 43–48.

[5]  V. B. Lobb, "26-Meter S-X Conversion Project," *The Deep Space Network Progress Report 42-39*, vol. March and April 1977, http://tmo.jpl.nasa.gov/progress_report/issues.html Accessed February 2002.

[6]  R. W. Hartop, "Dual Frequency Feed System for 26-meter Antenna Conversion," *Deep Space Network Progress Report 42-40*, vol. May and June 1977, http://tmo.jpl.nasa.gov/progress_report/issues.html Accessed February 2002.

[7]  R. W. Hartop, "Dual Frequency Feed Cone Assemblies for 34-Meter Antennas," *Deep Space Network Progress Report 42-47*, vol. July and August 1978, http://tmo.jpl.nasa.gov/progress_report/issues.html Accessed February 2002.

[8]  D. L. Nixon and D. Bathker, "S/X-Band Microwave Optics Design and Analysis for DSN 34-Meter Diameter Antenna," *Deep Space Network Progress Report 42-41*, http://tmo.jpl.nasa.gov/progress_report/issues.html Accessed February 2002.

[9]  P. D. Potter, *S- and X-Band RF Feed System*, vol. VIII, JPL TR 32-1526, Jet Propulsion Laboratory, Pasadena, California, pp. 53–60, April 15, 1972.

[10] A. C. Ludwig, *Calculation of Scattered Patterns From Asymmetrical Reflectors*, JPL TR 32-1430, Jet Propulsion Laboratory, Pasadena, California, February 15, 1970.

# Chapter 4[1]
# Deep Space Station 13: Venus

The Venus site began operation in Goldstone, California, in 1962 as the Deep Space Network (DSN) research and development (R&D) station and is named for its first operational activity, a successful radar detection of the planet Venus. The 26-m Venus antenna was originally located at the Echo site (see Chapter 3 of this monograph), where it was erected to support Project Echo, an experiment that transmitted voice communications coast to coast by bouncing signals off the surface of a passive balloon-type satellite. In 1962, the antenna was moved, en masse, by truck, to its present location, a shielded site where research on and development of high-power transmitters could be carried out without causing radio interference at the other stations, and where the electromagnetic radiation danger to personnel could be minimized by the station layout. The Venus antenna is equipped with an azimuth-elevation-type mount. Its hydraulic drive system is designed for relatively fast angular movement and can be operated at 2 deg/s in elevation and azimuth. See Fig. 4-1 for a photograph of the Venus antenna in 1968.

The 2400-MHz planetary radar feed system configuration [1] is shown in Fig. 4-2. The Pioneer antenna, the Venus antenna, and all subsequent 26-m Cassegrain antennas were built with the same $F/D$ ratio so that support cones could be interchangeable from one antenna to another. This cone interchangeability philosophy was used throughout the DSN for new antenna designs (the 64-m [2] and the 34-m high-efficiency [HEF] antennas) and has had an impact on the design parameters available for new designs. This philosophy was ulti-

---

[1] Based on "Evolution of the Deep Space Network 34-M Diameter Antennas," by William A. Imbriale, which appeared in *Proceedings of the IEEE Aerospace Conference*, Snowmass, Colorado, March 21–28, 1998. (© 1998 IEEE)

**Fig. 4-1. DSS-13, the Venus antenna.**

mately abandoned when beam-waveguide (BWG) antennas were introduced into the DSN.

The subreflector of the planetary radar feed system consists of a 96-in. (2.438-m)-diameter truncated hyperboloid with a 120-in. (3.048-m)-diameter nonoptical flange to reduce antenna noise temperature [3,4]. The flange is smaller than that of the 140-in. (3.556-m) Pioneer antenna system because of the higher frequency. A type of vertex plate and subreflector remote control similar to those of the Pioneer antenna were used.

A block diagram of the support-cone equipment is shown in Fig. 4-3. The polarizer is a turnstile-junction type [5]. This type of junction is a six-port device; two ports are spatially orthogonal (hybrid) $H_{11}$ circular waveguide modes, two are $H_{10}$ rectangular waveguide outputs, and the final two are short-circuit terminated $H_{10}$ rectangular waveguide ports. By choosing the appropriate short-circuit lengths, it is possible to excite the feed horn with any type of polarization. Normally, circular polarization is used for radar experiments; two continuously rotatable spatially orthogonal modes of linear polarization may be obtained by manual change of the short circuits. The polarization switch allows remotely controlled selection of either right- or left-hand circular polarization, or two orthogonal linear polarizations.

As shown in Fig. 4-3, a second waveguide switch is used to switch the polarizer output either to the high-power transmitter or the receiving system. The third switch allows the receiver (maser) input to be switched between the

 

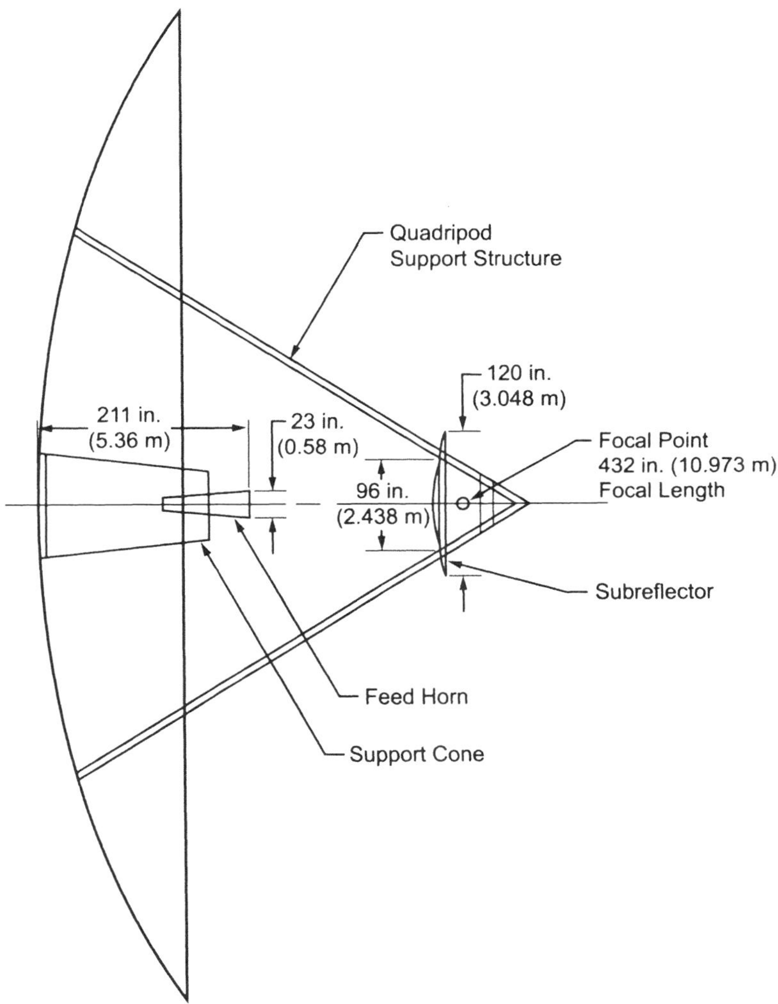


**Fig. 4-2. Planetary radar antenna system.**

antenna or either of two calibrating cryogenic terminations. During the normal transmit mode of radar operation, the polarization switch is in the right circular position, the transmit–receive switch is in the transmit position, and the calibration switch is in the nitrogen load position (the latter to provide additional isolation between transmitter output and receiver input). In the normal receiving configuration, the polarization and transmit–receive switch positions are reversed and the calibrate switch positioned to the antenna port. During radar operation, the transmitter drive and switch positions are changed remotely and automatically at time intervals that correspond to the round-trip propagation time between Earth and the planetary target.

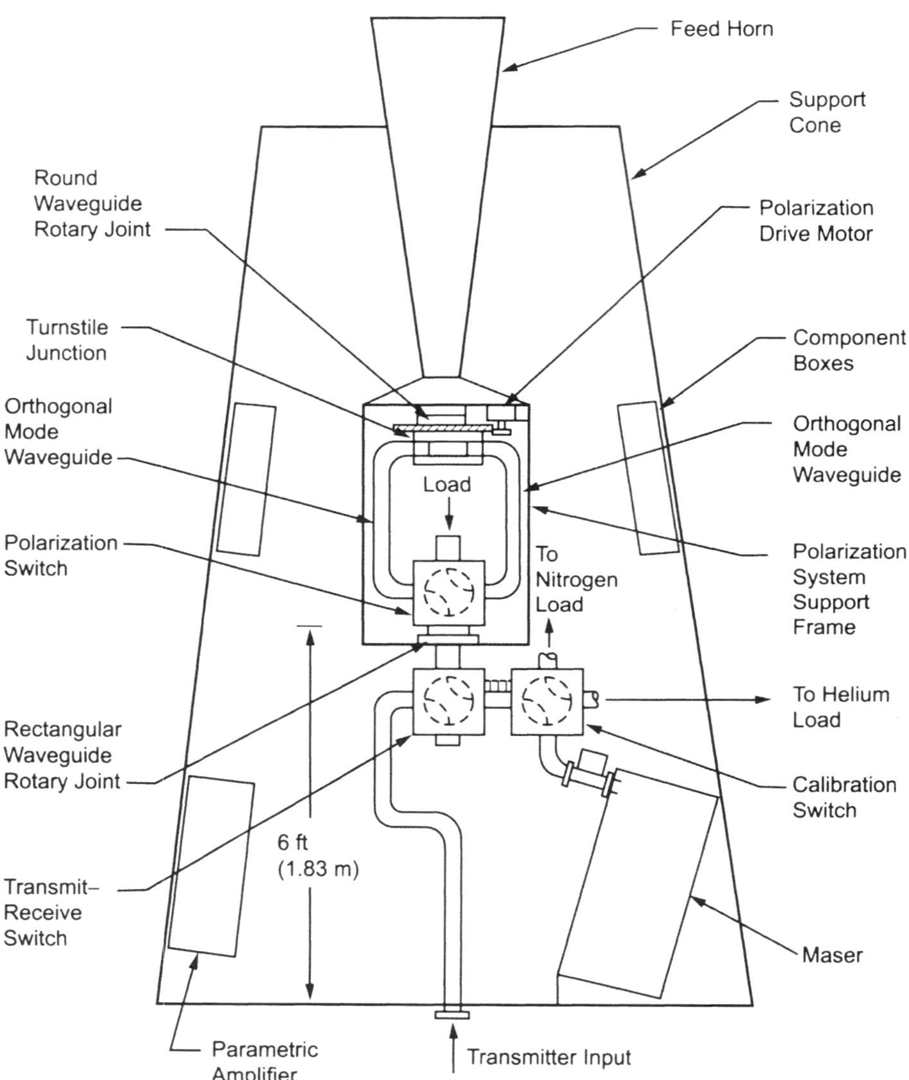



**Fig. 4-3. Support-cone experiment layout.**

The polarization flexibility and excellent axial ratio of the overall antenna system have been employed by Schuster and Levy [6,7] to perform a number of interesting polarization experiments with the planet Venus as a radar target.

## 4.1 The Dual-Mode Conical Feed Horn

The most significant improvement in the design of the Venus antenna was the change to a dual-mode conical feed horn [8,9], sometimes referred to as a Potter feed horn. It is well known that the conical feed horn, operating in the dominant circular (transverse-electric) $TE_{11}$ mode, effectively has a tapered aperture distribution in the electric plane. For this reason, its beamwidths in the electric and magnetic planes are more nearly equal than those with a square pyramidal feed horn; this is a valuable feature for polarization diversity applications. An additional result of this tapered electric-plane distribution is a more favorable sidelobe structure than with the square feed horn. The dual-mode conical feed horn utilizes a conical feed horn excited at the throat region in both the dominant circular $TE_{11}$ mode and the higher-order (transverse-magnetic) $TM_{11}$ mode. These two modes are then excited in the feed-horn aperture with the appropriate relative amplitude and phase to effect complete beamwidth equalization in all planes, complete phase center coincidence, and at least 30-dB sidelobe suppression in the E-plane. Because of the desired presence of the higher-order $TM_{11}$ mode in the feed horn, it is necessary to maintain extreme mechanical precision (of the order of $10^{-3}$ wavelength) on the inner surfaces to prevent mode conversion, with attendant pattern degradation. For this reason, very conservative mechanical design and fabrication techniques are used, and, as a result, the entire feed horn was machined from three solid aluminum billets.

## 4.2 Gain Calibration

The accurate gain calibration of large antenna systems poses a special problem because of the large distance involved in far-field tests (10.6 km for a 26-m [85-ft] antenna at 2.4 GHz). Consequently, a calibration signal source was installed on Mt. Tiefort, 20.4 km from the antenna. The technique used for gain calibration involved direct measurement of the signal attenuation between the antenna and the Mt. Tiefort signal source, using suitable corrections for atmospheric loss [10,11]. An accurate standard horn calibration of an identical dual-mode feed horn as the feed was performed, yielding an expected feed-horn gain value of $22.02 \pm 0.09$ dB. Using the above results, the 26-m antenna gain at the feed-horn output was found to be $54.40 \pm 0.15$ dBi; the corresponding aperture efficiency was 0.65. This was the most accurate gain measurement made on the 26-m antennas and was used for calibration of radio source absolute flux density.

A computer program was developed for secondary pattern and gain calculations, using feed radiation pattern data and calculated aperture blockage. The quadripod was included in the calculation as four wedge-shaped regions corresponding to the physical outline of the trusswork with an opaqueness factor to account for the fact that trusswork is partially transparent. Table 4-1 compares the predicted and measured gain. The 63 percent opaqueness factor is the number derived from equating the measured and predicted gain. Table 4-1 indicates possibilities for future improvement. As was found in later developments, about a 1-dB improvement in performance was achieved by increasing the illumination efficiency by dual-reflector shaping and by decreasing effective quadripod blockage through more efficient structural design.

General techniques for determining effective antenna temperature have been derived by Schuster et al. [12]. The method used for evaluating the planetary radar system basically consists of comparing the noise power received by the antenna with that emitted by the liquid nitrogen and liquid helium terminations. Corrections for the small (0.1-dB) insertion losses in the various transmission paths were made, resulting in an overall standard deviation for the zenith antenna temperature of approximately 0.75 K. The mean value for the antenna temperature at the feed-horn output is 10.5 K. Figure 4-4 is a plot of the antenna temperature as a function of elevation angle, together with the temperature that would be observed if the only elevation angle dependence were that predicted by Hogg [10] for the atmosphere. Note that the forward sidelobe contribution to antenna noise temperature is scarcely discernible.

**Table 4-1. Predicted and measured gain.**

| Item | Loss Factor (dB) | Associated Gain (dB) |
|---|---|---|
| Prediction factors | | |
| Theoretical maximum gain | Not applicable | 56.24 |
| Illumination factor | –1.06 | Not applicable |
| Gain for perfect surface and no quadripod | Not applicable | 55.18 |
| Surface tolerance loss (rms = 0.81 mm) | –0.05 | Not applicable |
| Gain for no quadripod | Not applicable | 55.13 |
| Loss for 100 percent opaque quadripod (machine computed) | –1.19 | Not applicable |
| Loss for 63 percent opaque quadripod | –0.73 | Not applicable |
| Predicted gain for 63 percent opaque quadripod | Not applicable | 54.40 |
| Measured gain | Not applicable | 54.50 ± 0.15 |

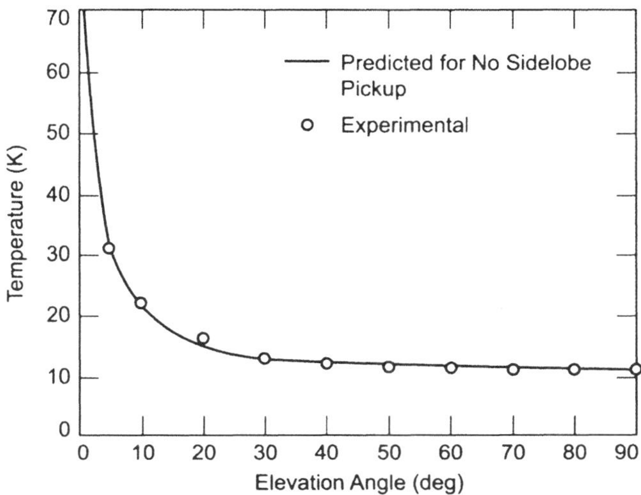


**Fig. 4-4. Antenna temperature versus
elevation angle.**

Predicted and measured zenith antenna noise temperature performances are shown in Table 4-2; they demonstrate that zenith noise temperature may be predicted to good accuracy from a knowledge of the feed system patterns and the antenna physical characteristics.

The Venus antenna remained 26 m and was last equipped with the prototype common-aperture feed horn developed for the 34-m HEF antenna. With the advent of the new research and development 34-m BWG antenna built at the Venus site, the original 26-m antenna was removed from service.

**Table 4-2. Predicted and measured zenith noise temperature.**

| Budget Elements | Predicted Values |
| --- | --- |
| Feed spillover (0.5%) | 1.0 K (predicted from scale model tests) |
| Quadripod scattering | 5.5 K (predicted from 9% blocking, energy assumed to scatter isotropically, averaged 240 K ground) |
| Surface leakage between panels | 0.5 K (extrapolated from a measured value at a different frequency) |
| Atmosphere and extra atmospheric noise | 3.0 K (measured) |
| Totals | |
| Predicted | 10.0 K |
| Measured | 10.5 ± 0.75 K standard deviation |

# References

[1]  P. D. Potter, *The Design of a Very High Power, Very Low Noise Cassegrain Feed System for a Planetary Radar*, JPL TR 32-653, Jet Propulsion Laboratory, Pasadena, California, August 1964.

[2]  Technical Staff, Tracking and Data Acquisition Organization, *The NASA/ JPL 64-Meter-Diameter Antenna at Goldstone, California: Project Report*, JPL TM 33-671; Jet Propulsion Laboratory, Pasadena, California, July 1974.

[3]  P. D. Potter, "Unique Feed System Improves Space Antennas," *Electronics*, vol. 35, pp. 36–40, June 1962.

[4]  P. D. Potter, *A Simple Beamshaping Device for Cassegrainian Antennas*, JPL Technical Report TR 32-214; Jet Propulsion Laboratory, Pasadena, California, January 1962.

[5]  R. S. Potter, "The Analysis and Matching of the Trimode Turnstile Waveguide Junction," NRL Report 4670, Naval Research Laboratory, Washington, D.C., December 1955.

[6]  D. Schuster and G. S. Levy, "Faraday Rotation of Venus Radar Echoes," *Astronomical Journal*, vol. 69, no. 1, pp. 42–48, February 1964.

[7]  G. S. Levy and D. Schuster, "Further Venus Radar Depolarization Experiments," *Astronomical Journal*, vol. 69, no. 1, pp. 29–33, February 1964.

[8]  P. D. Potter, "A New Horn Antenna with Suppressed Sidelobes and Equal Beamwidth," *Microwave Journal*, vol. VI, no. 6, pp. 71–78, June 1963.

[9]  P. D. Potter, and Ludwig, A. C., "Beamshaping by Use of Higher Order Modes in Conical Horns," *Nerem Record*, pp. 92–93, 1963.

[10] D. C. Hogg, "Effective Antenna Temperature Due to Oxygen and Water Vapor in the Atmosphere," *Journal of Applied Physics*, vol. 30, September 1959.

[11] J. H. Van Vleck, "Absorption of Microwaves by Oxygen," *Physical Review*, vol. 71, no. 6, pp. 413–433, March 1947.

[12] D. Schuster et al., "The Determination of Noise Temperature of Large Paraboloidal Antennas," *IRE Transactions on Antennas and Propagation*, vol. AP-10, no. 3, pp. 286–291, May 1962.

# Chapter 5[1]
# Deep Space Station 14: Mars

The need for large-aperture, ground-based antennas with greater capability than the previously existing 26-m-diameter antennas was foreseen and documented in early 1960. An outgrowth of this need was the Advanced Antenna System (AAS) Project [1] to design, build, and initially operate a 64-m antenna at the Deep Space Communications Complex (DSCC) in Goldstone, California. The antenna was inaugurated with the tracking of the Mariner 4 spacecraft as it was occulted by the Sun in April 1966 and, as there was a tradition to name Deep Space Network (DSN) antennas for their first operational activity, the new antenna was designated the Mars antenna. Ultimately, two more antennas of the same design were built for Spain and Australia.

Whereas the 64-m antenna was a significant advancement mechanically over existing antennas, the radio frequency (RF) design was initially a simple scaling of the highly successful systems that had been previously utilized on the DSN 26-m antennas. For the 64-m antenna, the Cassegrain geometry was chosen to be a physical scale of the 26-m Cassegrain design, allowing interchangeability of the feed-horn structures and permitting test of the feed horns on a 26-m antenna prior to commitment for use on the 64-m antenna.

In Fig. 5-1, the 64-m antenna, which featured a fully steerable azimuth–elevation mount, is shown at the time of its initial operation; Fig. 5-2 details the major components of the antenna system.

---

[1] Based on "Evolution of the Large Deep Space Network Antennas," by William A. Imbriale, which appeared in *IEEE Antennas & Propagation Magazine,* vol. 33, no. 6, December 1991. (© 1991 IEEE)

**Fig. 5-1. The 64-m antenna at the time of initial operation.**

The tipping structure contains the significant RF components: the primary reflector surface, the secondary hyperboloidal reflector surface, a supporting quadripod structure, and a Cassegrain feed-horn support.

This chapter describes the evolution in performance, from 1966 to 2001, of the largest DSN antenna: the size increase from 64 to 70 m; enhanced efficiencies through dual-reflector shaping; operational frequency upgrade from S-band (2.110–2.120 GHz transmit and 2.290–2.300 receive) to X-band (7.145–7.190 transmit and 8.400–8.450 receive), with a plan for future Ka-band (34.200–34.700 transmit and 31.800–32.300 receive); and growth from single-frequency to multifrequency operation. Design, performance analysis, and measurement techniques are highlighted.

## 5.1 Antenna Structure

The primary reflector structure was a 64-m-diameter paraboloidal space frame supported by a truss-type backup structure. The space frame was a network of 48 rib trusses radiating from the center hub and interconnected by a rectangular girder, intermediate rib trusses, and ten circular hoop trusses. The backup structure was formed by two elevation-wheel trusses, supported and braced by a tie truss. Elevation-wheel trusses supported the elevation-gear seg-

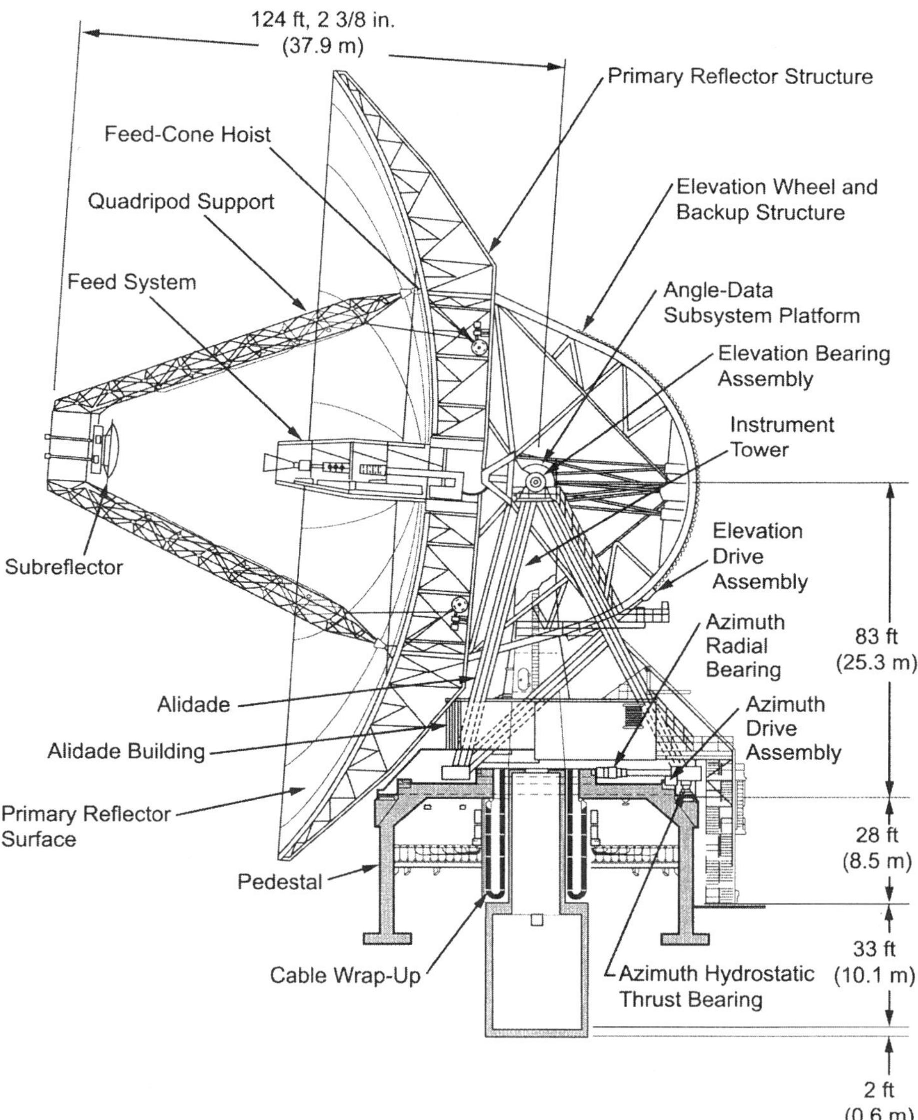


**Fig. 5-2. Tipping structure of the 64-m antenna, showing major components.**

ments, and counterweights were placed to statically balance the tipping assembly about the elevation axis.

The surface of the primary reflector comprised 552 individual panels contoured from aluminum sheet and riveted to a precisely formed aluminum frame. The panels were affixed to the reflector backup structure with adjusting screws that were accessible from the exposed surface of the reflector; the design is such that loads in the backup structure were not transmitted into the surface panels. Together, the panels formed a solid reflecting surface over the inner half-radius of the reflector and a perforated surface over the outer half-radius.

The quadripod that supported the subreflector was a tubular space-frame structure of four trapezoidally shaped legs meeting in a large-apex space frame. The four legs were supported at the hard points of the rectangular girder in the primary reflector structure. The quadripod also supported pulleys and/or hoists for handling the Cassegrain feed horns, the subreflector, or other heavy equipment being brought to or from the reflector, with the elevation motion of the quadripod legs acting as a crane.

The subreflector was a precision hyperboloidal reflecting surface 20 ft (6.1 m) in diameter, with a 12-in. (305-mm) radial flange for improving the microwave feed system. A hub and backup space-frame structure were designed with linear adjustment capability provided by remotely controlled electric motors and screwjack assemblies.

The Cassegrain feed-horn support (Fig. 5-3) was a large, three-module structure used to mount the actual feed horn and to house the receiver and transmitter equipment. The lower module was an open-framed steel truss (this

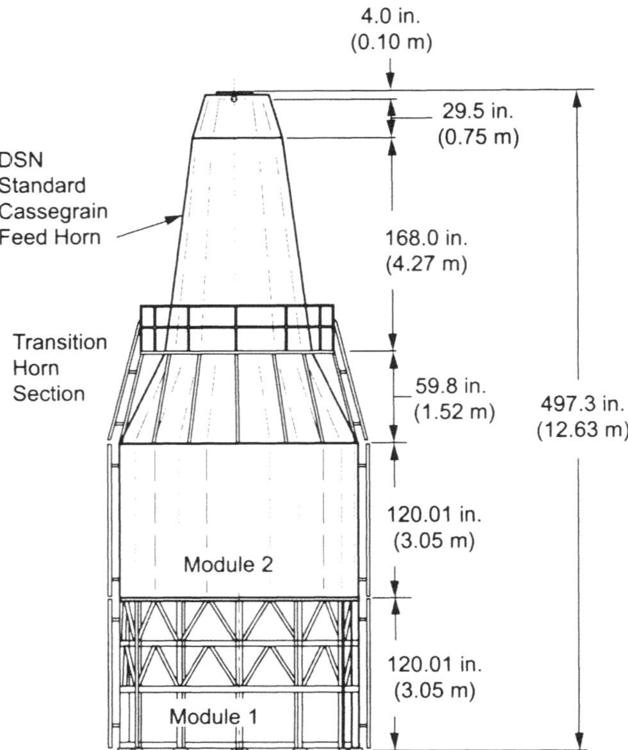


**Fig. 5-3. Feed-horn configuration during initial antenna operations.**

allowed optical surveys of the reflector surface), attached to the center hub of the primary reflector structure. The center module was a framed, stressed-skin cylinder that provided a controlled environment area for microwave and electronics equipment. The upper module was a framed, stressed-skin truncated cone that provided a more controlled environment for equipment. The upper module also contained the Cassegrain feed-horn adapter, designed to permit use of any standard DSN Cassegrain cone on the 64-m-diameter antenna. Air conditioning equipment to control the environment in upper modules and the Cassegrain cone was located in the lower module.

## 5.2  S-Band, 1966

The RF design of the original S-band feed system that was installed on the antenna was a physical scale of the 26-m antenna [2,3]. It consisted of the dual-mode [4] feed horn, a subreflector consisting of a 20-ft (6.090-m) hyperboloid together with a beam-shaping extension flange [5], a vertex plate in the central region [6], and a main paraboloid with $F/D = 0.4235$. The basic design philosophy for the subreflector was to (a) choose the largest subreflector possible, based on analysis of the blocking sidelobe level and (b) use the smallest horn-aperture-to-hyperboloid spacing that was structurally practical and did not further increase the blockage by feed-horn shadow on the main reflector (caused by the central rays from the subreflector). This design philosophy minimized forward spillover while maintaining a modest feed-horn size.

Potter [5] suggested an additional improvement in the subreflector design, involving the use of a flange that would both reduce forward spillover (less noise from interfering sources near the main beam) and rear spillover (less noise from the ground) while simultaneously increasing the aperture efficiency. The geometry of the beam-shaping flange is shown in Fig. 5-4. The forward spillover was reduced because of the larger extended angle, with consequent lower feed-horn pattern level. The rear spillover was reduced because the energy radiated from the flange near the edge of the main reflector was subtracted from the energy radiated from the hyperboloid, effectively steepening the slope at the edge of the main reflector and consequently reducing the rear spillover. Reduced forward and rear spillover were also aided by the increased size of the subreflector. A one-tenth scale model was built and tested, and the flange angle was experimentally determined to be 18.4 deg as a suitable compromise between aperture efficiency and low-noise performance.

A block diagram of the equipment used in the initial feed horn is shown in Fig. 5-5; the feed horn provided a gain of 61.47 ± 0.4 dB and a noise temperature at zenith of 27 K at 2.295 GHz. (The gain was measured using the high signal level (−100 dBw) received from the Surveyor 1 spacecraft, using a gain

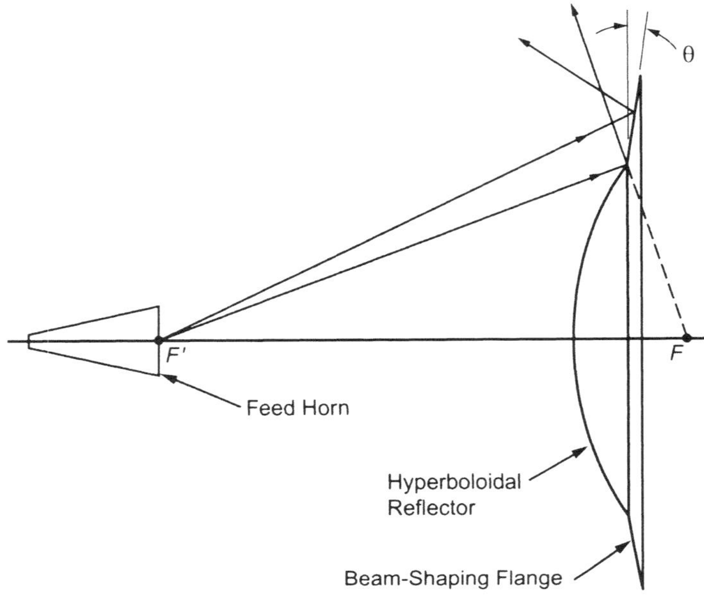


**Fig. 5-4. Geometry of shaped-beam subreflector.**

comparison method.) This feed system used a dual-mode horn, polarizer, and orthogonal-mode transducer.

Later, an experimental S-band feed system (called the ultracone) was designed as a minimum system-temperature, listen-only feed horn suitable for use on the Venus 26-m antenna (see Chapter 4) during the Mariner 5 Venus encounter in 1967. This feed horn was modified to include a diplexing capability and was installed on the Deep Space Station 14 (DSS-14) 64-m antenna in March 1968.

Based on the advanced capability demonstrated in terms of low total (zenith) operating temperatures of 17 and 28 K (listen-only and diplexed, respectively), a commitment was possible to support a high-data-rate (16.2 kb/s) experimental video data system experiment for the 1969 Mariner 6 and 7 project.

Further improvements in the feed-horn system were completed in time for the Mariner 6 and 7 Mars encounter video playback operation, which provided total (zenith) operating temperatures of 16 and 22 K, primarily a result of an improved diplexing filter. The 16-K system-noise temperature, in the listen-only mode, consisted of 9.6, 2.2, and 4.2 K for the antenna, dissipative loss, and preamplifier input noise contributions, respectively.

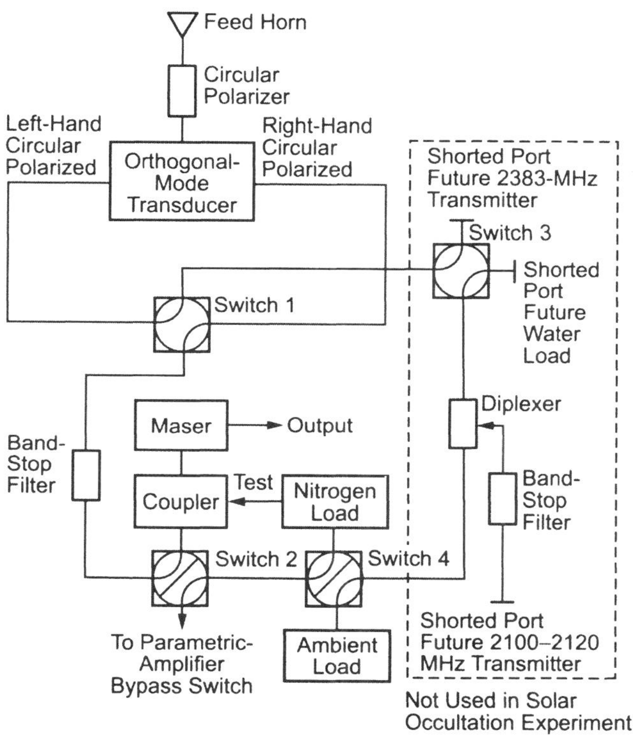



**Fig. 5-5. Original feed-horn block diagram.**

## 5.3  Performance at X-Band

Late in 1966, an X-band (8.448-GHz) Cassegrain experimental feed cone (XCE) was constructed. This feed cone was initially used for detailed evaluation of the 26-m Venus antenna [7]. In February 1968, it was used to evaluate the X-band performance of the 64-m antenna [8,9]. This new system gave the DSN the ability to clearly define surface tolerance. Previously, RF loss at S-band was sufficiently low as to make difficult an accurate RF measure of surface tolerance. The XCE cone contained a feed that was a scale model of the Venus radar feed (a dual-mode Potter horn), an X-band receive maser, and a total power radiometer. The subreflector flange, which was nominally 2λ wide at S-band, was considerably larger at X-band. Because of this, an analysis of the scattering from the Cassegrain system was performed using Rusch's scattering programs [10,11]. There was a slight pattern distortion due to the large wavelength flange, which degraded the system somewhat, but did not necessitate the removal of the flange.

The predicted value for the RF optics efficiency of the 64-m antenna fed with the XCE feed-horn system [8] was

$$\eta_0 = 0.629 \pm 0.024$$

where the surface tolerance quoted below in Section 5.3.1 is an estimated 1-sigma value, primarily due to quadripod blocking uncertainty.

## 5.3.1  Surface Tolerance

The RF surface comprised the 552 individual surface panels having factory-measured distortions with respect to their mounting legs. The distortions of the surface-panel mounting points were measured at the 45-deg elevation angle, and the gravity distortion as a function of elevation angle was computed using the STAIR structural computing program [12]. The gravity-load-deflection vectors as output from STAIR were evaluated by calculating one-half the RF path lengths of the residuals following best-fitting by a perfect paraboloid, assuming a new focal length as a best-fitting parameter.

Best-fit paraboloid translations, x-axis rotations, quadripod, and subreflector deflections were also computed. An equivalent root-mean-square (rms) distortion was compiled from a power sum of the key components (each expressed in rms millimeters): (a) the lateral and (b) axial misalignments occurring at the paraboloid focus, (c) the paraboloid structure gravity distortion, and (d) the component due to panel manufacturing error.

The errors associated with each of these components were taken in two ways: one allowed for axial focusing of the subreflector and a second constrained the system to a fixed-focus mode. The fixed-focus mode was adequate at S-band, but the axial focusing mode was required at X-band. The measured surface errors at 45 deg are shown in Table 5-1, and the rms surface tolerance as a function of elevation is shown in Fig. 5-6 compared to an RF-derived mea-

**Table 5-1. Measured surface distortion of the 64-m antenna at a 45-deg elevation angle.**

| Error Source | Error (rms, in mm) |
|---|---|
| Surface-panel setting | 0.48 |
| Primary surface-panel manufacturing | 0.89 |
| Subreflector manufacturing | 0.69 |
| Total RF half-path-length error | 1.22 |

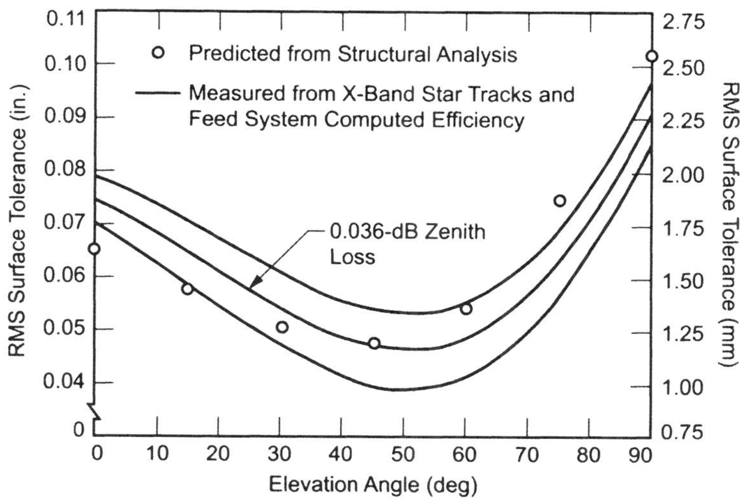



**Fig. 5-6.  Surface tolerance as a function of
elevation angle, 8.448 GHz.**

surement of the rms. The gain measurement was converted to rms, using the standard Ruze [13] technique.

### 5.3.2    Measured X-Band Performance

Operating on the ground in a checkout mode, the XCE feed horn, pointing to the zenith, exhibited a 33.5-K total operating temperature. The individual noise sources were thought to be $T_M = 13.7$ K at the maser input, $T_L = 13.2$ K, $T_F = 0.1$ K at the maser input, and $T_A = 6.8$ K at the horn aperture, where $T_M$, $T_L$, $T_F$, and $T_A$ are the maser, the loss, the follow-up receiver, and the antenna temperature contributions, respectively. When operated on the 64-m antenna, an additional 3.3-K noise temperature at zenith was observed. This additional noise was caused primarily from feed spillover and quadripod scatter towards the warm earth. A total X-band Cassegrain antenna temperature defined at the feed-horn aperture of 10.1 K appears a reasonable estimate. Comparable S-band values were an incremental 3.6 K and a total of 9.3 K.

The 64-m antenna was equipped with subreflector axial drives and, as indicated earlier, axial focusing as a function of elevation angle was used. The RF data was found to be in good agreement with the structural predictions and was used throughout the X-band evaluation.

The peak measured system gain and efficiency were 72.3 dB and 52 percent, respectively. The inaccuracy of radio star flux at these frequencies yields an error in the measurement of ± 0.8 dB. Accepting a system surface tolerance maximum of 1.65-mm rms as applicable over a very wide range of elevation angles, the predicted gain limit should have occurred in the 2-cm band.


## 5.4  Tricone Multiple Cassegrain Feed System

Several major modifications were made during the early years of operating the 64-m antenna, to improve the capability of the antenna. The most significant of these was to provide a method for rapidly changing the RF feed horns. The unicone system not only required valuable operating time to change cones, but it also presented risks of possible damage during handling procedures. The solution was to replace the standard feed-horn support structure with a structure capable of supporting three fixed feed horns clustered about the antenna axis and pointed toward the antenna subreflector. By suitable choice of Cassegrain geometry, a focal ring was realized, with a given cone brought into focus by rotating an asymmetrically truncated subreflector about its symmetric axis and pointing it toward the feed.

The principle of the focal ring is illustrated in Fig. 5-7. In a symmetric Cassegrain system [Fig. 5-7(a)], the feed is located on the centerline of the paraboloid at one of the two foci (point B) of a hyperboloid, and the other focal point (A) is located at the focal point of the paraboloid. The line AB thus defines the axis of the hyperboloid, and the hyperboloid transforms a spherical wave from the feed-phase center, B, into a spherical wave from the vertical focus at point A. For the asymmetric case [Fig. 5-7(b)] the axis (line AB) is rotated through an angle β around point A. Since A is also located at the paraboloid focus, the paraboloid is illuminated by a spherical wave from its focus, and no boresight error or aperture phase error is introduced by the tricone geometry. However, since the geometry is asymmetric with respect to the center line,

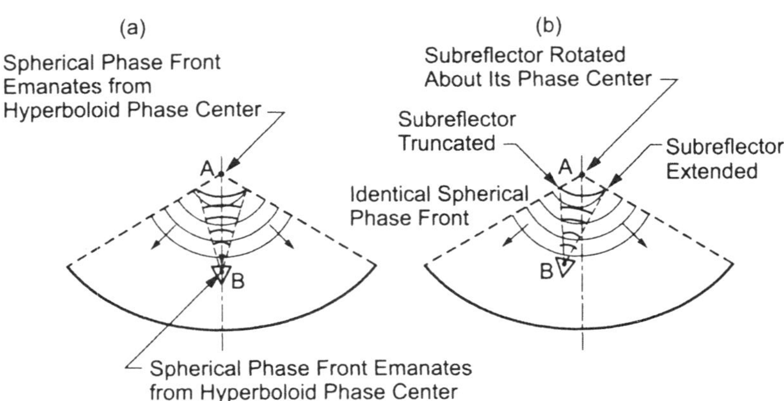


**Fig. 5-7.  Comparison of symmetric and asymmetric feed systems.**

there is an amplitude taper, which is controlled by two methods. The first is by truncating the hyperboloid asymmetrically so that the subtended half-angle at the edge as seen from the paraboloid focus A is uniformly 60 deg—the same value utilized in the unicone design. This asymmetric truncation is necessary to control the rear spillover. The second is by rotating the feed about its phase center B so that the axis of radiation intersects the geometric center of the hyperboloid's area.

The geometry of the tricone is further illustrated in Fig. 5-8. The feed-illumination angle is sufficiently close to the unicone values so that the same feed design can be used. For the 64-m geometry, the rotation angle β was 4.517 deg, and the feed phase center was 42.523 in. (1.08 m) from the center axis. The concept of the subreflector flange was maintained in the asymmetric design. Figure 5-9 shows the tricone installed on the 64-m antenna.

Rotation of the subreflector about its axis of symmetry was achieved by means of a large bearing within the torus assembly. Five rotational positions were provided by an indexing pin system. Three of these positions corresponded to the centers of the tops of the three feed horns; this provision was made in anticipation of using multiple high-frequency feed horns.

## 5.4.1 Radio Frequency Performance

During the design of the tricone system, a series of scale model tests of the subreflector and horn assembly was made, and additional analysis was performed to verify the expected performance. Analysis based upon both the calculated and the scale-model-measured scatter patterns from the subreflector showed a gain loss of 0.06 to 0.09 dB over the unicone, with the loss almost

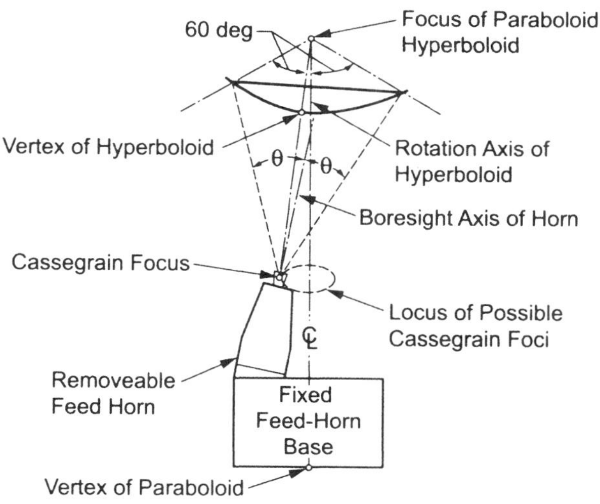


**Fig. 5-8. Tricone geometry.**

**Fig. 5-9. Tricone installed on the 64-m antenna.**

entirely due to the increased central blockage in the tricone configuration rather than to a difference in illumination function. In fact, the portion due to difference in illumination function was only 0.008 dB.

After the successful completion of the tricone installation, verification tests of S- and X-band performance were conducted. At S-band, in order to perform a direct test of the tricone effect on performance, the same feed horn was utilized as in prior unicone tests. Contrary to the predicted small degradation, the tricone gain was higher than the unicone gain by 0.1 dB. This discrepancy has never been fully explained. At X-band, a larger improvement, 0.5 dB, was measured. However, in this case, a different feed horn was utilized that had 0.1-dB less insertion loss. Thus, the improvement of the tricone was only 0.4 dB. This discrepancy has also not been fully explained but is thought to be due to an improved surface tolerance caused by the fact that the reflector panels were originally aligned for an estimated central weight that more closely resembled that of the tricone rather than that of the unicone.

### 5.4.2    New Wideband Feed Horns

Around the same time as the tricone geometry modification, a new wideband feed horn was developed [14] that was designed around the principles described by Minnett [15]. As previously described, the Potter dual-mode horn [4] achieved equal beamwidths, suppressed sidelobes, and identical E- and H-plane phase centers by the use of multiple waveguide modes in the feed horn. However, Potter's technique of generating and phasing the modes had the disadvantage of being relatively narrowband (about 9 percent). As described by Minnett, the use of a corrugated waveguide structure achieves the same results with a much wider bandwidth (28 percent in the development model). A feed horn was designed with equal E- and H-plane beamwidths, sidelobes more than 30-dB below the peak of the beam, and a voltage standing-wave ratio (VSWR) of less than 1.1 over two frequency bands of 7.767–7.801 GHz and 8.429–8.466 GHz and at two distinct frequencies of 7.84 and 8.79 GHz. This bandwidth would be sufficient to cover the S-band uplink and downlink frequencies as well as projected experimental frequencies. The use of such a horn at S-band increased the gain of the DSN antennas by more than 1 dB at the uplink frequency. There was no appreciable change at the downlink frequencies since the prior dual-mode horns were already optimized for them. Physically, the developmental mode horn had a 7.077-in. (17.98-cm)-diameter aperture, a 6.2542-deg half-flare angle, and an input section diameter of 1.369 in. (3.48 cm). A single, large circumferential slot was used for impedance matching. The remaining corrugations were the wideband mode generators, which continued for the entire length of the horn.

The developmental feed horn was accepted as the JPL standard 22-dB horn, and all subsequent 22-dB horns have utilized this design, even though better methods of matching corrugated horns have been developed. The S-band horn was a scale of 8.448/2.295 times the X-band horn. Both the S-band and X-band tricones were fitted with wideband corrugated horns.

### 5.4.3    Dual-Hybrid-Mode Feed Horn

Prior to 1970, most DSN feeds were based upon the use of two waveguide modes propagating within a smooth conical waveguide—the so-called Potter horn [4]. Due to the length of high-gain feed horns, the proper phase synchronism at the feed-horn aperture between the waveguide modes was achieved only over a restricted bandwidth (5 to 10 percent). In the DSN, this had previously meant that enhanced horn performance was achieved at the downlink band while the transmit performance was equal to or less than single-waveguide-mode feed horns.

Work to realize better feed-horn performance (pattern shaping) was only partially successful at that time. Three- and four-mode smooth conical waveguide feed horns were devised by Ludwig [16]. However, the difficulties of controlling mode amplitudes and phases for specific use on DSN antennas were great. It was concluded that only marginal performance was practically available beyond the two-mode feed horns.

By 1970, the principles described by Minnett were applied to feeds for the DSN, with excellent results. Minnett's type of feed horn is based upon one hybrid mode (transverse-electric [TE] and transverse-magnetic [TM] component) conducted along a corrugated conical waveguide. This lowest-order hybrid mode, $HE_{11}$, used within a waveguide, maintains itself nearly distortion free over a very broad bandwidth, since no phase asynchronism problems arise. It is then possible to provide superior performance over both the uplink and downlink bands.

In a manner analogous to adding individual TE and TM modes within a smooth-surface conical waveguide, it is possible to add hybrid modes within a corrugated waveguide. By so doing, it is possible to approximate a radiation pattern that puts additional energy near the outer edge of the parabolic reflector, approximating a more uniform aperture illumination taper and, hence, yielding a higher overall gain.

Figure 5-10 shows the throat region of the 8.45-GHz horn described in the preceding paragraph [17]. The $TE_{11}$ input waveguide remained the same as in previous X-band horns. The abrupt step, in addition to exciting the $HE_{12}$ mode, also played a major role in controlling the amplitude ratio between the two hybrid modes. The horn flare angle was the same as for existing (at the time) JPL feed horns. The aperture size of 12.563 in. (31.91 cm) was approximately optimum for the 14.784-deg half-aperture angle of the 64-m antenna with tricone feed subreflector. By experiment, the phase center was found to be 7.88 in. (20 cm) inside the aperture.

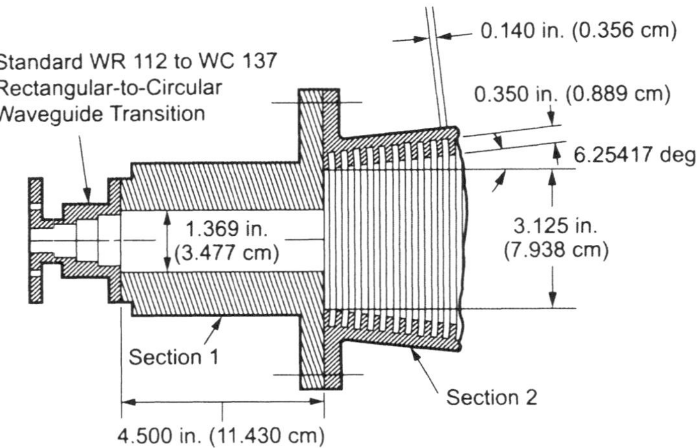


Fig. 5-10. Throat region of dual-hybrid-mode feed horn.

Figure 5-11 shows the feed-horn amplitude patterns of the single-hybrid and dual-hybrid-mode feed horns at 8.45 GHz, and Fig. 5-12 the E- and H-plane amplitude patterns scattered from the symmetric equivalent subreflector of the existing 64-m tricone feed system. The more uniform amplitude illumination produced by the dual-hybrid feed horn was clear. Table 5-2 gives the results of efficiency calculations on the two patterns of Fig. 5-12. For the symmetric case, a performance increase of 0.36 dB was available. It was estimated that a loss of 0.07 dB would occur if the feed were used in the 64-m asymmetric system, due to energy converted into the $m \neq 1$ modes. The net available improvement was, therefore, +0.29 dB.

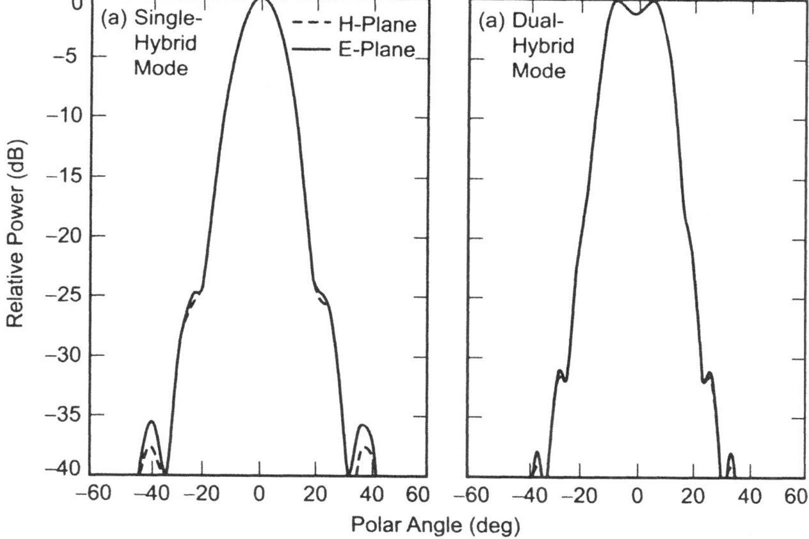


**Fig. 5-11. Feed-horn amplitude patterns (8.45 GHz): (a) single-hybrid-mode feed and (b) dual-hybrid-mode feed.**

**Table 5-2. Subreflector scattered pattern efficiencies, tricone symmetric equivalent subreflector (8.45 GHz).**

| Efficiency Terms | Single-Hybrid-Mode Feed | Dual Hybrid-Mode Feed | Dual-Hybrid-Mode Improvement (dB) |
|---|---|---|---|
| Spillover | 0.9447 | 0.9529 | +0.037 |
| Illumination | 0.8474 | 0.9008 | +0.266 |
| Cross-polarization | 0.9999 | 0.9999 | 0.000 |
| Phase | 0.9718 | 0.9670 | −0.021 |
| Blocking | 0.9441 | 0.9616 | +0.080 |
| Total | 0.7344 | 0.7981 | +0.361 |

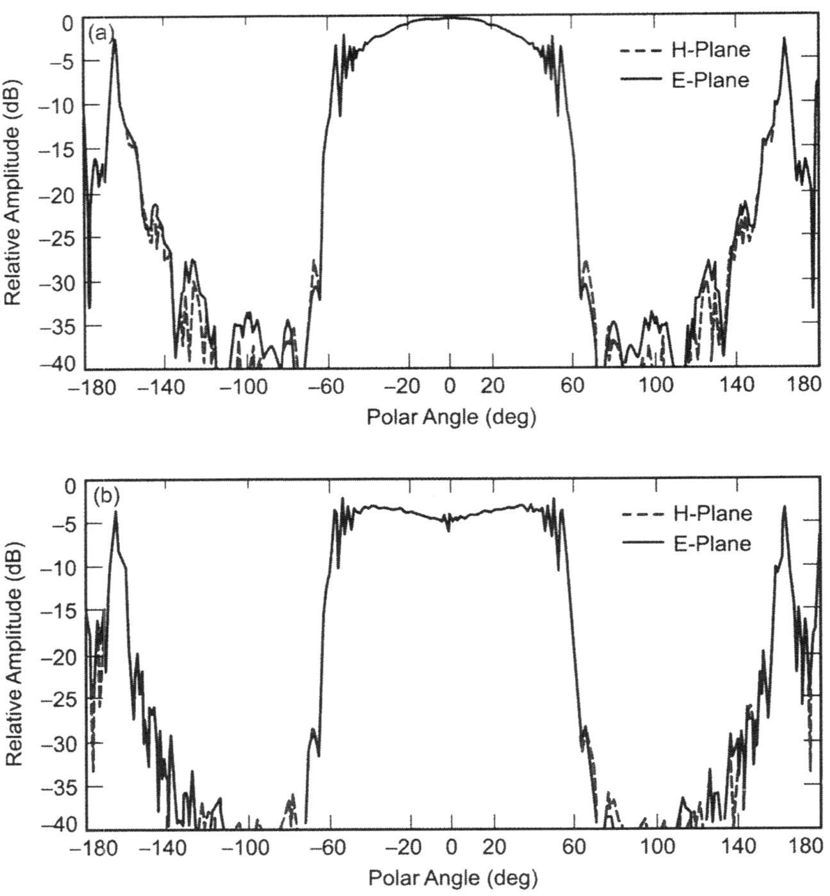



Fig. 5-12. Symmetric subreflector amplitude patterns: (a) single-hybrid-mode feed and (b) dual-hybrid-mode feed.

The X-band feed-horn assembly [18] on the DSN 64-m antennas originally provided the capability for selectable right-hand circular polarization (RCP) or left-hand circular polarization (LCP) [19]. Designed to work with a signal traveling-wave maser (TWM), the X-band receive-only (XRO) cone assembly was then upgraded to include dual TWMs [19]. As part of that reconfiguration, provision was made for a new feed assembly, the Mod III XRO, that would include two basic improvements over the feeds that had been in use since early 1977. The first improvement was the development of an orthogonal-mode transducer that permits simultaneous reception of two different polarizations. The second improvement was the incorporation of a dual-hybrid-mode feed horn to increase the antenna gain.

The new feed assembly is shown in Fig. 5-13. It was designed for mounting within the XRO cone assembly on the same circular mounting plate as the previous feeds (Mod I and Mod II), so that replacement in the field was expedited [20,21]. Each output arm of the orthogonal-mode transducer has its own waveguide switch for TWM calibration.

The 64-m antenna X-band system at DSS-14 was evaluated to determine the performance with the new dual-hybrid-mode feed. The peak system efficiency increased from 42.0 to 45.6 percent, resulting in a 0.36-dB increase in antenna gain (see Fig. 5-14). The new measured gain was 71.6 dB. Antenna pointing, beamwidth, optimum subreflector focusing, and operating system temperature were unchanged from the previous feed. Some evidence of antenna aging was apparent (from the 1973 measurements reported earlier) both in peak gain and in the pointing angle at which the peak occurs. Further data can be found in [22].

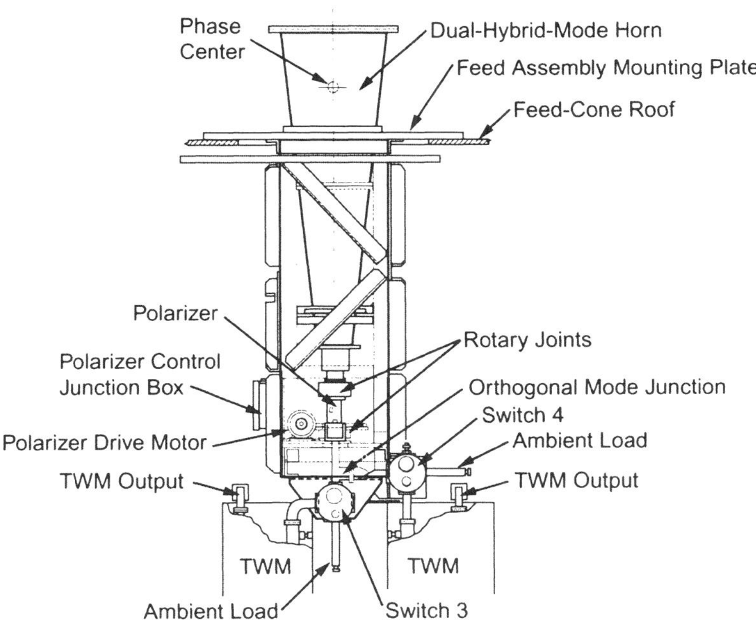


**Fig. 5-13. Mod III XRO feed assembly.**

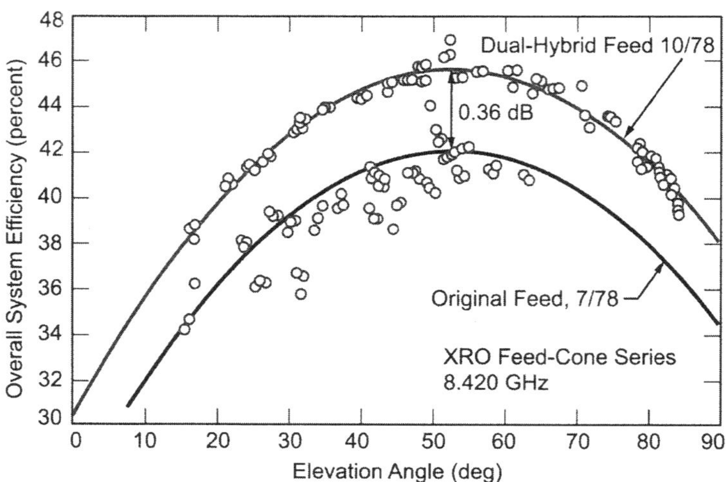


**Fig. 5-14.  Antenna system efficiency at X-band with
dual-hybrid-mode feed horn.**

## 5.5   Reflex–Dichroic Feed System

The tricone geometry was adequate to support single-band communication, but was unable to handle simultaneous multifrequency operations. This need arose in support of the Mariner 1973 X-band experiment, where a system was required to provide simultaneous low-noise reception at S- and X-bands and high-power transmission at S-band. Since S-band was still the primary frequency, the emphasis of the design was for minimal degradation at the S-band receive frequency.

A cross-sectional view of the reflex feed system [23,24] is shown in Fig. 5-15. The feed system made simultaneous use of both an X- and S-band feed horn. By using two reflectors—one an ellipse and the second a planar dichroic reflective at S-band—the effective S-band phase center was translated from its normal position in the S-band feed horn to a new point that nearly coincided with the X-band feed-horn phase center. This translation occurred as follows: since the ellipsoidal reflector was small in terms of wavelengths, the radiated energy focused short of the ellipsoid $F_2$ geometric focus, point C in Fig. 5-15.

The energy was then redirected by the planar reflector to the antenna subreflector. By the image principle, this redirected energy appeared to emanate from the point $F_x$, which is both the far-field phase center of the X-band feed horn and the subreflector focal point when it is aligned with the X-band feed-

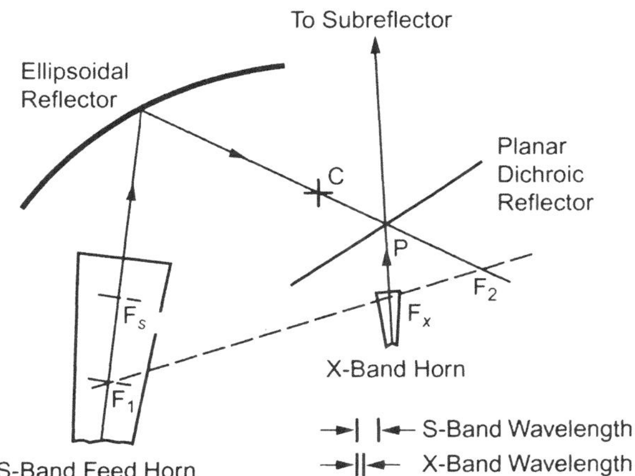


**Fig. 5-15. Reflex–dichroic feed system.**

horn position. To permit simultaneous X-band operation, the central region of the planar reflector was a thick plate filled with circular holes designed to make the reflector essentially transparent to X-band while still reflective to S-band.

The basic performance of the reflex–dichroic feed system is given in [25], from which the following explanation is abstracted.

Because of the asymmetric geometry, the S-band reflex feed patterns did not exhibit as high a level of symmetry as the corrugated waveguide feed horn taken alone. Spherical-wave techniques were used [26–28], to predict overall S-band performance. Because this feed design was based on a series of computer programs, it was considered necessary to experimentally check the predictions. A 1/7 scale-model program produced measured patterns in both polarizations essentially identical to those shown in Fig. 5-16. Also shown in Fig. 5-16 are the predicted scattered patterns of both the feed-horn ellipsoidal reflector, referenced to $F_s'$, and the horn ellipsoid and flat plate (simulating an opaque dichroic filter) taken together, referenced to $F_x$.

Table 5-3 summarizes the predicted gain performance of the S-band reflex feed and compares that with previous tricone performance, itself a slightly asymmetric system. Table 5-3 illustrates the familar trade-off in forward spillover and amplitude illumination loss, which is a function of feed-horn gain. In this case, the reduced forward spillover was a result of slightly higher gain for the horn/ellipsoid/flat-plate patterns than for the horn alone. Table 5-3 also shows the major penalty of asymmetric systems: radiated energy not contributing to forward antenna gain ($m \neq 1$) and resultant cross-polarization. In summary, the reflex system was expected to degrade S-band gain performance less than 0.06 dB, and negligibly impact the S-band noise temperature.

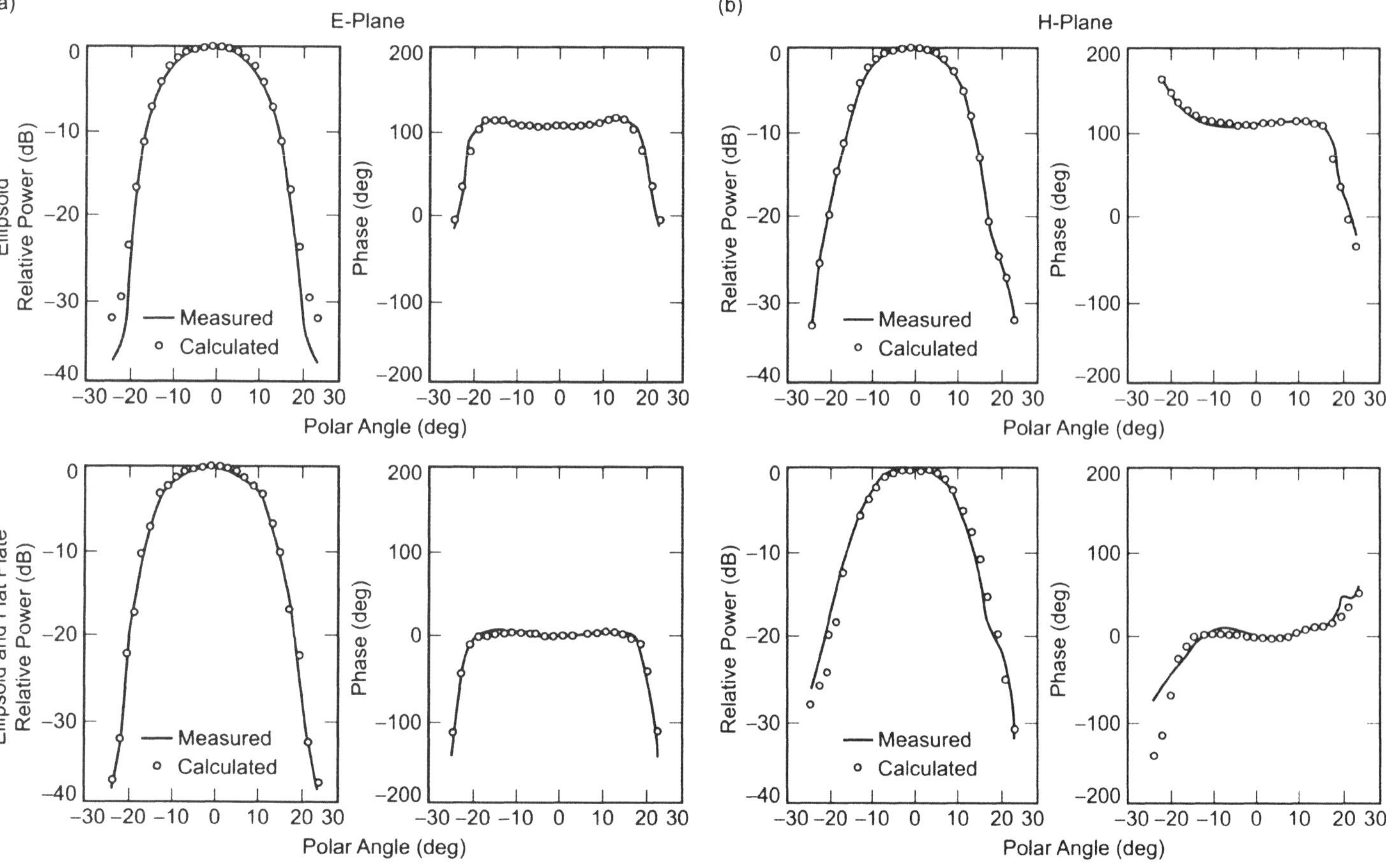

Fig. 5-16.  Calculated and measured low-band patterns: (a) E-plane and (b) H-plane.

**Table 5-3. Predicted S-band gain performance with tricone and reflex–dichroic feeds.**

| Gain Factors | Predicted Gain (dB) | | Notes |
| --- | --- | --- | --- |
| | Tricone | Reflex–Dichroic | |
| Forward spillover | −0.247 | −0.133 | None |
| Rear spillover | −0.011 | −0.010 | None |
| Amplitude illumination | −0.747 | −0.840 | None |
| Phase illumination | −0.088 | −0.070 | None |
| Cross-polarization | −0.001 | −0.005 | None |
| $m \neq 1$ energy | −0.009 | −0.089 | Due to asymmetric geometry |
| Central blockage | −0.262 | −0.260 | None |
| Quadripod blockage | −0.580 | −0.580 | None |
| Surface measurement | −0.093 | −0.093 | Main and subreflector (rms = 1.52 mm) |
| Surface tolerance | 0 | −0.010 | Ellipsoid (rms = 0.51 mm) |
| Surface tolerance | 0 | Negligible | Dichroic |
| Resistivity | 0 | −0.003 | None |
| Total | −2.038 | −2.093 | None |

For the X-band dichroic plate, preliminary designs were based on discussions with and previous work by individuals at The Ohio State University. That work was extended for circular polarization [29]. These filters were physically ultrathin (0.4 mm) and perforated with an array of reactively loaded close spaced slots. For this application, the original filters were found inadequate in the dissipation loss sense, providing 0.3-dB loss at X-band for aluminum construction. It should be noted that 0.3 dB of ambient-temperature dissipation, when added ahead of a 21-K total operating-noise-temperature receive system, results in a gain-over-temperature $(G/T)$ degradation of 3 dB due to doubling of the temperature term.

Intermediate filter designs increased bandwidth with attendant lower dissipation losses (2 to 4 K) and thicker construction (3 mm). These intermediate designs also suffered from mechanical tolerance sensitivity problems as well as probable severe degradation due to paint, dirt, and water droplet accumulation. Finally, due to the aspect ratio (3-mm thickness, 1-m diameter), and the geometrical constraints, a difficult thermal distortion problem was identified. When illuminated with 400-kW S-band, with a beam diameter of approxi-

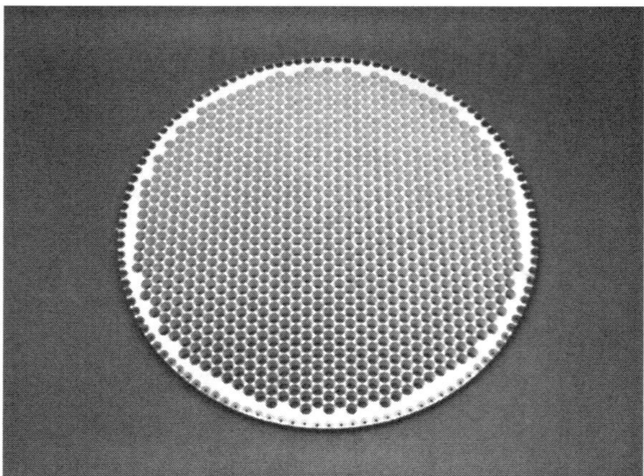

**Fig. 5-17. Half-wave-thick dichroic filter.**

mately 40 cm, the thin filters invariably heated and became bowed relative to S-band flatness requirements.

An excellent alternative to the thin, thermal-dependent-type dichroic filters was proposed and developed by Potter [30]. The solution was to obtain resonance by the simple means of a half-wave-thick array of dominant-mode cylindrical waveguide apertures. At X-band, this construction is 1.42-in. (3.6-cm) thick, with a calculated dissipation loss of 0.012 dB, using aluminum (Fig. 5-17). Measurements showed an operating system-noise-temperature increase of 1.2 K (0.01 dB) due to dissipation, with virtually no X-band pattern degradation. Bandwidth was approximately 100 MHz for the specific tilt angle and horn beamwidth (angular scan) employed.

Because the geometry shown in Fig. 5-15 tilts the dichroic filter at 30 deg off normal to the X-band feed-horn boresight, the E- and H-plane dichroic resonant frequencies were slightly split. Use of circular polarization resulted in different transmission phase shifts for orthogonal waves transversing the filter. The differential phase shift resulted in an on-axis X-band ellipticity of 1.8 dB, which, accepting a perfectly circularly polarized antenna at the other end of a link, produced an additional 0.04-dB transmission loss. Another result of differential resonant frequencies for the tilted array was an approximate −18-dB reflection. This had two effects: an additional 0.04-dB reflection loss and a slowly varying noise-temperature term for the complete antenna, due to scatter, which is a function of elevation angle. The magnitude of this additional noise was 0.4 to 1.2 K.

The total gain degradation at X-band was, therefore, 0.10 dB for a link employing circular polarization. The total noise-temperature degradation was 1.6 to 2.4 K. Accepting 2.0 K as an average, and referenced to a base operating noise temperature of 21 K, 0.4 dB was lost to noise, for a total $G/T$ reduction of 0.5 dB.

Tests were conducted on the 64-m antenna to include measurement of $G/T$ at both bands; boresight coincidence or beam coaxiality; and high-power transmission, at frequencies of 2.295, 8.415, and 2.115 GHz, respectively. Table 5-4 summarizes the measured performance of the full-scale prototype dual-band feed horn as installed at Goldstone, California, in January 1973.

**Table 5-4. Predicted and measured performance summary, reflex–dichroic feed.**

| Band and Gain or Temperature | Predicted | Measured | Notes |
| --- | --- | --- | --- |
| S-band gain | –0.055 dB | –0.03 dB | Measurement includes 0.045-dB pointing loss |
| S-band temperature | | | |
|     Above 30-deg elevation angle | 0.0 K | 0.0 K | |
|     Below 30-deg elevation angle | 0.0 K | 2.0 K | Improvement |
| X-band gain | –0.10 dB | –0.10 dB | Circular polarization |
| | –0.06 dB | –0.06 dB | Random polarization |
| X-band temperature | +1.6 to 2.4 K | +1.5 to 2.3 K | None / None |

Measured S-band gain indicated improved performance (+0.01 dB) when pointing on the S-band beam peak. When allowance for a 0.04-dB pointing loss was made (discussed below), S-band gain performance difference between reflex and non-reflex modes (0.03 dB) was difficult to distinguish. Because of the reduced forward spillover and the tricone geometry, S-band noise-temperature performance was improved. However, most of this improvement was due to relative feed-horn positions upon the tricone with respect to the ground and was not attributed to the new feed. This was so because the S-band radiation, in the reflex mode, effectively emanated from $F_x$ rather than $F_s$. X-band measured performance was in perfect agreement with predictions. The standard deviations in the gain difference determinations were approximately 0.02 dB and 0.10 dB for S- and X-bands, respectively. No deleterious thermal effects were observed with use of full uplink power.

Figure 5-18 is a view, facing the parabolic reflector relative to local vertical, of the RF beam positions in space. Four beams are shown: the singular S- and X-band nonreflex beams (subreflector individually focused and dual-band reflectors stowed) and the simultaneous reflex beams (subreflector

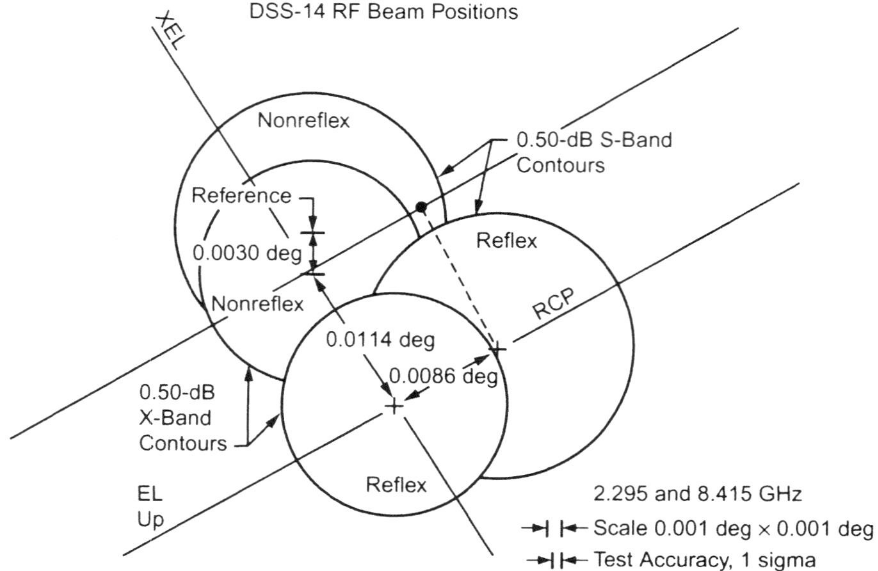


**Fig. 5-18. RF beam positions.**

focused to $F_x$ and dual-band reflectors functional). Several observations are worth noting. The singular S- and X-band beams were offset 0.003 deg, which is typical of DSN attained accuracy in feed-horn placement within the feed horns (±6 mm) and tricone subreflector repeatability. Use of the dichroic reflector produced a noticeably lower X-band reflex beam (0.011 deg); this resulted from refraction through the relatively thick dichroic plate. The 30- to 60-deg characteristic seen in Fig. 5-18 was traceable to the tricone geometry with respect to local vertical. Upon first installation, the S-band reflex beam was found to be displaced 0.009 deg to the right of the nonreflex beam. This effect resulted from use of circular polarization in the presence of the asymmetric geometry-induced cross-polarization mentioned above. Although not verified, it was predicted that use of the opposite-hand circular polarization would produce an S-band reflex-beam peak displaced an equal amount to the left of the nonreflex beam.

Figure 5-19 shows the full-scale reflex–dichroic feed on the 64-m antenna.


## 5.6  L-Band

In support of several international space exploration projects—the French/Soviet Vega mission to Venus (June 1985) and comet Halley flybys (March

**Fig. 5-19.  Reflex–dichroic feed on the 64-m antenna.**

1986)—JPL was asked in late 1983 to modify the DSN to receive L-band (1.668-GHz) telemetry used by the Soviet space program.

An extensive description of the Venus Balloon Project and L-band system requirements is given in [31]. The following requirements affected the design of the microwave system:

- Antennas must receive 1.668 GHz ± 5 MHz

- Antenna gain must be at least 58 dBi, or 50 percent efficiency on a 64-m antenna

- System-noise temperature $(T_{op})$ must be <35 K at zenith.

## 5.6.1    Design Approach

The required 58-dBi gain precluded the use of all DSN antennas except the three 64-m antennas. Since there was insufficient room for the feed horn in any of the existing feed cones, the solution was to suspend the feed horn in the area between the X-band receive-only (XRO) and the host-country feed cone (Fig. 5-20). The feed horn was cantilevered on a bracket bolted to the top of the XRO feed cone. As weight was then a critical concern, a sheet metal smooth-wall dual-mode horn 3.4 m long, with a 0.97-m aperture, was designed to be a light, low-loss, but narrowband (approximately ±40-MHz) solution with the illumination efficiency necessary to meet system requirements.

Due to the large aperture of the L-band feed horn, the phase center of the L-band feed lies some 60 cm (3.4 wavelengths) radially outward from the focus ring, which produces a small scan loss in beam-peak gain and beam-pointing

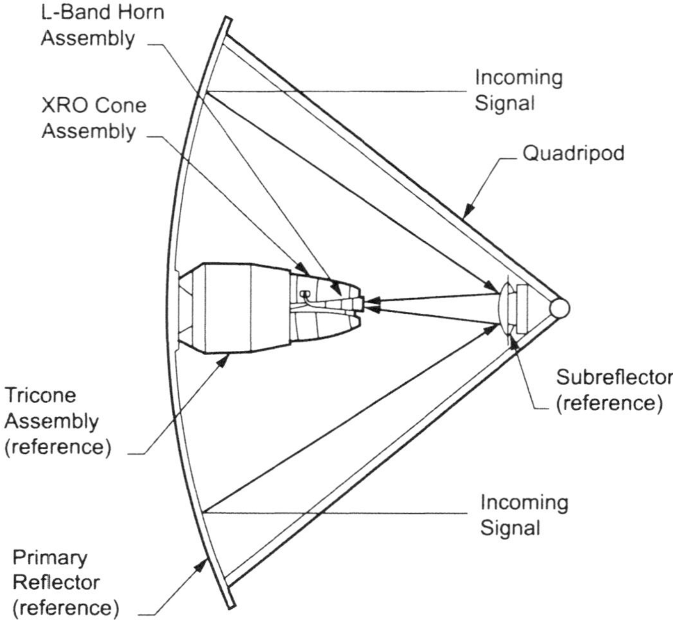



**Fig. 5-20. L-band feed-horn antenna position.**

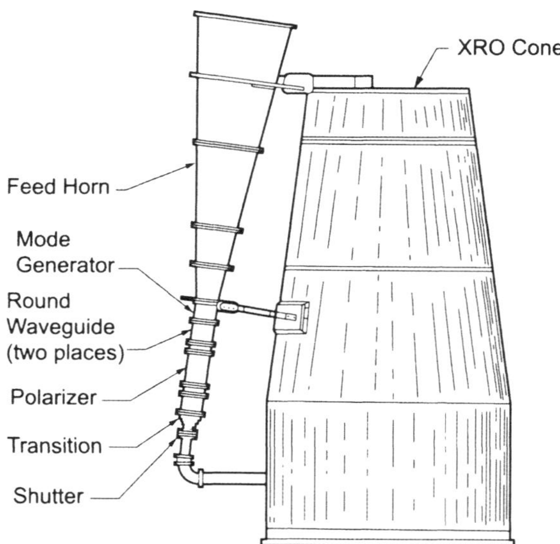



**Fig. 5-21. L-band feed-horn geometry.**

squint. Figure 5-21 shows the basic horn and antenna geometry used on the three 64-m antennas.

A narrowband (1.690 GHz ± 50 MHz) quarter-wave plate polarizer was used to meet the circular polarization requirement. Spare DSN WR 430 waveguide components and a WR 430 switch completed the microwave feed system.

Two completely redundant, cryogenically cooled (physically cooled to 14 K) L-band field-effect-transistor (FET) low-noise amplifiers (LNAs) provide the necessary preamplification. The FET LNAs were designed with 38 dB of gain and a usable bandwidth of about 200 MHz. A bandpass filter in front of the amplifiers is used to limit the bandwidth response of the FET LNA to about 100 MHz (1.668 GHz ± 50 MHz). This was done to prevent out-of-band noise (at 2.1 GHz) from the S-band transmitters of co-situated antennas in Spain and Australia and to prevent known radio frequency interference (RFI) threats from saturating the preamplifiers. (A signal level approaching –40 dBm may be enough to saturate these FETs.)

An L-band to S-band upconverter was used to convert the output of the L-band FETs to S-band. This allowed use of all station S-band receiver equipment necessary to meet Vega telemetry-processing requirements. The upconverter further limits the bandwidth of the L-band system to 10 MHz, fixing the total bandwidth of the overall L-band receive system at 1.668 GHz ± 5 MHz.

### 5.6.2 Performance Predictions and Measurements

The complete cooled FET system, with horn, polarizer, and waveguide, was assembled and tested to determine the microwave temperature contribution to the overall system $T_{op}$. The temperature contribution was determined to be approximately 10 K for the hardware and 14 K for the L-band FET LNA.

The measured values for gain, $T_{op}$, and scan offset on the three 64-m antennas are shown in Table 5-5. The $T_{op}$ values were obtained using the Y-factor method. The predicted noise-temperature component values at zenith, given in kelvins, are shown in Table 5-6.

**Table 5-5. Calculated and measured L-band microwave system efficiency, $T_{op}$, and scan offset.**

| Source | Efficiency at Approximately 45 deg of Elevation [percent (gain, dBi)] | $T_{op}$ (K) | Scan Offset (deg) |
|---|---|---|---|
| Calculated (PO) | 57 (58.5) | 33 | 0.260 |
| Measured at DSS-14 | 51 (58.2) | 33 | 0.260 |
| Measured at DSS-43 | 52 (58.2) | 36 | 0.260 |
| Measured at DSS-63 | 55 (58.4) | 34 | 0.260 |

**Table 5-6. Predicted noise temperature at zenith.**

| Parameter | Value (K) |
|---|---|
| Antenna temperature | 8.5 (cosmic plus sky plus spillover) |
| Feed components | 10.0 |
| FET LNA | 14.0 |
| Receiver follow-on contribution | 1.0 |
| Total | 33.5 |

A PO analysis predicted a scan offset angle of 0.26 deg, a half-power beamwidth of 0.19 ± 0.01 deg, with a slightly elliptically shaped beam (0.01-deg difference), and a peak gain of 59.46 dBi, or 70.6 percent efficiency. Additional antenna losses not included in the PO calculation, expressed in decibels and shown in Table 5-7, must be subtracted from the PO result.

Adding all losses, the PO-based prediction was that the scan axis gain peak should be 58.5 dBi for an efficiency of 57 percent. This is compatible with the measured data. The lower efficiency at DSS-14 may be due to saturation by RFI as these measurements were made before the full extent of FET saturation by RFI was understood. Some gain nonlinearity caused by saturation may account for the lower efficiencies. Further details on design and performance may be found in [32], along with a block diagram of the L-band receive-only system.

### 5.6.3  L-Band System Modifications

The L-band system successfully supported the French/Soviet Vega mission to Venus as well as the comet Halley flybys. The 64-m antennas were upgraded to 70-m, and the L-band system was modified to include a C-band transmit capability to support the Russian Phobos missions. The C-band uplink was

**Table 5-7. Additional antenna losses.**

| Source | Loss (dB) |
|---|---|
| Surface rms (97 percent) | 0.13 |
| Spar and subreflector blockage (88 percent) | 0.56 |
| Feed dissipation losses (98 percent) | 0.09 |
| Feed-mode losses (96 percent) | 0.18 |
| Total additional loss | 0.96 |

designed to handle a maximum power of 15 kW, while the L-band channel is receive-only. The C-band feed consists of a disc-on-rod antenna, which is located in the center of the existing L-band horn. The design of the feed is given in [33] and the performance measurements in [34]. After Phobos mission support, the C-band transmitter was removed from the antenna and the original L-band–only feed restored. The L-band receive-only system is currently in use to support radio science.

## 5.7   The Upgrade from 64 Meters to 70 Meters

The Voyager spacecraft flyby of Neptune in 1989 required a substantial enhancement in DSN performance, and a portion of that performance enhancement was accomplished by the 70-m upgrade project. The project goal was to improve performance by 1.9 dB at X-band (8.4 GHz). This improvement was to be obtained by increasing the antenna diameter to 70 m, increasing the RF efficiency through shaping of the reflective surfaces, improving the main reflector surface accuracy through the use of high-precision panels, and redesigning the subreflector support structure (quadripod) to reduce shadowing on the main reflector.

The increase in diameter involved reinforcement of the elevation wheel and the inner 34-m-diameter reflector backup structure, replacement of the reflector outer radial ribs beyond this inner 34-m segment, and replacement of all the surface panels.

The new surface panels were fabricated and installed to within a 0.65-mm rms deviation from the theoretically desired shape. These panels were fabricated using an adhesive bonding technique that achieved individual panel surface tolerance of better than 0.12 mm. In addition to high performance at X-band, these precision panels will support effective operation at Ka-band (32 GHz)—the next higher frequency band allocated for future deep-space communications.

The new subreflector measured 8 m in diameter and weighed approximately 6000 kg. It was machined as a single piece of aluminum to within a surface tolerance of 0.15 mm.

The mechanical installation and erection work at all three antennas was followed by a period of measurement and adjustments of the surface panels to reduce the overall dish-surface alignment tolerance to a 0.5-mm goal. The initial surface setting resulted from installing the panels on adjustable standoffs, using optical theodolite measurements. In accordance with the project plan, this provided an effective performance slightly in excess of that of the original 64-m antenna. Then, a series of microwave holography measurements were obtained in conjunction with Eikontech Ltd., which "mapped" the antenna surface, precisely determining deviations from theoretical. The deviation calculations were subsequently used to adjust the panel positions [35].

### 5.7.1    Design and Performance Predictions

The basic elements of the 64- to 70-m upgrade are shown in Fig. 5-22. The upgrade consisted of both mechanical and RF improvements aimed at providing a 1.9-dB gain improvement at X-band and a 1.4-dB gain improvement at S-band, while suffering no adverse effects due to wind and gravity deformations [36]. A discussion of each element of the upgrade follows.

**5.7.1.1  The 70-Meter Extension.** The 64-m main reflector radial ribs were removed, beginning at a radius of approximately 15 m, and replaced with entirely new ribs to a final radius of 35 m. This nearly total replacement was driven by preassembly and field erection requirements to minimize time out of service. The physical area increase of the resulting instrument provides a +0.8-dB frequency-independent performance increase.

**5.7.1.2  Structural Stiffenings.** The original 64-m structural design, accomplished in the very early 1960s, was modeled on a computer and optimized based on a quarter-section analysis. This necessary limitation reflects machine capacity capabilities of that time. For the design of the upgrade, it was possible to model a half-section. Because of the twin elevation wheels necessary in a design employing a separately founded master equatorial pointing reference and control platform (rising to the intersection of azimuth and elevation axes), the symmetric half-section computer model was superior to the quarter model. Accordingly, some structural bracketing within the inner 15-m radius zone (so-called box girder) was identified as beneficial in further limiting gravitational

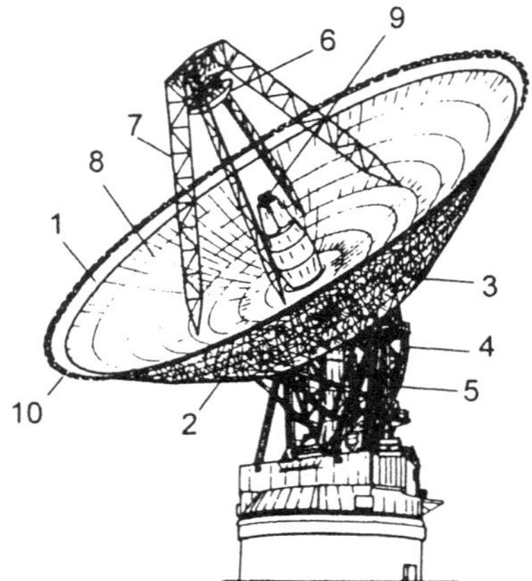


1)   Extend from 64 m to 70 m
2)   Replace and extend radial ribs
3)   Reinforce reflector backup structure
4)   Reinforce elevation wheel
5)   Add more counterweights
6)   Install new shaped subreflector
7)   Install new quadripod
8)   Install new shaped-surface panels
9)   Modify microwave feed
10)  Add 1- to 2-m S-band noise shield
     (never implemented)

**Fig. 5-22.  64-m to 70-m antenna extension.**

distortions as a function of elevation angle. At X-band and at 10- and 80-deg elevation angles, the improvement was approximately +0.3 dB.

**5.7.1.3  Improved Reflecting Panels.** The existing 64-m reflecting panels were measured to be approximately 0.9-mm rms, leading to an X-band loss factor of nearly 0.5 dB. Considerably better manufacturing and installation accuracies were possible, with 0.25 mm or better achievable for the manufacturing portion. Although conventional civil engineering (theodolite) techniques were used initially to provide an X-band-quality surface, a microwave holographic technique was later applied, using strong available sources in geosynchronous Earth orbit, to fine-tune reflecting-panel settings. The combination of improved panel manufacturing and setting accuracies increased X-band gain by +0.5 dB.

**5.7.1.4  Automated Two-Axis Focusing.** For many years, DSN 64-m X-band operations were conducted by first manually activating the axial (z) focus, as elevation angle was changed, and later by automating this function. Conceptually, this maintained the subreflector rear focus in necessary juxtaposition with the more limber and moving main reflector focus, as a function of elevation angle. A careful structural analysis, involving main reflector best-fit paraboloid focus together with quadripod feed support and subreflector unit deflections, identified the value of lateral (y-axis or gravity-vector) focusing as well as axial focusing. However, lateral focusing produced a beam shift of the final ensemble radiation pattern, necessitating a pointing feedback correction. A new microprocessor-based table lookup/compare/feedback machine was developed for this purpose. The performance benefits at X-band were similar in nature and magnitude and additive to those achieved by the structural stiffening activity mentioned above.

**5.7.1.5  Improved Quadripod Feed Support.** In order to accommodate a different (shaped) subreflector, significant changes were necessary in the apex region of the existing quadripod structure. Again, based on the important constraint for short out-of-service time, it became optimal to have a completed all-new structure that could be rapidly installed. Further, having the system on the ground enabled preassembly and test of the automated two-axis focusing subsystem. Because of the upgrade, advantage was taken to provide a reduced shadow (blocking) design. An anticipated improvement of 0.3 dB was calculated for this change.

**5.7.1.6  Optional Main Reflector Noise Shield.** In the initial design tradeoffs, a noise shield to improve S-band performance was considered but deleted due to economic considerations.

**5.7.1.7 Shaped Dual-Reflector Design.** Techniques have been available since the mid-1960s to synthesize by geometrical optics (GO) uniform aperture illumination of a two-reflector antenna. However, the 70-m antenna presented a difficult shaping problem because of the asymmetric geometry and the RF design constraints required by economic considerations.

To minimize feed costs, it was necessary to leave the basic feed geometry unchanged; that is, both the cone tilt angle and the height of the feed had to remain the same. The solution had to best-fit to the existing 64-m contour so the basic inner structure would remain the same. The main reflector was to be symmetrical and 70 m in diameter. The basic design trade-offs considered the subreflector size, a subreflector flange versus a noise shield, and an asymmetric versus a symmetric subreflector [37]. Furthermore, since shaping uses GO, which is independent of frequency, it was necessary that all the feeds have the same radiation pattern. All the feeds on the antenna matched the standard 22-dB horn pattern, except the X-band feed, which had been converted to a dual multimode horn design to obtain 0.36 dB over the standard design. However, since the shaping converted any radiation pattern into uniform illumination, it was decided to convert the X-band feed horn back to a standard 22-dB horn pattern and base the shaping upon the standard pattern.

The first step of the basic synthesis technique was to determine the main reflector shape using main and subreflector symmetric synthesis. This design then was optimized for X-band gain and noise temperature [38]. An important parameter was the subreflector illumination angle, which determined the subreflector size. Table 5-8 delineates the subreflector size trade-off. An illumination angle of

**Table 5-8. Theoretical 22-dB feed-horn pattern used to shape and scatter. Option selection is shaded in gray.**

| Parameter | Options Studied | | | | |
|---|---|---|---|---|---|
| Subreflector illumination angle (deg) | 15 | 16 | 17 | 18 | 19 |
| Subreflector diameter (m) | 7.1 | 7.68 | 8.26 | 8.84 | 9.45 |
| S-band (2.295 GHz) | | | | | |
|     Efficiency (percent) | Not calculated | 90.2 | 90.4 | 91.4 | 90.4 |
|     Noise temperature (K) | Not calculated | 3.6 | 3.6 | 4.5 | 5.3 |
| X-band (8.45 GHz) | | | | | |
|     Efficiency (percent) | 91.1 | 92.4 | 92.5 | 90.6 | 85.9 |
|     Noise temperature (K) | Not calculated | 0.67 | 0.52 | 0.6 | 1.8 |

16 deg was chosen (versus 17 deg for the 34-m antenna design) to minimize the subreflector size while achieving nearly optimum performance. Second, a new subreflector design was computed by putting the feed point in the offset position and synthesizing an asymmetric subreflector by ray tracing for perfect phase [39], using the previously determined main reflector. The basic geometry of the synthesis is shown in Fig. 5-23. Since only phase synthesis was used, the amplitude illumination was slightly tilted, but the loss due to the tilt was negligible [40].

To minimize the cost of the implementation, the X-band dual-hybrid-mode horn was not directly replaced with a standard 22-dB feed horn because the phase center location of the standard feed horn would require moving the input flange and associated equipment. Rather, a standard-design feed horn was mated with an input flange and the phase center moved to the proper location through the use of a feed-horn extension termed a stovepipe design [41]. There was a slight degradation in gain, but the loss was deemed acceptable to reduce the cost. As shown in Table 5-9, detailed design expectations were 75.8 percent aperture efficiency at S-band and 71.1 percent at X-band.

### 5.7.2  S- and X-Band Performance

The measured S-band efficiency was 75.6 percent ± 0.8 percent. Compared with previous performance, a 1.91-dB improvement was obtained at S-band. For X-band, the three antennas had an efficiency of 68.7 percent ± 1.5 percent [42]. Compared with previous 64-m performance, an average improvement of 2.13 dB at X-band was obtained. Figure 5-24 shows the measured X-band 70-m gain performance for the three antennas, together with the nominal design expectation.

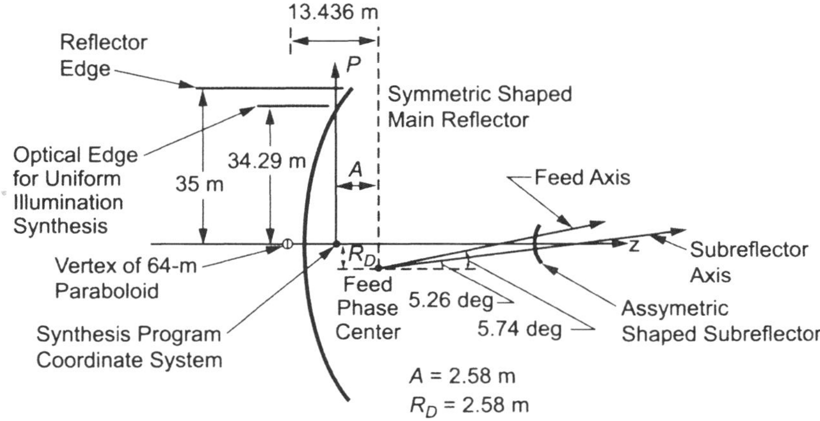


**Fig. 5-23. 70-m antenna geometry used in physical optics analysis.**

**Table 5-9.  Design-expected S-/X-band 70-m efficiency performance.**

| Gain Factors | S-Band | X-Band | Notes |
|---|---|---|---|
| Rear spillover | 0.994 | 0.997 | None |
| Forward spillover | 0.959 | 0.964 | None |
| Illumination amplitude | 0.959 | 0.982 | Standard feed horn |
| Illumination phase | 0.994 | 0.989 | None |
| Cross-polarization | 1.00 | 1.00 | None |
| $m \neq 1$ modes | 0.980 | 0.996 | None |
| Central blockage | 0.983 | 0.988 | None |
| Quadripod blockage | 0.9008 | 0.9008 | 5.1 percent effective |
| Dichroic reflectivity | 0.9993 | N/A[a] | Below plane reflector |
| Surface reflectivities | 0.998 | 0.998 | None |
| Surface tolerances | 0.9946 | 0.9527 | None |
| Pointing squint | 0.996 | 1.000 | X-band beam peaked |
| Waveguide dissipation | 0.9795 | 0.984 | None |
| Waveguide VSWR | 0.9908 | 0.9908 | None |
| Dichroic VSWR | N/A | 0.990 | None |
| Compromise feed horn | N/A | 0.975 | Stovepipe extension |
| Total | 0.758 (+63.28 dBi at 2.295 GHz) | 0.711 (+74.34 dBi at 8.42 GHz) | None |

[a] Not applicable.

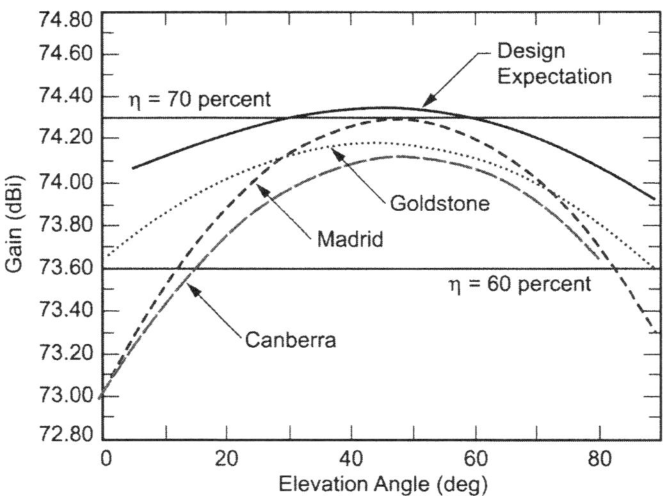


**Fig. 5-24.  X-band measured gain performance
for DSN 70-m antennas.**

In addition to the large improvement in aperture efficiency, somewhat lower total system operating noise levels were obtained: approximately 1.5- and 0.7-K improvements at S- and X-bands, respectively. The best measurements at S-band show a 18.3-K total at zenith and about a 21-K total at 30-deg elevation. At X-band, all three antennas are quite uniform at 20.9 K ± 0.3 K at zenith and about 25.5 K at 30-deg elevation. Figure 5-25 shows system-noise levels with elevation angle. The two upper curves are total operating system-noise levels (with atmosphere) while the lower curves have the nominally clear–dry atmosphere numerically removed. The lower curves are indicative of acceptably small antenna spillover and scattering-noise components.

A very important element of this project was the application of the microwave holographic technique. Holographic imaging was achieved by use of 12-GHz-band geostationary satellite beacon signals providing approximately 75-dB signal-to-noise ratio (SNR) on beam peak, with over 40-dB SNR in the reference channel. Imaging was used for adjustment of the nearly 1300 individual reflector panels on each antenna. Additionally, holographic imaging provided data for quantifying the aperture illumination uniformity, the blockage and diffraction phenomena, the rotating subreflector alignment quality, gravity loading, and excellent estimates of the achieved surface tolerance following adjustment. At high lateral resolution (0.42 m), the surface tolerance achievable, presuming infinite ability to adjust, is found to be 0.3-mm rms. This is essentially the panel manufacturing precision together with a holography data system noise component. It was believed that the achievable surface tolerance was about 0.4-mm rms for these antennas. However, due to time constraints,

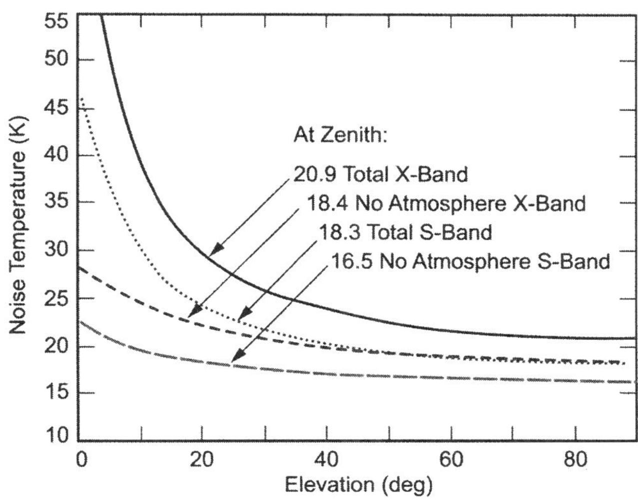



**Fig. 5-25. 70-m antenna system-noise level.**

the goal was to meet only the project specification, which was 0.65-mm rms at 0.42-m resolution. Effectively, surface adjustments were not refined for the last available 0.1 dB at X-band. A photograph of the 70-m antenna at Goldstone is shown in Fig. 5-26.

### 5.7.3    Ka-Band Performance

In 1989, a 32-GHz receiver was installed on the 70-m antenna at Goldstone. The primary purpose of the installation was to measure the high-frequency performance of the antenna after it was upgraded and enlarged from 64 m to 70 m. System calibrations were performed, and several radio sources were measured between March and July of 1989. The measurements were highly repeatable, which suggests that the precision of the data is about 2.5 percent. The peak of aperture efficiency function at 32 GHz is $0.35 \pm 0.05$ centered around a 45-deg elevation (Fig. 5-27). The estimated error (2 sigma) of the efficiency number was dominated by the uncertainty in the absolute flux-density calibration. Further details can be found in [43].

### 5.7.4    Adding X-Band Uplink

Prior to 1999, the X-band system on the 70-m antenna was receive-only. X-band uplink (7.145–7.190 GHz) had been available on DSN 34-m antennas since the mid-1980s (starting with the 34-m high-efficiency antenna described in Chapter 6 of this monograph). However, in 1995, due to the advent of

**Fig. 5-26. The 70-m antenna after the 64- to 70-m upgrade.**

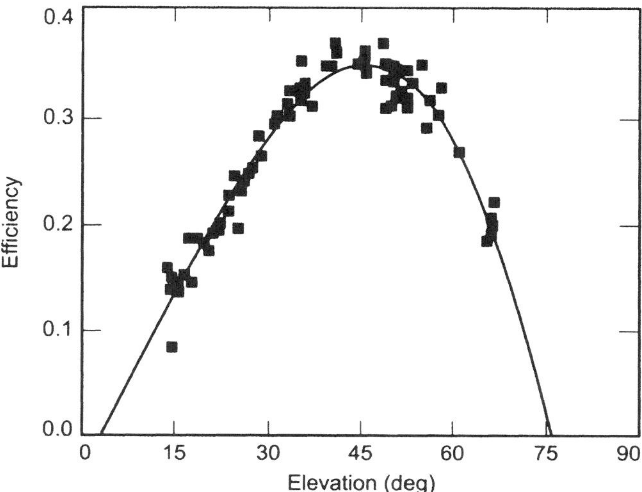



**Fig. 5-27. The 70-m antenna's 32-GHz efficiency
as a function of elevation angle.**

smaller spacecraft requiring higher data rates and the need for high-power
X-band for emergency situations, the National Aeronautics and Space Administration (NASA) funded JPL to implement a 20-kW X-band uplink on all 70-m
antennas. There were two difficulties with X-band uplink on the 70-m antenna:
(a) the existing dichroic plate could only pass the downlink frequency, so a new
dichroic plate design was required that could pass both the uplink and downlink
frequencies, and (b) the additional noise temperature from a traditional diplexer
implementation might be as large as 9 K. To circumvent the additional noise
temperature caused by a traditional diplexer, a new ultralow-noise–diplexed
X-band microwave system [44], called the X-/X-band (X/X) diplexing feed
was developed for the 70-m antenna uplink.

**5.7.4.1 New Deep Space Network X/X Diplexing Microwave Feed.** An excellent discussion of the benefits of the new feed versus a traditional DSN
diplexer can be found in [44]. The significant advantage comes from the ability
to cryogenically cool most of the components in the downlink signal path.

Figure 5-28 shows a simplified block diagram of the X/X diplexing microwave feed. A diplexing junction is used for combining the uplink and downlink
signals. This diplexing junction is a six-port device. Two of the ports use circular waveguides. One is connected to a 22-dB feed horn through a coupler, and
the other is connected to the LNA package through a transmitter reject filter
and an ambient-load sliding switch. The filter provides 40 dB of rejection for
the fundamental transmitter frequency. The ambient-load switch is used for calibration of the LNAs. The other four ports, which are in rectangular

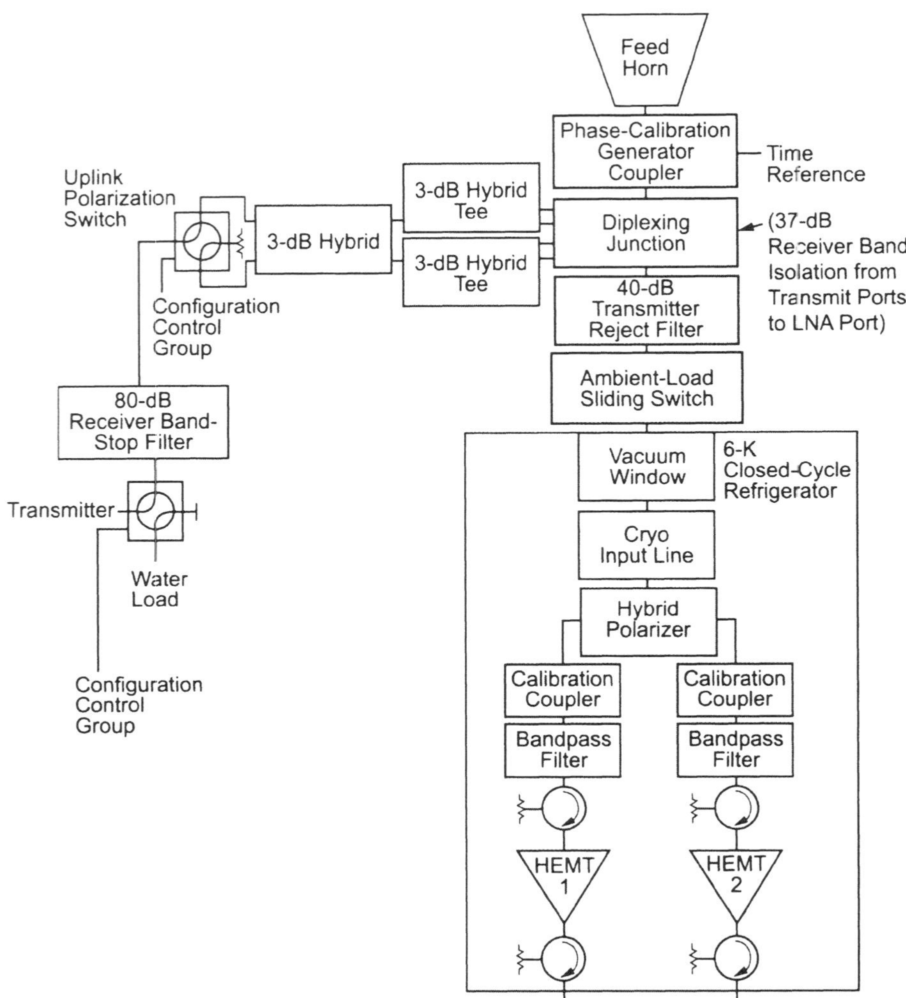



**Fig. 5-28. Simplified block diagram of the X/X diplexing feed.**

waveguides, are used for injection of the uplink signal. These ports are combined into one port with a network of hybrids, tees, and a polarization selection switch. For additional protection against the spurious signal generated by the transmitter, an 80-dB absorptive filter is used between the polarization switch and the transmitter.

In the downlink path, the signal coming out of the ambient load switch is guided to the LNA package. The LNA package consists of a cryogenically cooled container that holds the rest of the downlink path components. These components include a polarizer and orthomode junction to convert the circularly polarized signal to linear polarization, additional bandpass filters for protection of the super-sensitive high-mobility-electron transistor (HEMT) LNAs, and circulators for improving the match and reducing the standing wave generated between the reflective filters in the downlink path. One LNA is used for the RCP signal, the other for the LCP signal.

The noise contribution of the microwave circuit in this feed is approximately 2.7 K for diplexed operations. Table 5-10 shows the breakdown of the predicted and measured noise temperature of the individual components of this feed at 8.45 GHz.

**Table 5-10. Noise-temperature breakdown of X/X diplexing feed components, in kelvins.**

| Item | Predicted Noise Temperature | Measured Noise Temperature |
| --- | --- | --- |
| Feed horn | 1.0 | 0.8 |
| Phase-calibration-generator coupler | 0.3 | 0.3 |
| Diplexing junction and filter | 1.7 | 1.4 |
| Ambient load switch | 0.2 | 0.2 |
| LNA | 6.0 | 8.4 |
| Total | 9.20 | 11.1 |

**5.7.4.2 S-/X-Band Dichroic Mirror.** Prior to 1999, the 70-m antennas supported downlink only. Therefore the dichroic mirror utilized in the reflex–dichroic feed system was a narrowband design capable of reflecting both the S-band transmit and receive frequencies while only passing the X-band downlink band. With the new diplexed microwave feed, the antenna needed a wideband dichroic plate that would pass the X-band uplink frequencies (7.145–7.190 GHz) as well as the downlink frequencies, with very little loss.

A new dichroic plate that uses cross-shaped holes to provide low loss and minimizes grating lobes for both the uplink and downlink bands was designed [45]. Figure 5-29 shows the perforated section of this dichroic plate. It was fabricated in hard, oxygen-free copper, using a wire electric discharge machine.

Cross-shaped holes can be packed tighter than circular holes and, therefore, it is possible to design a wideband dichroic plate that can support X-band uplink and downlink bands without undesired grating lobes. Figure 5-30 shows the theoretical and measured power reflection coefficient of this mirror for a circularly polarized signal incident at a 30-deg angle with respect to the normal to the plane of the mirror [45]. As can be seen, the mirror is designed to have two resonant frequencies: one at the center of the uplink band, the other at the center of the downlink band. There is excellent agreement between the theoret-

**Fig. 5-29. The wideband S-/X-band dichroic mirror.**

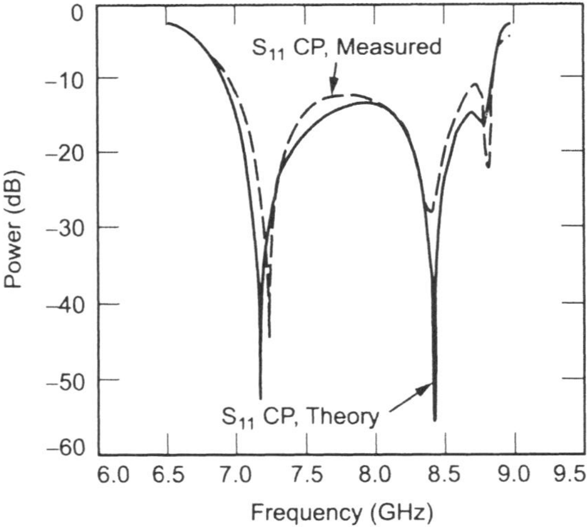

**Fig. 5-30. Power reflection coefficient: comparison of measurement and theory for circular polarization (CP).**

ical and measured values. The error between the theoretical and measured resonant frequencies is less than 1 percent. However, when the feed cone containing the dichroic plate was tested on the ground prior to installation on the antenna, there was a significantly higher noise temperature than initially anticipated. This higher noise temperature was later confirmed by measurements on the dichroic plate alone [46].

The noise-temperature contribution of the dichroic plate is the result of two effects: (a) ohmic loss as the signal propagates through the holes in the plates and (b) the mismatch or scattering loss of energy that is reflected from the plate surface. The ohmic loss tends to be small, contributing less than 1 K.

The scattering loss from the rear surface of the plate is quite large and would contribute 12 to 18 K of noise if the scattered energy were to see ambient temperature. Fortunately, when located on the 70-m antenna, the rear scatter is largely reflected to the sky and causes a small (1-K) contribution. The noise-temperature contribution can be used to estimate the antenna efficiency loss due to the plate. A 12 to 18 K contribution corresponds to an efficiency loss of 4 to 6 percent (0.2 dB). This is consistent with the measurements of the antenna efficiency with and without the plate in place.

Since the dichroic plate is only required for simultaneous S- and X-band operation, the plate was made retractable, as shown in Fig. 5-31, to provide for high-performance X-band-only operation.

### 5.7.4.3 Radio Frequency Performance.

A 20-kW X-band uplink capability was installed on all three 70-m antennas: at Goldstone and Canberra in 2000 and at Madrid in 2001. The intent was to have minimal impact at all other frequencies, and this was accomplished utilizing the new retractable dichroic plate and X-/X-band feed. The critical X-band receive performance is summarized in Table 5-11 for noise temperature and Table 5-12 for gain. All three antennas are capable of transmitting 20 kW on the uplink.

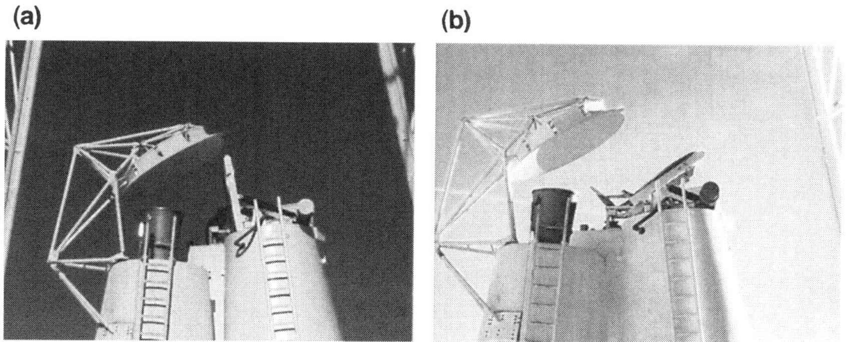


**Fig. 5-31. The dichroic plate retraction mechanism:
(a) retracted and (b) extended.**

**Table 5-11. X-band downlink performance post-installation: noise temperature, in kelvins.[a]**

| Operation Mode ($T_{op}$ [SNT[b], in vac, at zenith, K]) | Predicted | DSS-14 | | | | DSS-43 | | | | DSS-63 | | | |
|---|---|---|---|---|---|---|---|---|---|---|---|---|---|
| | | Measured at LNA Input | | Measured at Horn Aperture | | Measured at LNA Input | | Measured at Horn Aperture | | Measured at LNA Input | | Measured at Horn Aperture | |
| | | LNA-1 | LNA-2 | LNA-1 | LNA-2 | LNA-1 | LNA-2 | LNA-1 | LNA-2 | LNA-1 | LNA-2 | LNA-1 | LNA-2 |
| X-band-only mode, transmitter off | 14.2 | 14.22 | 14.22 | 14.35 | 14.35 | 14.57 | 14.57 | 14.7 | 14.7 | 13.76 | 14.05 | 13.89 | 14.18 |
| S-/X-band mode, transmitter off | 15.4 | 15.12 | 15.12 | 15.26 | 15.26 | 15.77 | 15.78 | 15.92 | 15.93 | 15.03 | 15.11 | 15.18 | 15.26 |
| Noise temperature increase, transmitter on | <0.1 | <0.2 | <0.2 | N/A[c] | N/A | <0.2 | <0.2 | N/A | N/A | <0.2 | <0.2 | N/A | N/A |

[a]Courtesy of Manuel Franco.
[b]SNT = system noise temperature, vac = vacuum.
[c]Not available.

Table 5-12. X-band downlink performance at LNA-1 post-installation: gain, in dBi.[a]

| Elevation Angle (deg) | Predicted | DSS-14 | | | | DSS-43 | | | | DSS-63 | | | |
| | | Measured at LNA Input | | Measured at Horn Aperture | | Measured at LNA Input | | Measured at Horn Aperture | | Measured at LNA Input | | Measured at Horn Aperture | |
| | | Dichroic Plate | | Dichroic Plate | | Dichroic Plate | | Dichroic Plate | | Dichroic Plate | | Dichroic Plate | |
| | | Retracted | Extended | Retracted | Extended | Retracted | Extended | Retracted | Extended | Retracted | Extended | Retracted | Extended |
|---|---|---|---|---|---|---|---|---|---|---|---|---|---|
| 10 | 74.0 | 74.04 | 73.84 | 74.08 | 73.88 | 74.25 | 73.98 | 74.29 | 74.02 | 74.01 | 73.57 | 74.05 | 73.61 |
| 25 | 74.2 | 74.22 | 74.02 | 74.26 | 74.06 | 74.47 | 74.20 | 74.51 | 74.24 | 74.46 | 73.98 | 74.50 | 74.02 |
| 45 | 74.2 | 74.30 | 74.10 | 74.34 | 74.14 | 74.55 | 74.28 | 74.59 | 74.32 | 74.67 | 74.19 | 74.71 | 74.23 |
| 80 | 73.9 | 74.04 | 73.84 | 74.08 | 73.88 | 74.11 | 73.84 | 74.15 | 73.88 | 73.96 | 73.61 | 74.00 | 73.65 |

[a]Courtesy of Manuel Franco.

## 5.8   Distortion Compensation

As is well illustrated in Fig. 5-27, a problem with Ka-band is the significant loss in efficiency at low and high elevation angles due to the gravity-induced deformations of the primary antenna surface.

There are a number of techniques that can be employed to compensate for large-scale surface errors in large reflector antennas. One approach to gravity compensation is to install actuators at the corners of each main reflector panel or at the panel junctions. The actuators can either be controlled open loop if there is a prior knowledge of the surface errors or in a closed-loop system if the surface is actively measured. This approach has been implemented on the clear-aperture 100-m antenna at Greenbank, West Virginia [47].

A second approach for gravity compensation is to install a deformable sub-reflector. This approach has been implemented at MIT's Haystack antenna and is one of the proposed solutions for the 25-m ARISE (Advanced Interferometry between Space and Earth) antenna operating at 86 GHz [48].

Since the early 1990s, extensive work [49–53] has been performed at JPL on the use of a deformable mirror to correct for the gravity-induced distortions on a large reflector antenna. This work culminated in a demonstration of a deformable mirror on the NASA/JPL 70-m antenna in early 1999 [54,55]. The deformable mirror, nominally a flat plate, is placed in the beam path and deformed in order to compensate for the gravity-induced distortions as the antenna moves in elevation. Actuators on the mirror are driven via a lookup table. Values in the lookup table are derived using the distortions, ray tracing, and a structural finite element model of the mirror system.

During the same period, an array-feed compensation system (AFCS) was also studied extensively at JPL as an alternative to the deformable mirror approach for gravity compensation [56–58]. The system consists of a small array of feed horns, LNAs, downconverters, and digital-signal-processing hardware and software for optimally combining the signals received by the feed horns.

A combined system consisting of a deformable flat plate (see below) and the AFCS was also demonstrated on the 70-m antenna. The combined system worked better than either one of the systems acting alone [59,60].

A comprehensive study was undertaken [61,62] that considered each of the possible options for implementing Ka-band on the existing 70-m antennas. The study described the advantages and disadvantages of each approach and estimated the relative cost and RF performance.

### 5.8.1   Deformable Flat Plate

The deformable flat plat (DFP) is a deformable mirror designed to correct gravity-induced antenna distortions on the main reflector surface by correcting wavefront phase errors. An experimental DFP (Fig. 5-32) was built with actuators laid out to enable gravity correction for either the 70-m or the DSS-13 34-m DSN antennas [51]. The DFP surface was constructed from a 6000-series aluminum sheet, 0.04-in. (1-mm) thick and 27 in. (68.6 cm) in diameter, nominally producing a flat surface. The plate is driven by 16 actuators, via a lookup table, that change as a function of antenna elevation angles. Open-loop calibration of the DFP is achieved by following these steps:

1) Model the RF distortions of the main reflector (base this model upon measured holography maps of the 70-m antenna obtained at three elevation angles); process the holography maps, using an appropriate mechanical model of the gravity distortions of the main reflector to provide surface-distortion maps at all elevation angles (this technique is further described in [63])

2) From the surface-distortion maps, use ray optics to determine the theoretical shape of the DFP that will exactly phase-compensate the distortions

3) Using the theoretical shape and a NASTRAN mechanical model of the plate, select the actuator positions that generate a surface that provides the best rms fit to the theoretical model (using the actuator positions and the NASTRAN model provides an accurate description of the actual mirror shape, permitting the performance of the DFP to be determined [55]).

## 5.8.2 Array-Feed Compensation System

The AFCS comprises a feed array and cryogenic dewar, downconverter assembly, and digital-signal-processing assembly (DSPA). The feed array con-

**Fig. 5-32. Experimental DFP.**

sists of seven identical, smooth-walled Potter horns, each 1.732 in. (4.4 cm) in diameter, providing 22-dBi gain at 32 GHz. The wall of each feed horn is tapered near the top to facilitate close packing, minimizing the loss of signal energy between horns. The cryogenic front end contains three-stage HEMT LNAs that provide 25 dBi of total gain, thus establishing high SNR without adding significant receiver noise to the signal. Each HEMT LNA outputs an amplified RF signal at 32-GHz center frequency, which is then translated to 300-MHz intermediate frequency (IF) by the seven-channel downconverter assembly and transmitted to the DSPA. The DSPA estimates the combining weights in real time and combines the weighted digital samples to produce a "combined output" with improved SNR. A photograph of the feed array prior to installation in the cryogenic dewar is shown in Fig. 5-33.

### 5.8.3    The Array-Feed Compensation System–Deformable Flat-Plate Experiment

During the period from November 1998 through February 1999, a series of measurements was made on the 70-m antenna at DSS-14 to determine the performance characteristics of two systems designed to compensate for the effects on antenna gain of elevation-dependent gravity distortion of the main reflector. The AFCS and DFP were mounted in the same feed cone, and each was used

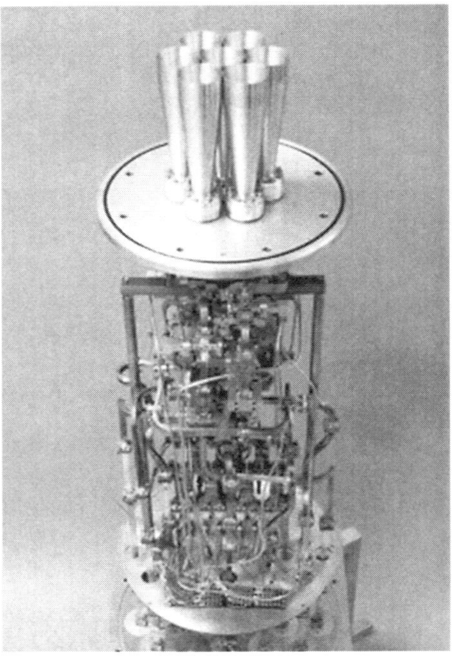

**Fig. 5-33. Feed array for the AFCS.**

independently as well as jointly to measure and improve the antenna aperture efficiency as a function of elevation angle. The experimental data is presented in [54,58] and the theoretical formulation in [55].

The experiment geometry is shown in Fig. 5-34. The main elements are the 70-m main reflector and subreflector, a refocusing ellipse, the DFP, and the receive feed system. The focal point of the dual-reflector 70-m system is labeled $F_1$, and the focal point where the feed is placed is labeled $F_2$. The antenna Cassegrain focus was 0.6 in. (1.5 cm) above $F_1$, which was corrected by z-axis alignment of the subreflector. Looking at the system in the transmit mode, the output of the feed system is refocused at $F_1$, the input to the dual-reflector system. The parameters of the ellipse are chosen to map the fields (with no magnification) from $F_2$ to $F_1$. Hence, the performance of the 70-m system would be the same if the same feed were placed at either $F_1$ or $F_2$. Experiments were done with the AFCS placed at both $F_1$ and $F_2$. When the AFCS was placed at $F_2$, combined AFCS–DFP performance measurements were possible.

#### 5.8.3.1 AFCS Antenna Deformation Compensation.

The New Millennium Program spacecraft, Deep Space 1 (DS1), provided a stable Ka-band signal throughout the holography cone experiments. During a typical DS1 track, it

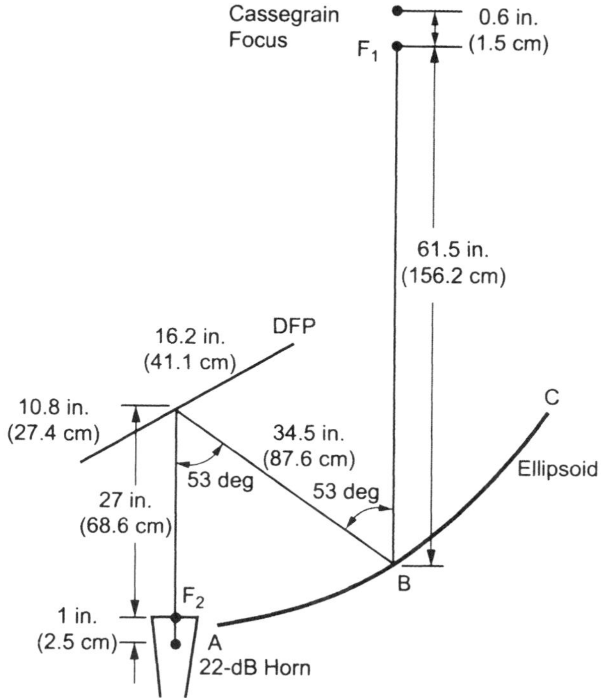


**Fig. 5-34. Experimental geometry for the 70-m experiment.**

was necessary to update antenna pointing in real time, using the coherent version of the array-feed boresighting algorithm [58]. In addition to providing pointing information, this algorithm records the complex signal voltages at the end of each integration time in all seven channels as well as in the combined channel, measures rms noise voltages, and updates optimum combining weights. This data can be used to compute the combining gain, $G_c$, which is defined as the ratio of the SNR of the combined channel to the SNR of the central channel, and is an important quantity for characterizing AFCS performance.

A representation of combining gain data recorded during the DS1 holography cone experiment is shown in Fig. 5-35. Note that above 60-deg elevation, the data appears to split into two branches, indicating an azimuth dependence in the combining gain: the upper branch corresponds to the ascending part of the track (before transit), and the lower branch corresponds to its descent towards the horizon (after transit). This asymmetrical behavior of the antenna has been observed on all occasions, but the exact cause is not well understood at this time. The third-order trend line indicates average gain at a given elevation, without taking into account the azimuth dependence of the gain curve at higher elevations. Also shown on the curve is the predicted combining gain, using the holography-determined distortions and the computed feed-horn pattern.

**5.8.3.2 DFP Compensation.** As indicated above, the computation of the required deformable-mirror surface begins with a map of the main-reflector distortions to be corrected for at the elevation angle of interest. The final output

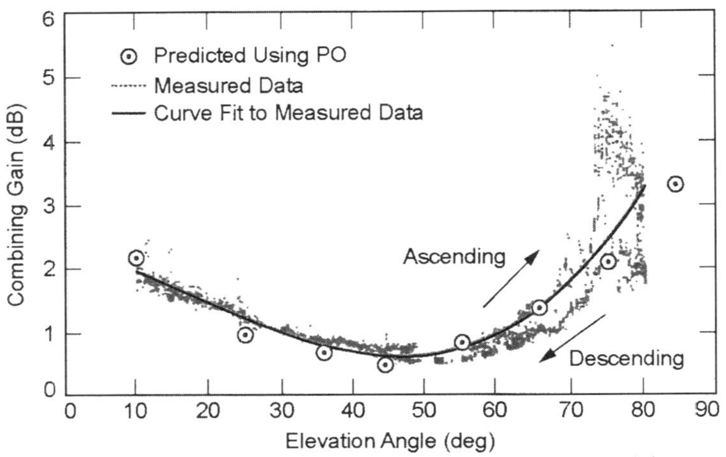


**Fig. 5-35. Computed and measured performance of the AFCS on the 70-m DSS-14 antenna at 32 GHz.**

of the design process is the actuator settings required to correct for these distortions.

The gravity deformations of the antenna at all elevation angles are derived from holography measurements taken at only three elevations: 47.2, 36.7, and 12.7 deg [63]. The computations assume that

- The antenna structural response due to gravity loading is linear
- The antenna response to gravity loading is symmetric relative to the antenna elevation-rigging angle
- The elevation-rigging angle (maximum gain) of the antenna can be accurately inferred from total power radiometry (TPR) efficiency measurements.

Note that in the absence of measurement noise, three is the minimum number of angle observations needed to solve for the three unknowns in the linear model used.

In the usual case where noise is present, it is desirable to acquire measurements at more than three elevation angles so that a least-squares solution can be obtained. Conversely, if only three measurements can be made, making two of them near the extreme values of 0 and 90 deg will produce the most accurate results. Due to the constraints of time and the highest viewable angle (47 deg) available from this antenna location, neither of these conditions could be met at the time of the holography cone experiment at DSS-14; therefore, the second and third conditions listed above were added to the derivation.

A typical holography distortion map is shown in Fig. 5-36(a). There is a significant random component in the surface that cannot be compensated for with a smoothly varying DFP surface. Consequently, a 91-term Zernike polynomial description of the main reflector antenna was used to design the DFP. The Zernike representation for the main reflector surfaces is shown in Fig. 5-36(b) for the 15-deg elevation. The difference from the full holography map representation of the dish and the Zernike representation is shown in Fig. 5-36(c). This component of distortion will not be corrected by the DFP.

The values in the actuator lookup table were computed by performing a ray trace from the deformed main reflector, represented by Zernike polynomials, to the DFP position and using the structural finite element model of the mirror [55]. This process was repeated at 5-deg-elevation increments to derive 17 sets of 16 actuator displacements. The performance of the DFP was assessed by tracking natural radio sources and obtaining antenna temperature using TPR [54]. Positive compensation was achieved over all elevation angles, with average compensations at Ka-band of 1.7 and 2.3 dB, at elevation angles of 10 and 80 deg, respectively. These results successfully demonstrate an open-loop calibration of the DFP (Fig. 5-37).

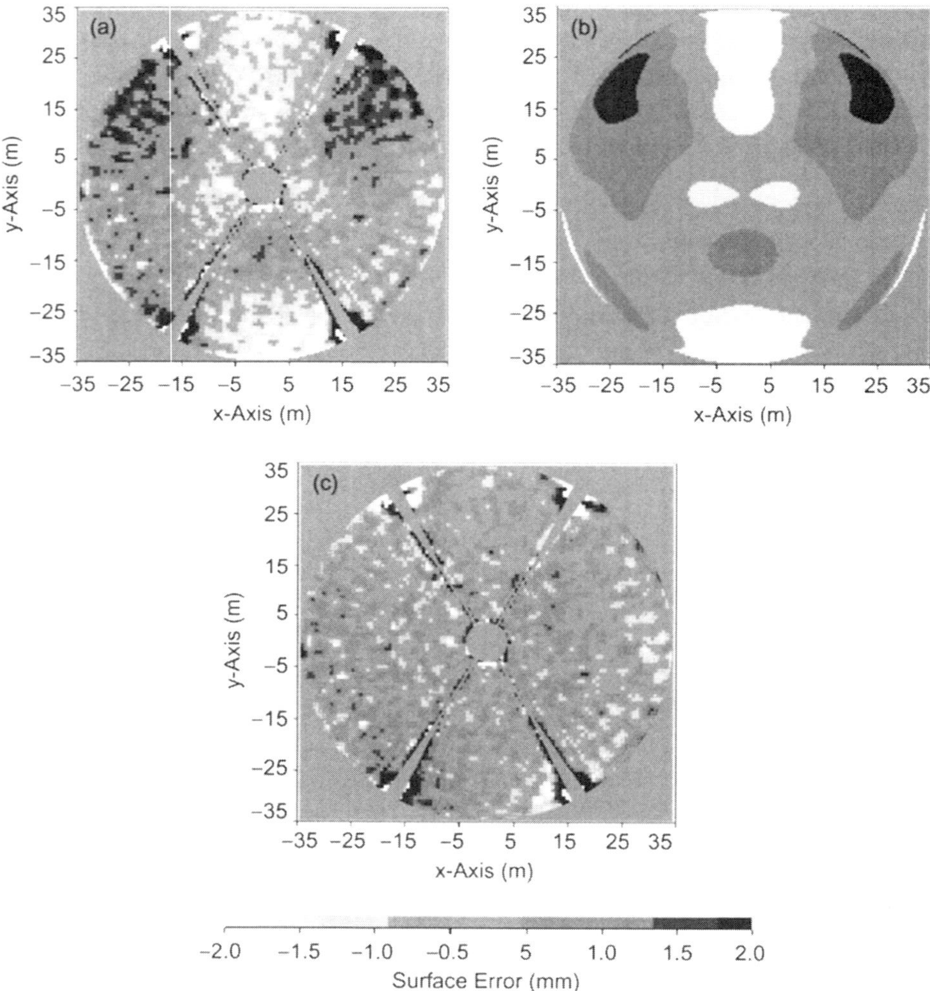

**Fig. 5-36. Main reflector distortion at an elevation of 15 deg: (a) full holography, (b) Zernike polynomial representation, and (c) the difference between the two surfaces.**

After completion of the AFCS–DFP experiment, another method (ranging theodolite) was used to measure the main-reflector distortion. This time, the measurements were made at widely scattered elevation angles (13, 30, 47, 68, and 88 deg), and thus it was not necessary to assume the second and third conditions listed above. These measurements also served as a verification of the assumptions. As might be expected, these holography-derived results at high elevation angles were less accurate than the results at angles below 47 deg. Although it was too late to use the theodolite data to recompute the DFP actua-

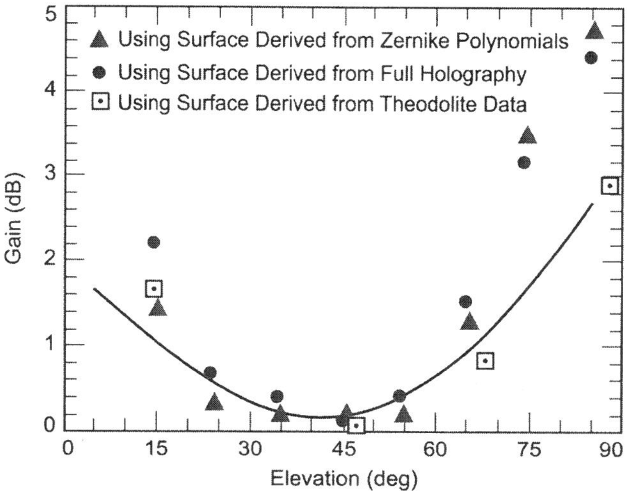


**Fig. 5-37. Computed and measured performance
of the DFP on the 70-m DSS-14 antenna at 32 GHz.**

tor positions, a more accurate assessment of the expected performance was made using the more accurate characterization of the distortion at high elevation angles. Also shown in Fig. 5-37 is the predicted performance using the main-reflector surface derived from the theodolite measurements. As can be inferred from the data, if the more accurate main-reflector distortion was used to compute the actuator positions, better performance would have been achieved at higher elevation angles, using the DFP.

### 5.8.3.3  Joint AFCS–DFP Performance Measurements.

As we have shown, operating either the AFCS or the DFP singly provides some gravity compensation over all elevation angles. However, the performance of a combined AFCS–DFP system was significantly better than each one performing individually. Analysis of the data obtained on day of year (DOY) 056 (Table 5-13), when DS1 was tracked continuously as it descended from an elevation of approximately 50 deg to 8.5 deg, clearly demonstrates that an AFCS–DFP compensation system working together in real time can recover most of the signal energy lost to gravitational deformations on the 70-m antenna at Ka-band. The various gains used in Table 5-13 are defined as follows:

$G_{AFCS} = G_c$ is the improvement over the central channel when the DFP is flat

$G_{DFP}$ = the increase in signal power in the central channel when the DFP is activated

$G_{joint}$ = the improvement over the uncorrected central channel when the DFP and the AFCS operate jointly to recover SNR losses

**Table 5-13. Average combining gain of AFCS, DFP,
and AFCS–DFP on DOY 056, in decibels.**

| Elevation (deg) | $G_{AFCS}$ | $G_{DFP}$ | $G_{joint}$ | $\Delta$AFCS | $\Delta$DFP |
|---|---|---|---|---|---|
| 8.5 | 2.1 | 1.8 | 2.9 | 1.1 | 0.8 |
| 13 | 1.8 | 1.6 | 2.5 | 0.9 | 0.7 |
| 16.5 | 1.6 | 1.4 | 2.2 | 0.8 | 0.6 |
| 23 | 1.2 | 1.1 | 1.8 | 0.7 | 0.7 |
| 28 | 1.1 | 0.9 | 1.5 | 0.6 | 0.4 |
| 33 | 0.75 | 0.6 | 1.0 | 0.4 | 0.25 |
| 38 | 0.5 | 0.3 | 0.6 | 0.3 | 0.1 |

$\Delta$AFCS = the contribution of the AFCS to $G_{joint}$ over that of the DFP acting alone

$\Delta$DFP = the contribution of the DFP to $G_{joint}$ over that of the AFCS acting alone.

The computed versus measured results for the combined DFP–AFCS are shown in Fig. 5-38. There are several curves shown on the figure. The lower solid curve is the measured baseline efficiency. The calculated results for the combined DFP–AFCS, using the full holography maps are shown in circles. The measured performance of the combined DFP–AFCS system is shown, but only data for lower elevation angles was obtained. Observe that the measured and calculated values agree to within a few percent. It should be noted that the measured efficiency curve includes a substantial random component. This random-component loss is estimated to be on the order of 3.4 dB. This estimate was derived by looking at the difference between the gain calculated by using a Zernike polynomial representation of the surface (smooth component) compared to the gain calculated using the surface derived from holography (containing smooth and random components). Also observe that the combined system recovers almost all of the gain lost due to the systematic distortion but does not recover the random component of the surface distortion. It is anticipated that a majority of the random component will be recovered by resetting the main reflector.

Also shown in Fig. 5-38 is the expected performance if a so-called "perfect DFP" were used. A perfect DFP means that the required shape of the plate as determined by the Zernike polynomial expansion of the main-reflector surface is replicated exactly. With an improved DFP, it should be possible to get close to the required surface and produce performance close to this prediction. The perfect DFP does not recover the random component of the distortion, as this distortion is not even included in its determination.

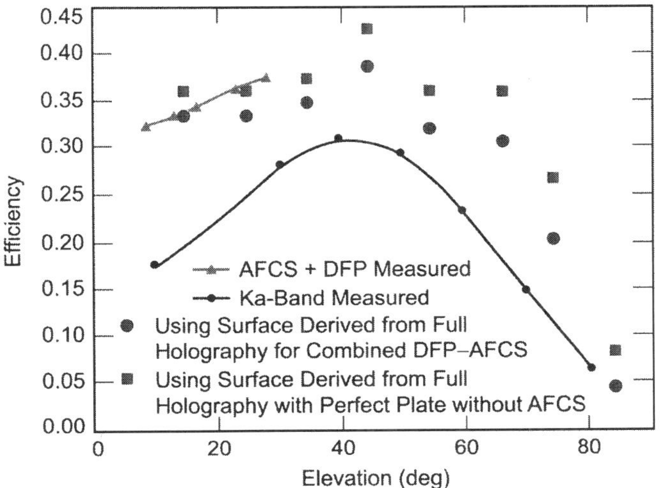



**Fig. 5-38. Computed and measured performance
of the combined AFCS–DFP system.**

### 5.8.4   Projected Ka-Band Performance

The present aperture-efficiency performance of the DSS-14 antenna at Ka-band is marginal as a result of two independent effects: loss at the rigging angle due to the rms surface error of the main reflector and losses due to gravity-induced deformation (see Fig. 5-39, the lower curve, "Ka-Band Current"). The marginal rigging-angle aperture efficiency of 31 percent is mostly due to the main reflector rms error of 0.73 mm. It is expected that this rms error can be reduced to between 0.35 and 0.45 mm, thus increasing the overall antenna aperture efficiency to between 55 and 60 percent at the rigging angle (see Fig. 5-39, the upper curve, "Ka-band + PNL Set"). The curve labeled "AFCS + DFP Current" in Fig. 5-39 shows the gravity performance of the antenna with joint AFCS–DFP compensation obtained from Table 5-12, while the curve "AFCS + DFP + PNL Set" shows the expected efficiency performance when compensation is applied to the antenna after panel setting. Thus, with appropriate panel setting and gravity compensation benefits from both systems, the DSS-14 antenna can achieve excellent performance at Ka-band, thereby enabling the predicted 6-dB communications improvement relative to X-band.

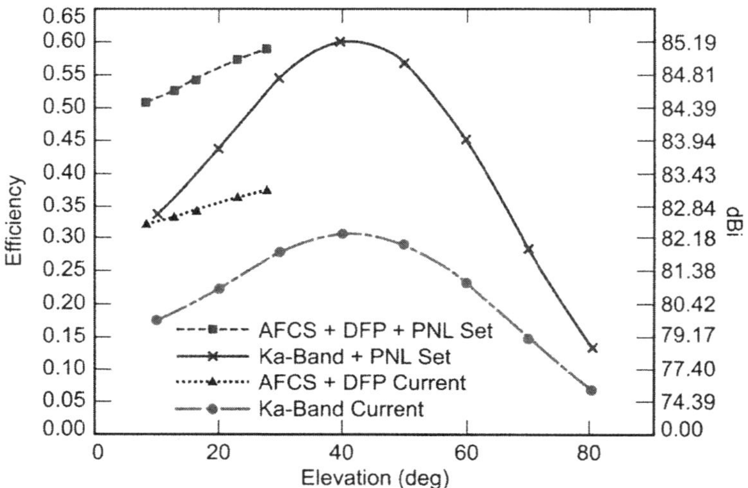

**Fig. 5-39.  Performance of the 70-m DSS-14
antenna, 32 GHz, present and future (predicted).**

## 5.9  Future Interests and Challenges

The large DSN antennas saw a significant evolution in performance during the last three decades of the 20th century. Size was increased from 64 to 70 m, efficiencies were enhanced through dual-reflector shaping, operational frequencies were upgraded from S-band to X-band, and a decision was made to add Ka-band. There was also a growth from single-frequency to multifrequency operation. On the other hand, as the designs have matured, it has become more and more difficult to enhance antenna performance. For example, it is unlikely that antennas will grow physically larger; rather, when needed, additional aperture will be achieved by arraying, that is, by combining the signals from several other DSN antennas or enlisting the aid of other non-DSN radio astronomy antennas. It is also unlikely that X-band efficiency will increase much beyond its present performance. The most significant improvement will likely be made by moving operational frequencies from X-band to Ka-band. The most significant Ka-band problem is with 70-m antenna gain performance versus elevation angle, where the drop-off is large (>5 dB). Several techniques for compensating for this drop-off have been described in this chapter. The next logical step will be to implement one of the compensation systems on the 70-m antenna, thus enabling its use at Ka-band.

# References

[1] Technical staff, Tracking and Data Acquisition Organization, *The NASA/ JPL 64-Meter-Diameter Antenna at Goldstone, California: Project Report*, Technical Memorandum 33-671, Jet Propulsion Laboratory, Pasadena, California, July 1974.

[2] P. D. Potter, "The Application of the Cassegrain Principle to Ground Antennas for Space Communications," *IRE Transactions on Space Electronics and Telemetry*, vol. SET-8, pp. 154–158, June 1962.

[3] P. D. Potter, *The Design of a Very High Power, Very Low Noise Cassegrain Feed System for Planetary Radar*, Technical Report 32-653, Jet Propulsion Laboratory, Pasadena, California, August 1964.

[4] P. D. Potter, "A New Horn Antenna with Suppressed Sidelobes and Equal Beamwidth," *Microwave Journal,* vol. VI, no. 6, pp. 71–78, June 1963.

[5] P. D. Potter, *A Simple Beamshaping Device for Cassegrainian Antennas*, Technical Report 32-214, Jet Propulsion Laboratory, Pasadena, California, January 1962.

[6] P. Foldes and S. G. Komlos, "Theoretical and Experimental Study of Wideband Paraboloid Antenna with Central Reflector Feed," *RCA Review,* vol. XXI, pp. 94–116, March 1960.

[7] D. A. Bathker, *Radio Frequency Performance of an 85-ft Ground Antenna: X-Band*, Technical Report 32-1300, Jet Propulsion Laboratory, Pasadena, California, July 1968.

[8] D. A. Bathker, *Radio Frequency Performance of a 210-ft Ground Antenna: X-Band*, Technical Report 32-1417, Jet Propulsion Laboratory, Pasadena, California, December 1969.

[9] A. J. Freiley, *Radio Frequency Performance of DSS-14 64-m Antenna at 3.56- and 1.96-cm Wavelengths*, Technical Report 32-1526, vol. XIX, November and December 1973, http://tmo.jpl.nasa.gov/progress_report/ issues.html Accessed August 2001.

[10] W. V. T. Rusch, "Scattering from a Hyperboloidal Reflector in a Cassegrainian Feed System," *IEEE Transactions on Antennas and Propagation,* vol. AP-11, no. 4, pp. 414–421, July 1963.

[11] W. V. T. Rusch, "Phase Error and Associated Cross-Polarization Effects in Cassegrainian-Fed Microwave Antennas," *IEEE Transactions on Antennas and Propagation,* vol. AP-14, no. 3, pp. 266–275, May 1966.

[12] STAIR (Structural Analysis Interpretive Routine), Lincoln Manual 40, Massachusetts Institute of Technology Lincoln Laboratory, Lexington, Massachusetts, March 1962.

[13] J. Ruze, *Physical Limitations on Antennas*, Technical Report 248, Research Laboratory of Electronics, Massachusetts Institute of Technology, Cambridge, Massachusetts, October 1952.

[14] S. A. Brunstein, "A New Wideband Feed Horn with Equal E- and H-plane Beamwidths and Suppressed Sidelobes," in *Space Programs Summary 37-58, Vol. II, The Deep Space Network*, pp. 61–64, Jet Propulsion Laboratory, Pasadena, California, July 1969.

[15] H. C. Minnett, B. Thomas, and A. Mac, "A Method of Synthesizing Radiation Patterns with Axial Symmetry," *IEEE Transactions on Antennas and Propagation,* vol. AP-14, no. 5, pp. 654–656, September 1966.

[16] A. C. Ludwig, "Antennas for Space Communication," in *Space Programs Summary 37-33, Vol. IV, Supporting Research and Advanced Development*, pp. 261–266, June 30, 1965.

[17] R. F. Thomas, D. A. Bathker, "A Dual Hybrid Mode Feedhorn for DSN Antenna Performance Enhancement," *The Deep Space Network Progress Report 42-22*, vol. May and June 1964, http://tmo.jpl.nasa.gov/progress_report/issues.html Accessed August 2001.

[18] R. Hartop, "X-Band Antenna Feed Cone Assembly," *The Deep Space Network Progress Report*, November and December 1973, Technical Report 32-1526, vol. XIX, http://tmo.jpl.nasa.gov/progress_report/issues.html Accessed August 2001.

[19] R. Hartop, "Selectable Polarization at X-Band," *The Deep Space Network Progress Report 42-39*, March and April 1977, http://tmo.jpl.nasa.gov/progress_report/issues.html Accessed August 2001.

[20] R. Hartop, "New X-Band Microwave Equipment at the DSN 64-meter Stations," *The Deep Space Network Progress Report 42-48*, vol. September and October 1978, http://tmo.jpl.nasa.gov/progress_report/issues.html Accessed August 2001.

[21] R. W. Hartop, "New X-Band Antenna Feed for DSN 64-meter Stations," *The Deep Space Network Progress Report 42-52*, vol. May and June 1979, http://tmo.jpl.nasa.gov/progress_report/issues.html Accessed August 2001.

[22] A. J. Freiley, "Radio Frequency Performance of DSS-14 64-m Antenna at X-Band Using a Dual Hybrid Mode Feed," *The Deep Space Network Progress Report 42-53*, July and August 1979, http://tmo.jpl.nasa.gov/progress_report/issues.html Accessed August 2001.

[23] P. D. Potter, *S- and X-Band RF Feed System*, Technical Report 32-1526, vol. VIII, November and December 1972, http://tmo.jpl.nasa.gov/progress_report/issues.html Accessed August 2001.

[24] A. C. Ludwig, *Calculation of Scattered Patterns From Asymmetrical Reflectors*, Technical Report 32-1430, Jet Propulsion Laboratory, Pasadena, California, February 15, 1970.

[25] D. A. Bathker, "Dual Frequency Dichroic Feed Performance," Proceedings 139, AGARD Conference on Antennas for Avionics, June 1974.

[26] A. C. Ludwig, "Spherical Wave Theory," in *The Handbook of Antenna Design, Volumes 1 and 2* (A. W. Rudge, K. Milne, A. D. Olver, and P. Knight, editors), London: Peter Peregrinus Ltd., pp. 101–123, 1986.

[27] P. D. Potter, "Application of Spherical Wave Theory to Cassegrainian-Fed Paraboloids," *IEEE Transactions on Antennas and Propagation*, vol. AP-15, no. 6, pp. 727–736, November 1967.

[28] A. C. Ludwig, "Near-Field Far-Field Transformations Using Spherical-Wave Expansions," *IEEE Transactions on Antennas and Propagation*, vol. AP-19, no. 2, pp. 214–220, March 1971.

[29] R. T. Woo, "A Low-Loss Circularly Polarized Dichroic Plate," IEEE G-AP 1971 *International Symposium Digest*, pp. 149-152.

[30] P. D. Potter, *S- and X-Band Feed System*, Technical Report 32-1526, vol. XV, Jet Propulsion Laboratory, Pasadena, California, pp. 54–62, June 15, 1973.

[31] C. T. Stelzried, "The Venus Balloon Project," *Tracking and Data Acquisition Progress Report 42-80*, vol. October–December 1984, http://tmo.jpl.nasa.gov/progress_report/issues.html   Accessed February 13, 2002.

[32] J. Withington, "DSN 64-meter Antenna L-band (1668-MHz) Microwave System Performance Overview," *Tracking and Data Acquisition Progress Report* 42-94, vol. April–June 1988, http://tmo.jpl.nasa.gov/progress_report/issues.html Accessed February 13, 2002.

[33] P. H. Stanton and H. F. Reilly, Jr., "The L-/C-band Feed Design for the DSS-14 70-meter Antenna (Phobos Mission)," *Tracking and Data Acquisition Progress Report 42-107*, vol. July–September 1991, http://tmo.jpl.nasa.gov/progress_report/issues.html   Accessed February 13, 2002.

[34] M. S. Gatti, A. J. Freiley, and D. Girdner, "RF Performance Measurement of the DSS-14 70-meter at C-band/L-band," *Tracking and Data Acquisition Progress Report 42-96*, vol. October–December 1988, http://tmo.jpl.nasa.gov/progress_report/issues.html   Accessed February 13, 2002.

[35] D. J. Rochblatt, "A Methodology for Diagnostic and Performance Improvement for Large Reflector and Beam-Waveguide Antennas Using

Microwave Holography," *Proceedings of 1991 Symposium of Antenna Mechanical Techniques Association*, pp. 1-7–1-12, October 7–11, 1991.

[36] D. A. Bathker, A. G. Cha, W. A. Imbriale, W. F. Williams, "Proposed Design and Performance Analysis of NASA/JPL 70-M Dual Reflector Antennas," *IEEE AP-S International Symposium*, Boston, Massachusetts, June 1984.

[37] T. Veruttipong, V. Galindo-Israel, and W. A. Imbriale, "Low-Loss Offset Feeds for Electrically Large Symmetric Dual-Reflector Antennas," *IEEE Transactions on Antennas and Propagation,* vol. AP-35, no. 7, July 1987.

[38] W. F. Williams, "Considerations for Determining the Shaped Main Reflector for the DSN 70-m Upgrade Program," Telecommunications and Data Acquisition document 890-170, JPL D-1875 (internal document), Jet Propulsion Laboratory, Pasadena, California, November 1984.

[39] A. G. Cha, and W. A. Imbriale, "70-m Antenna Reflector Surface Computer Programs," JPL D-1843 (internal document), Jet Propulsion Laboratory, Pasadena, California, November 1984.

[40] A. G. Cha, "Physical Optics Analysis of NASA/JPL Deep Space 70-m Antennas," JPL D-1853 (internal document), Jet Propulsion Laboratory, Pasadena, California, November 1984.

[41] F. Manshadi and R. Hartop, "Compound-Taper Feedhorn for NASA 70-m Antennas," *IEEE Transactions on Antennas and Propagation,* vol. AP-36, no. 9, September 1988.

[42] D. A. Bathker, A. G. Cha, D. J. Rochblatt, B. C. Seidel, and S. D. Slobin, "70-Meter Deep Space Network Antenna Upgrade Performance," *Proceedings of ISAP*, 1989.

[43] M. S. Gatti, M. J. Klein, T. B. H. Kuiper, "32-GHz Performance of the DSS-14 70-meter Antenna: 1989 Configuration," *Telecommunications and Data Acquisition Progress Report 42-99*, vol. July–September 1989, http://tmo.jpl.nasa.gov/progress_report/issues.html  Accessed August 2001.

[44] Manshadi, F., "The New Ultra Low Noise Diplexed X-band Microwave Feed for NASA 70-m Antennas," 2000 IEEE Aerospace Conference Proceedings, Big Sky, Montana, March 18–25, 2000.

[45] Epp, L. W., Stanton, P. H., Jorgenson, R. E., and Mittra, R., "Experimental Verification of an Integral Equation Solution for Thin-Walled Dichroic Plate with Cross-shaped Holes," *IEEE Trans. Antennas and Propagation*, vol. 42, pp. 878–882, June 1994.

[46] Britcliffe, M. J. and Fernandez, J. E., "Noise-Temperature Measurements of Deep Space Network Dichroic Plates at 8.4 GHz," *Telecommunications and Mission Operations Progress Report 42-145*, vol. January–March

2001, http://tmo.jpl.nasa.gov/progress_report/issues.html   Accessed August 2001.

[47] R. J. Lacasse, Green Bank Telescope Memo 184, http://info.gb.nrao.edu/ GBT/GBT.html  Accessed August 2001.

[48] M. Djobadze, J. Kibler, and M. Pryor, "ARISE Subreflector Demonstration Article Final Report," Composite Optics, Inc., internal memorandum, November 23, 1999.

[49] V. Galindo-Israel, S. R. Rengarajan, W. Veruttipong, and W. A. Imbriale, "Design of a Correcting Plate for Compensating the Main Reflector Distortions of a Dual Shaped System," *IEEE Antennas and Propagation Society International Symposium*, Ann Arbor, Michigan, pp. 246–249, June 1993.

[50] W. A. Imbriale, M. Moore, D. J. Rochblatt, and W. Veruttipong, "Compensation of Gravity-Induced Structural Deformations on a Beam-Waveguide Antenna Using a Deformable Mirror," *IEEE Antennas and Propagation Society International Symposium 1995*, Newport Beach, California, pp. 1680–1683, June 1995.

[51] R. Bruno, W. Imbriale, M. Moore, and S. Stewart, "Implementation of a gravity compensating mirror on a large aperture antenna," AIAA Multidisciplinary Analysis and Optimization, Bellevue, Washington, September 1996.

[52] S. R. Rengarajan, W. A. Imbriale, and P. W. Cramer, Jr., "Design of a Deformed Flat Plate to Compensate the Gain Loss Due to the Gravity-Induced Surface Distortion of Large Reflector Antennas," International Symposium on Electromagnetic Theory, Thessaloniki, Greece, May 1998, pp. 124–126.

[53] S. R. Rengarajan and W. A. Imbriale, "A Study of a Deformable Flat Plate for Compensating Reflector Distortions," USNC URSI Meeting, Orlando, Florida, pp. 181, July 1999.

[54] P. Richter, M. Franco, and D. Rochblatt, "Data Analysis and Results of the Ka-Band Array Feed Compensation System/Deformable Flat Plate Experiment at DSS 14," *Telecommunications and Mission Operations Progress Report 42-139*, vol. July–September 1999, Jet Propulsion Laboratory, Pasadena, California, pp. 1–29, http://tmo.jpl.nasa.gov/progress_report/ issues.html  Accessed August 2001.

[55] W. A. Imbriale, and D. J. Hoppe, "Computational Methods and Theoretical Results for the Ka-Band Array Feed Compensation System/Deformable Flat Plate Experiment at DSS 14," *Telecommunications and Mission Operations Progress Report 42-140*, vol. October–December 1999, http:// tmo.jpl.nasa.gov/progress_report/issues.html  Accessed August 2001.

[56] V. Vilnrotter, E. Rodemich, and S. Dolinar, Jr., "Real-time Combining of Residual Carrier Array Signals Using ML Weight Estimates," *IEEE Transactions on Communications*, vol. 40, pp. 604–615, 1992.

[57] V. Vilnrotter, D. Fort, and B. Iijima, "Real-time Array Feed System Demonstration at JPL," in *Multifeed Systems for Radio Telescopes*, Astronomical Society of the Pacific Conference Series, vol. 75, pp. 61–73, 1995.

[58] V. Vilnrotter and D. Fort "Demonstration and Evaluation of the Ka-Band Array Feed Compensation System on the 70-Meter Antenna at DSS 14," *Telecommunications and Mission Operations Progress Report 42-139*, vol. July to September 1999, http://tmo.jpl.nasa.gov/progress_report/issues.html  Accessed August 2001.

[59] D. Rochblatt and V. Vilnrotter, "Demonstration of a Ka-band Array Feed-Deformable Flat Plate Compensation System," *TMOD Technology and Science Program News*, Issue 12, Jet Propulsion Laboratory, Pasadena, California, June 2000.

[60] W. Imbriale, D. Hoppe, and D. Rochblatt, "Analysis of the DFP/AFCS System for Compensating Gravity Distortions on the 70-meter Antenna," International Symposium for Deep Space Communications and Navigation, Pasadena, California, September 21, 1999.

[61] D. Hoppe, "70-meter Ka-band Implementation Study," JPL D-19434, Jet Propulsion Laboratory, Pasadena, California, January 10, 2000.

[62] W. A. Imbriale, and D. J. Hoppe, "Advances in the Deep Space Network—Adding Ka-band to the 70-meter Antenna," URSI meeting, Boulder, Colorado, January 2000.

[63] D. J. Rochblatt, D. Hoppe, W. Imbriale, M. Franco, P. Richter, P. Withington, and H. Jackson, "A Methodology for the Open Loop Calibration of a Deformable Flat Plate on a 70-meter Antenna," *Proceedings of the Millennium Conference on Antennas Propagation AP2000*, Davos, Switzerland, 2000.

# Chapter 6[1]
# Deep Space Station 15: Uranus—The First 34-Meter High-Efficiency Antenna

During the mid-1980s, The National Aeronautics and Space Administration (NASA's) Deep Space Network (DSN) introduced a new 34-m high-aperture-efficiency dual-shaped reflector antenna (see Chapter 1, Section 1.2.4, of this monograph) into the DSN (Fig. 6-1). The initial requirements were to provide a simultaneous receive-only capability at X-band (8.400–8.500 GHz) and S-band (2.2–2.3 GHz), to be used operationally for the first time in support of the January 1986 Voyager 2 spacecraft flyby encounter with Uranus, more than 3 billion km away. The 34-m antenna was used as one leg of a receive array system that also included the DSN's 64-m and the 34-m standard antennas. The addition of the new 34-m antenna added a nominal 0.8 dB to the 64-m and 34-m standard antenna array used for the Voyager Saturn encounter.

The 34-m high-efficiency (HEF) antenna is equipped with an electrically driven azimuth–elevation type of mount. The antenna dish is supported by a steel space frame that rotates in azimuth on four self-aligning wheel assemblies that ride on a precisely leveled circular steel track. The track is held firmly in place by 16 tangential links that attach to a centrally reinforced concrete pedestal. The antenna dish structure is attached to the elevation gear wheel, which drives the antenna up and down. The operating speeds are 0.4 deg/s in azimuth and elevation.

Two significant radio frequency (RF) developments were the dual-frequency feed, which provides for simultaneous multifrequency operation

---

[1]Based on "Evolution of the Deep Space Network 34-M Diameter Antennas," by William A. Imbriale, which appeared in *Proceedings of the IEEE Aerospace Conference*, Snowmass, Colorado, March 21–28, 1998. (© 1998 IEEE)

**Fig. 6-1. The DSS-15 HEF antenna.**

without the use of the dichroic plate, and the dual-reflector shaping of the surface, which provides improved aperture efficiency.

## 6.1  The Common-Aperture Feed

In mid-1976, a program was initiated to develop an S- and X-band feed horn, with the objective to utilize a centerline (CL) symmetric unit to replace the then asymmetric, simultaneous S-/X-band reflex DSN feed systems, thereby eliminating the dichroic plate (see Chapter 1) and further optimizing X-band performance, with degradation of S-band performance allowed if necessary. This feed was also to be capable of high-power transmissions in both X- (7.145–7.235 GHz) and S- (2.110–2.120 GHz) bands.

The basic concept for the horn design came from a paper by Jueken and Vokurka [1]. In essence, that paper recalls that the corrugation depths for a corrugated horn need to be between $\lambda/4$ ($\lambda$ = wavelength) and $\lambda/2$ to support the proper (hybrid) $HE_{11}$ waveguide mode. It follows that any such corrugations would be odd multiples of these depths within certain other frequency bands as well. A careful choice of depth is then made to obtain operation within S-band and X-band, with depths greater than $\lambda/4$ and less than $\lambda/2$ in S-band and greater than $5\,\lambda/4$ and less than $3\,\lambda/2$ in X-band. Thus, a sort of "harmonic" operation of the corrugation depth is effected. As such, a feed horn with fixed flare angle is made longer; hence, with larger aperture, a point is reached when further increase in length does not increase horn gain or reduce an associated beamwidth. Details of pattern shape will differ with frequency, but not with gain or beamwidth. One might call this saturated gain operation. Further details of this approach may be found in [2–4].

Injecting or extracting S-band is accomplished by feeding the horn at a sufficiently large horn-diameter region (above waveguide cutoff) from a surrounding radial line. The signal is injected into this radial line from four orthogonally located peripheral feed points excited in a 90-deg phase progression to develop circular polarization (CP).

The radial line carries two radial chokes, which prevent X-band from propagating within the S-band injection device, now termed the X-/S-band combiner. The system works very well at X-band, since no noise-temperature increase was noted when this feed horn was compared to the DSN standard X-band 22-dB horn, which has no S-band operation. Several combiner designs were explored, including a first-generation combiner with only a narrow S-band bandwidth suitable in a receive-only system, a second-generation combiner that would enable low-power (20-kW) transmission at S-band, and a third-generation combiner aimed at high-power (400 kW) S-band. The third-generation combiner was sufficiently immature at the time of the HEF antenna construction that the second-generation combiner was used [5].

Figure 6-2 shows the feed horn that was installed and tested in the Deep Space Station 13 (DSS-13) 26-m parabolic reflector. This R&D feed was installed in a four-function feed cone (Fig. 6-3) and was the first simultaneous dual-band receive/dual-band transmit DSN system [6,7]. A version of the feed cone was first tested at DSS-13 before a contract to manufacture the feed cones to be installed on the operational antennas was given to industry.

## 6.2  Dual-Reflector Shaping

Dual-reflector shaping consists of creating slight distortions of the usual hyperboloid subreflector of a Cassegrain system so that the feed-horn pattern can be transformed into a nearly uniform illumination across the main aperture. When this occurs, the uniform phase pattern, normally present from the hyperboloid, is destroyed. This uniform phase is then recovered by compensating for the distortions by modifying the main paraboloid. The resulting uniform distribution of amplitude and phase gives the maximum possible illumination efficiency available for the given aperture size.

Since the shaping transforms a feed-horn pattern, it follows that a particular feed-horn pattern must be used to obtain the final reflector shape. Geometric optics (GO) is used to solve the problem, and uses only equal path lengths, Snell's law of reflection, and conservation of energy, so frequency of operation does not enter into the solution. Therefore, any feed-horn pattern at any frequency and reasonably uniform phase will be transformed to an illumination

**Fig. 6-2. Photograph of the R&D
common-aperture feed horn.**

that will be distorted to an extent dependent upon the similarity of its pattern to the pattern that was used in the basic shaping design. This property is especially important for the dual-frequency feed, since the frequencies of operation are so widely separated. It is only necessary that the feed-horn patterns at the two frequencies be nearly the same. It must be mentioned that the standard paraboloid-hyperboloid Cassegrain system is only a special case of the general shaped dual-reflector antenna. In that case, the transformation is 1 to 1, that is, the feed-horn pattern is unchanged. The illumination becomes that of the feed horn used in a prime-focus system of greater focal length, modified slightly by finite reflector diffraction detail.

Several modifications of a standard dual-reflector shaping algorithm were incorporated into the design. First, illumination in the central region was eliminated, reducing subreflector blockage and improving blockage efficiency by synthesizing what previously had been termed a vertex matching plate, but without the accompanying phase distortion. Second, the distance from aperture plane to main reflector (quasiparaboloid) vertex was chosen as a design parameter; this was done to match the geometries of the 26-m antennas of the DSN. Consequently, the special shape solutions closely approximate the 26-m paraboloid contour. The result was a feed-horn focal point location at 193.5 in. (4.915 m) from the quasiparaboloid vertex. Hence, the new feed horn is compatible with any other DSN antenna for maximum flexibility. Third, the geometric ray from the subreflector edge does not go to the main reflector edge but

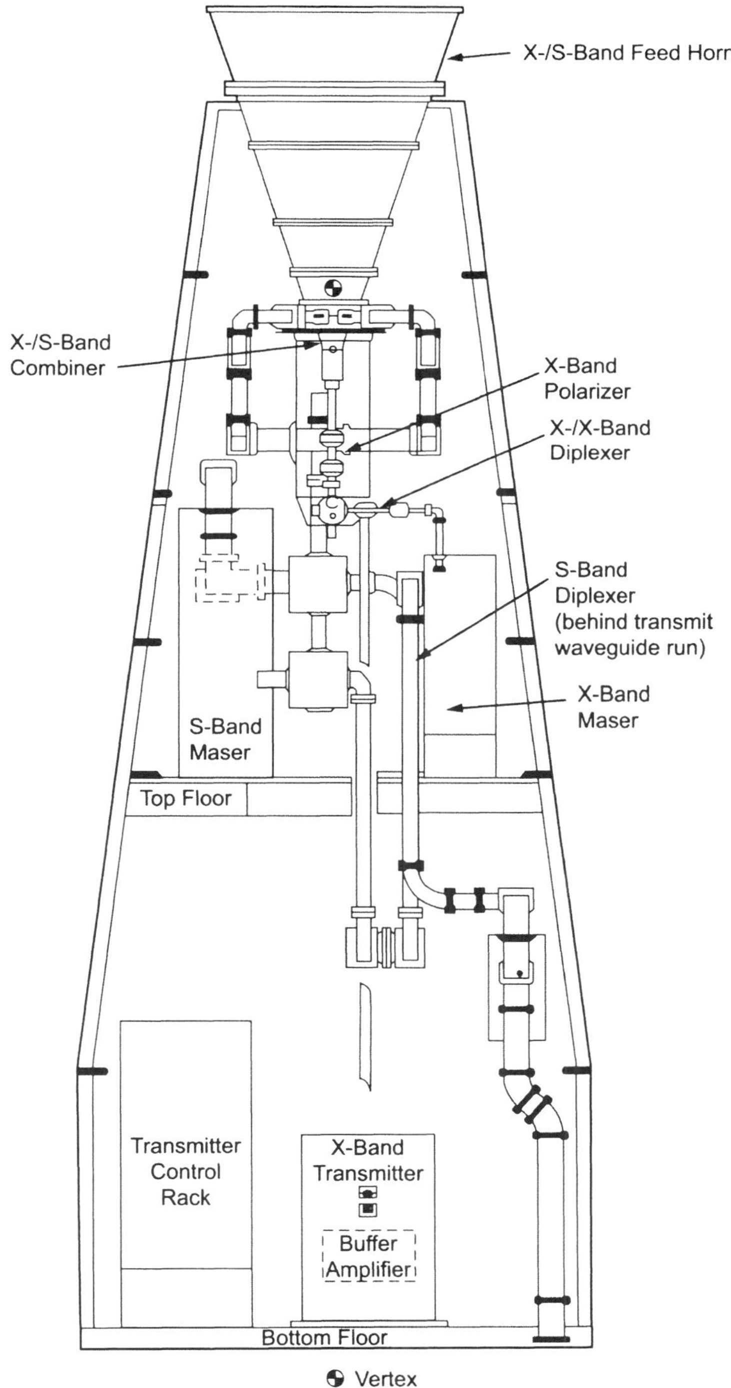


Fig. 6-3. Diagram of DSS-15 HEF antenna cone.

instead intercepts at 645 in. (16.383 m), as shown on the design control drawing (Fig. 6-4). The reason for this choice is as follows: The feed horn scattered
energy from the quasihyperboloid does not fall abruptly to zero at the angle $\theta_1$,
but instead tapers rapidly to a low level. The angle $\theta_1$, and, hence, the 16.393-m
dimension, is chosen so that the intensity at $\theta_2$ may be at a very low level relative to the central region of the main reflector, and the resulting rear spillover
noise contribution becomes acceptably small. This procedure results in a
slightly lower illumination efficiency, but the significant reduction in noise
from rear spillover results in an optimum gain over temperature ($G/T$) ratio.

Figure 6-5 gives a plot of $G/T$ versus edge illumination angle and justifies
the choice of 645 in. (16.393 m) for the optical shaping limit. The measured
pattern of the common-aperture feed horn at 8.45 GHz was used to calculate
the surface shapes. The angular pattern through 17 deg was used so that the forward spillover (past the subreflector) was limited to <2 percent at X-band. The
S-band pattern shape was similar, but nearly 9 percent of the energy was outside the 17-deg range. The improvement in gain from amplitude and phase distribution was nearly 1 dB.

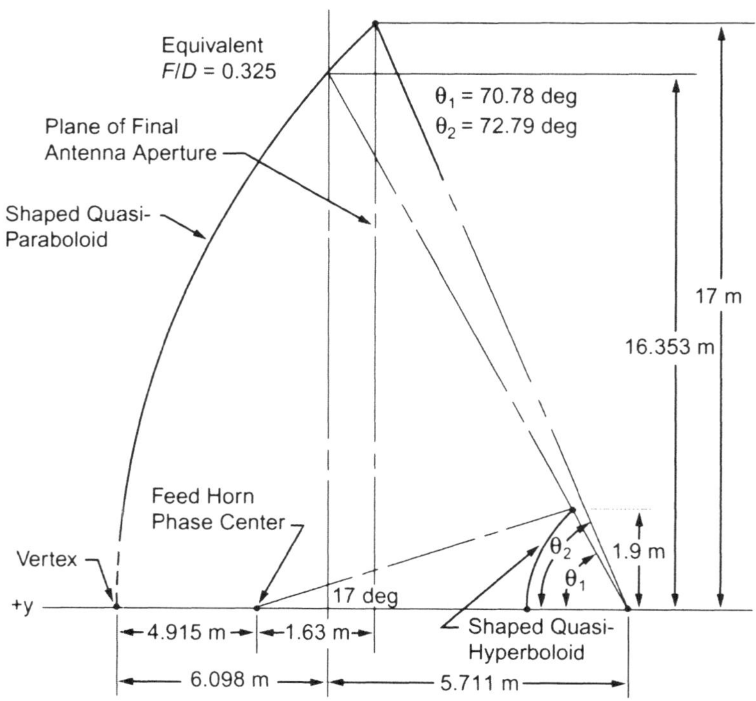


**Fig. 6-4. Synthesis coordinate system.**

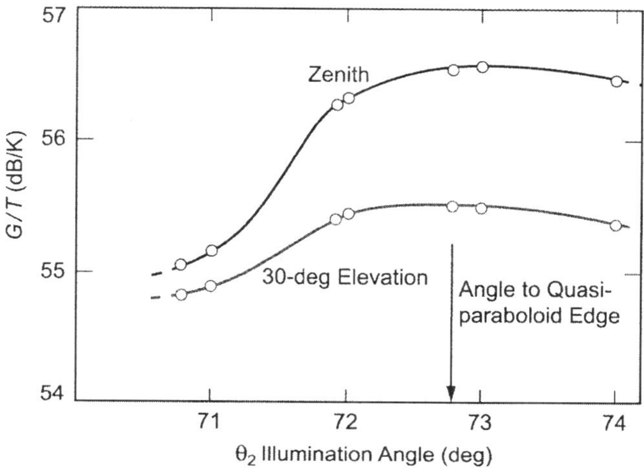


Fig. 6-5.  *G/T* of system versus edge illumination angle.

## 6.3   Computed versus Measured Performance

The RF performance of the new HEF antennas was calculated by scattering
the measured common-aperture X-/S-band horn patterns from the shaped
reflector surfaces using standard physical optics (PO) computer codes. The RF
efficiencies based upon the computer programs are given in Table 6-1 [8].

An estimate of noise contribution from the rear spillover is noted in
Table 6-1. This is the average blackbody radiation from the ground, estimated
to be 240 K multiplied by the percentage of energy in the rear spillover. Using
holographic techniques, the root-mean-square (rms) surface error was measured to be approximately 0.5 mm. The efficiency numbers in Table 6-1 were
derived using the Ruze formula. Area blockage due to the feed-support quadripod is estimated to be 6 percent, with the blockage efficiency determined from:

$$\text{Spar blockage efficiency} = [1 - 1.2(Ap)]^2$$

with *Ap* representing the percentage of geometric shadowing by the spars.
Using the estimated rms surface area and quadripod blockage, the resulting
expected efficiencies at the rigging angle are 75 percent (68.30 dB) at X-band
and 67 percent (56.5 dB) at S-band. Measured X-band gain was 68.26 dB.

**Table 6-1. DSS-15 efficiencies at S-band and X-band.**

| Element | S-Band | X-Band |
|---|---|---|
| Rear spillover efficiency | 0.9746 | 0.9982 |
|    Resulting noise temperature contribution | (6.1 K) | (0.4 K) |
| Forward spillover efficiency | 0.9110 | 0.9839 |
| Illumination efficiency | 0.9874 | 0.9823 |
| Cross-polarization efficiency | 0.9995 | 0.9988 |
| Phase efficiency | 0.9383[a] | 0.9744 |
| Central blockage efficiency | 0.9765 | 0.9826 |
| Dissipation, VSWR efficiency | 0.9700 | 0.9700 |
| RF efficiency | 0.7779 | 0.8949 |
|    Spar blockage efficiency | 0.8610 | 0.8610 |
|    Surface efficiency | 0.9977 | 0.9692 |
| Total efficiency | 0.6682 | 0.7468 |
| Resulting gain (dB) | 56.4900 | 68.300 |

[a] The S-band phase center does not coincide with the 8.450-GHz phase center, resulting in somewhat poorer phase efficiency at S-band. (The feed-horn location and shaping are based upon the 8.450-GHz phase center.)

# References

[1] E. J. Jeuken, and V. J. Vokurka, "Multi-Frequency Band Corrugated Conical Horn Antenna," *1973 European Microwave Conference Proceedings, Volume 2*, Brussels University, Brussels, Belgium, September 4–7, 1973.

[2] W. F. Williams, "A Prototype DSN X-S Band Feed: DSS 13 First Application Status," *Deep Space Network Progress Report 42-44*, vol. January and February 1978, http://tmo.jpl.nasa.gov/progress_report/issues.html Accessed April 9, 2001.

[3] W. F. Williams, "A Prototype DSN X- and S-Band Feed: DSS 13 Application Status (Second Report)," *Deep Space Network Progress Report 42-47*, vol. July and August 1978, http://tmo.jpl.nasa.gov/progress_report/issues.html Accessed April 9, 2001.

[4] W. F. Williams and S. B. Cohn, "Dual Band Combiner for Horn Antenna," U.S. patent No. 4199764, April 22, 1980.

[5] W. F. Williams and J. R. Withington, "A Common Aperture S- and X-Band Feed for the Deep Space Network," presented at the 1979

Antenna Applications Symposium, University of Illinois, Allerton Park, September 1979.

[6]  W. F. Williams, and H. Reilly, "A Prototype DSN X/S Band Feed: DSS 13 Application Status (Fourth Report)," *Telecommunications and Data Acquisition Progress Report 42-60*, vol. September and October 1980, http://tmo.jpl.nasa.gov/progress_report/issues.html  Accessed April 9, 2001.

[7]  J. R. Withington and W. F. Williams, "A Common Aperture X- and S-Band Four Function Feedhorn," presented at the 1981 Antenna Applications Symposium, University of Illinois, Allerton Park, September 1980.

[8]  W. F. Williams, "RF Design and Predicted Performance for a Future 34-Meter Shaped Dual Reflector System Using the Common Aperture X-S Feedhorn," *Telecommunications and Data Acquisition Progress Report 42-73*, January–March 1983, http://tmo.jpl.nasa.gov/progress_report/issues.html Accessed April 9, 2001.

# The 34-Meter Research and Development Beam-Waveguide Antenna

There are a number of advantages to feeding a large ground station antenna via a beam-waveguide (BWG) system rather than directly placing the feed at the focal point of a dual-reflector antenna. In a BWG system, the feed horn and support equipment are placed in a stationary room below the antenna, and the energy is guided from the feed horn to the subreflector, using a system of reflecting mirrors. Thus, significant simplifications are possible in the design of high-power water-cooled transmitters and low-noise cryogenic amplifiers, since these systems do not have to tilt as in normally fed dual-reflector antennas. Furthermore, these systems and other components can be placed in a more accessible location, enabling easier servicing and repair. In addition, the losses associated with rain on the feed-horn cover are eliminated because the feed horn is sheltered from weather.

Consequently, in the late 1980s, the National Aeronautics and Space Administration (NASA) Deep Space Network (DSN) undertook a comprehensive research program aimed at introducing BWG-fed antennas into the operational network. The research encompassed (a) new analytical techniques for predicting the performance of BWGs; (b) a model test facility to experimentally verify the analytical tools; and (c) the design, construction, and test of a new 34-m research and development (R&D) antenna.

Around the same time, various studies pointed to the advantage of using Ka-band (31.8–32.2 GHz receive and 34.2–34.7 GHz transmit) for the telecom-

---

[1] Based on "Evolution of the Deep Space Network 34-M Diameter Antennas," by William A. Imbriale, which appeared in *Proceedings of the IEEE Aerospace Conference*, Snowmass, Colorado, March 21–28, 1998. (© 1998 IEEE)

munications link [1]. Therefore, the R&D antenna was designed to operate at Ka-band and thus be the testbed for introducing Ka-band frequencies into the operational network.

## 7.1  New Analytical Techniques

A BWG system consists of a number of conic-section mirrors enclosed in a metal tube. There is a feed horn on the input, and the output irradiates the sub-reflector of the dual-reflector system. The commonly used analysis of this system ignores the presence of the metallic tube enclosing the BWG mirrors and uses either physical optics (PO), the Geometrical Theory of Diffraction, or Gaussian-mode analysis of the diffracted field calculations. However, the basic weakness of these analyses is that they do not shed any light on the effect of the metal tube. Therefore, a new and fundamentally more correct BWG analysis that considers the presence of the metal tube was developed [2,3]. The basic concept is to use a Green's function appropriate to the circular-waveguide geometry to compute the scattered field. With the new analysis, an accurate assessment of the effects of the tube, including noise-temperature increase due to conduction losses in the tube, can be factored into the design. The technique is described in detail in Chapter 1 of this monograph.

## 7.2  Beam-Waveguide Test Facility

In support of analytical and test activities, a flexible test facility [4] was constructed to study BWG performance parameters. The objectives of the test facility were to (a) measure and characterize multiple mirror systems used in BWG antennas, (b) verify computer software and software models, (c) characterize BWG components not easily modeled by software, and (d) predict performance of BWG antenna designs.

The BWG test structure was installed in a microwave anechoic chamber 6 m wide by 6 m high and 18 m long. The test facility setup consists of (a) a structure to hold the BWG elements under test; (b) a test probe, mounted and independent of the BWG structure support, to provide the radiating patterning sampling device; and (c) an instrument control and its acquisition software. The test probe assembly consists of an open-ended circular-waveguide feed horn mounted at the end of a long radial arm that is rotated in a circular arc by a 20-cm optical-grade rotating table. The feed/arm/rotating table element is itself mounted in an azimuthal positioner. This design allows complete tangential field sampling over a hemispheric surface at a given radius from the center of

**Fig. 7-1. The BWG test facility
with two-mirror configuration.**

the radial arm rotation point, thus producing a spherical near-field range measurement setup. A picture of the test facility is shown in Fig. 7-1.

One-quarter-scale mirrors (of those used in the full-scale 34-m R&D antenna) were machined from solid aluminum blocks and used in one-, two-, and three-mirror test configurations (see Figs. 7-1 through 7-3). The one-mirror beam-magnifier configuration results are shown in Fig. 7-4(a) for the offset plane (orthogonal to the plane of symmetry) at Ka-band. The beam-magnifier configuration consists of an ellipse and a 22-dB-gain input feed horn at one focal point that produces a 29-dB-gain output feed pattern at the second ellipse focal point. Compared are the computed and measured near-field results. Computations were made using a near-field PO program. The two-mirror results are shown in Fig. 7-4(b) for X-band (8.4 GHz) and Fig. 7-4(c) for Ka-band (32 GHz). The two-mirror system is a standard two-parabola configuration that replicates the input feed-horn pattern at the second parabola focal point. Three-mirror results are shown in Fig. 7-4(d) and consist of a 22-dB-gain input horn at the focal point of a beam-magnifier ellipse that produces a 29-dB output pattern for input to the standard two-mirror system. The excellent correlation that was achieved between measured and predicted results enabled the full-scale 34-m antenna system implementation to proceed with confidence.

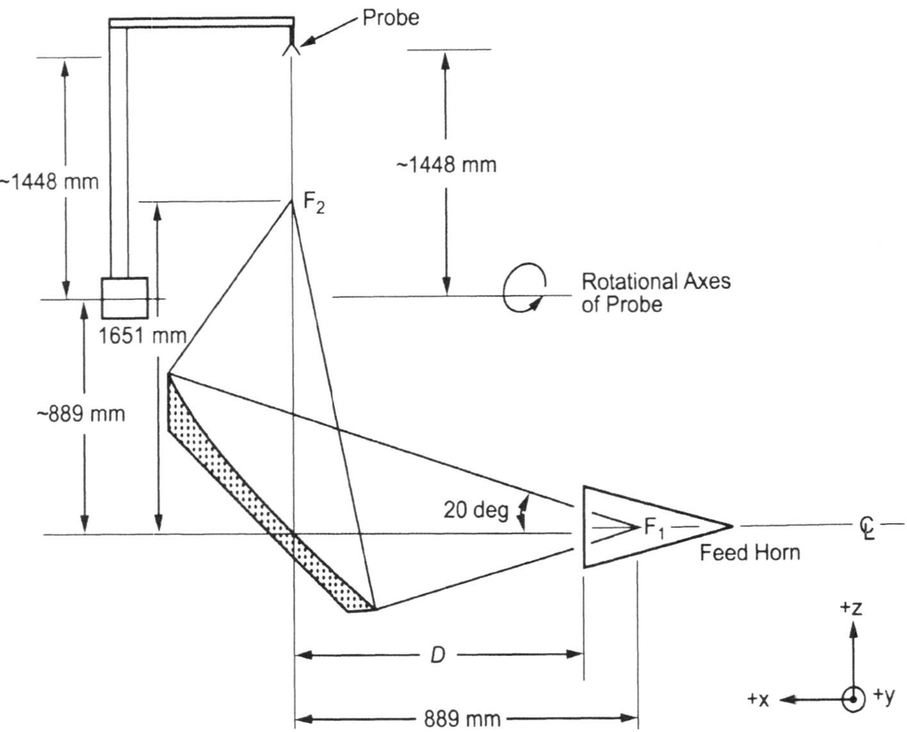



**Fig. 7-2. Schematic of the one-mirror configuration.**

## 7.3 The New Antenna

A 34-m R&D antenna [5–7] was constructed at the DSN's Deep Space Station 13 (DSS-13) in Goldstone, California, as a precursor to introducing BWG antennas and Ka-band frequencies into the operational network. Construction was completed in May 1990 (see Fig. 7-5), and the antenna has been continuously used for the DSN's R&D program since that time. Phase I of the project was for independent X- and Ka-band operation. Phase II of the project introduced simultaneous S- and X-band or X- and Ka-band operation.

The design of the upper portion of the BWG is based upon a geometrical optics (GO) criterion introduced by Mizusawa and Kitsuregawa in 1973 [8], which guarantees a perfect image from a reflector pair. Since it is desirable to retrofit existing antennas with BWGs as well as to construct new antennas, there are actually two independent designs built into the R&D antenna (Fig. 7-6). The first, termed a bypass design, places the BWG outside the existing elevation bearing on the rotating azimuth platform. The second, a center

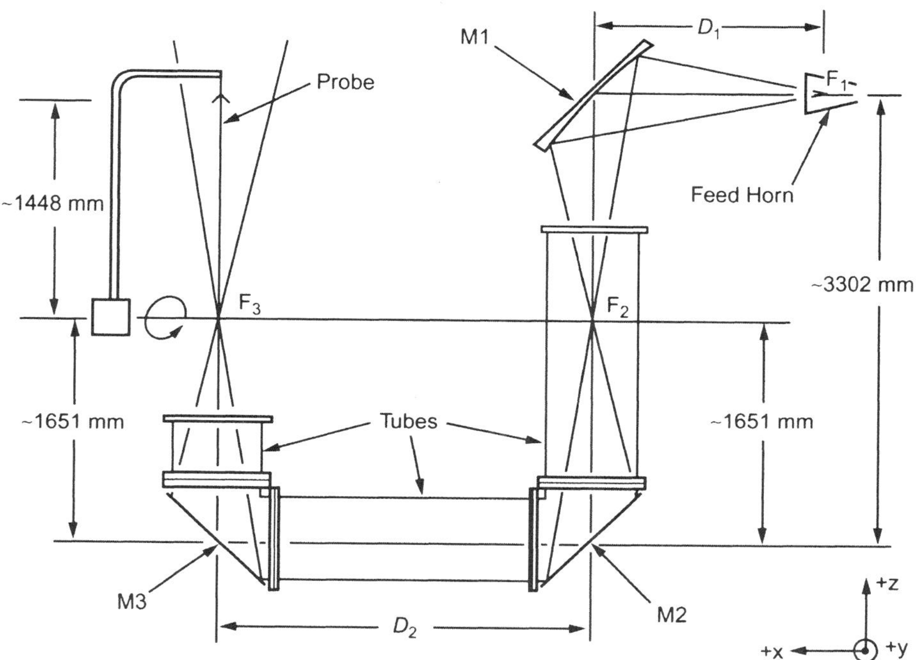



**Fig. 7-3. Schematic of the three-mirror configuration.**

design (see Fig. 7-7), places the BWG through the center of the dish, inside the elevation bearing and through the azimuth axis, into a pedestal room located below the antenna. The bypass design uses a single pair of paraboloids and two flat mirrors whereas the center design uses the same design (although not physically the same mirrors) above the ground, and a flat plate and beam-magnifier ellipse in the pedestal room. A beam magnifier is required, since the pair of paraboloids requires the use of a 29-dB-gain feed horn that at lower frequencies would be too large to fit into the pedestal room. The ellipse design allows the use of a smaller-gain feed horns in the pedestal room.

### 7.3.1    Antenna Design Considerations

At present, the DSN operates three 34-m high-efficiency (HEF) dual-shaped reflector antennas, each with a dual-band (2.3- and 8.4-GHz) feed horn having a far-field gain of 22.4 dBi (see Chapter 6 of this monograph for a full discussion of the first HEF antenna, DSS-15). The feed horn is conventionally located at the Cassegrain focal point. The structures were designed prior to BWG requirements, and therefore feature a continuous elevation axle and a carefully designed elevation-wheel substructure. The elevation wheel is supported by an alidade, which rotates on a circular azimuth track. To minimize

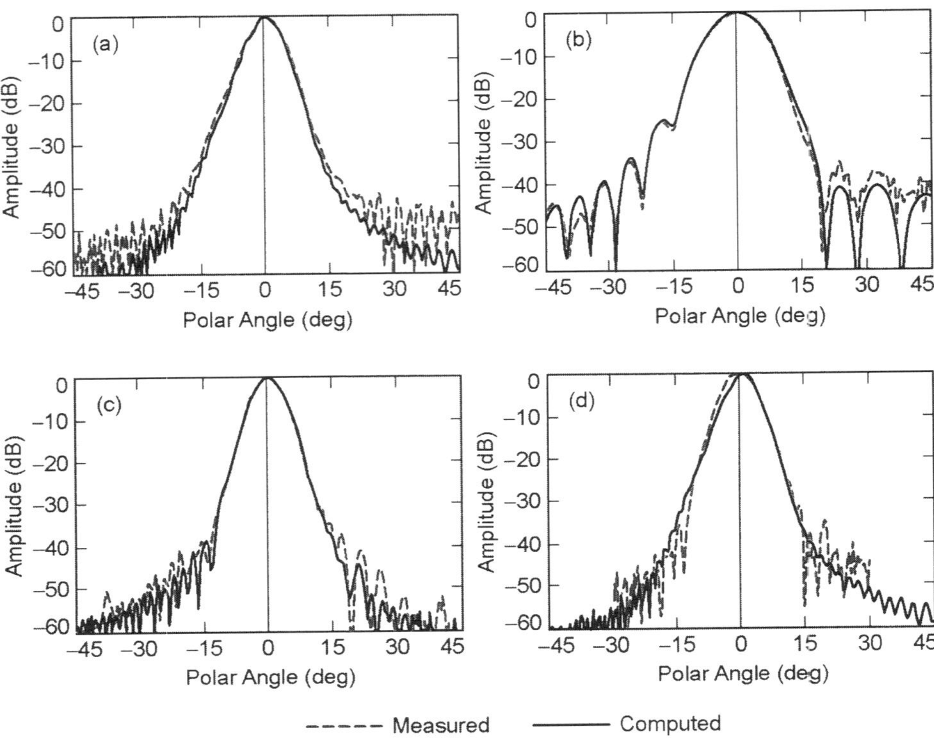



**Fig. 7-4. Measured and computed data for the offset plane: (a) one-mirror, Ka-band; (b) two-mirror, X-band; (c) two-mirror, Ka-band; and (d) three-mirror, Ka-band.**

the cost of developing a new 34-m BWG antenna, as much of the existing structure design as possible was used. Through the use of a clever mechanical design, the elevation tipping structure was modified to accommodate a central BWG inside the elevation bearing. To provide clear access for an 8-ft-(2.438-m)-diameter, center-fed BWG, the main reflector backup trusses are connected to the elevation wheel via the integral ring girder, or IRG. The IRG is a toroidal structure—an octagonal space truss with a square cross-section, approximately 290 in. (7.366 m) in maximum radius and 80 in. (2.032 m) high. It is interwoven with, but separate from, the antenna backup structure. In order to minimize the distortion of the reflector's surface under gravity loading, the reflector connections to the elevation-wheel structure were selected to provide equal stiffness supports. This is achieved by grouping either equally spaced reflector radial ribs into four pairs and connecting each pair to the IRG top plane at alternate vertices of the octagon. The vertices lying on the elevation axis, however, are reserved for supporting the IRG at the two elevation bearing points. The counterweight and single elevation bullgear lie in a plane orthogo-

**Fig. 7-5. The beam-waveguide antenna
after construction in 1990.**

nal to the elevation axis. The entire tipping structure, including the main reflector, elevation wheel, subreflector and its support, is weight balanced about the elevation axis.

By selecting the existing HEF reflector structure design, the focal length over diameter *(F/D)* of the main reflector surface is fixed. The reflector shape could be different than that for the original HEF design but would have to be within an adjustable tolerance ~1 in. (~2.5 cm) of the existing surface.

## 7.3.2   Upper-Mirror Optics Design

Geometrical optics is used to design the upper portion of the centerline BWG system (mirrors M1 to M4). As shown in Fig. 7-7, the first mirror, M1, has azimuth and elevation rotations together with the main and subreflectors. A plane surface is used for M1 to ensure an imaged feed pattern that is independent of the elevation angle of the antenna. Mirrors M2 and M3 are parabolas, and the system is designed such that a feed horn placed at $F_2$ is perfectly imaged at $F_1$ [8]. Mirrors M2 through M4 rotate in azimuth around the centerline axis.

An imaged feed pattern at $F_1$ is used to illuminate a subreflector with a narrow-angle, high-gain (~30-dBi) pattern. This configuration is chosen because of the large distance between the subreflector and the first BWG mirror (M1), and also because the size of M1 (as well as M2) is smaller than the subreflector. The position of focal point $F_1$ in Fig. 7-7 has to be closer to M1 in order to

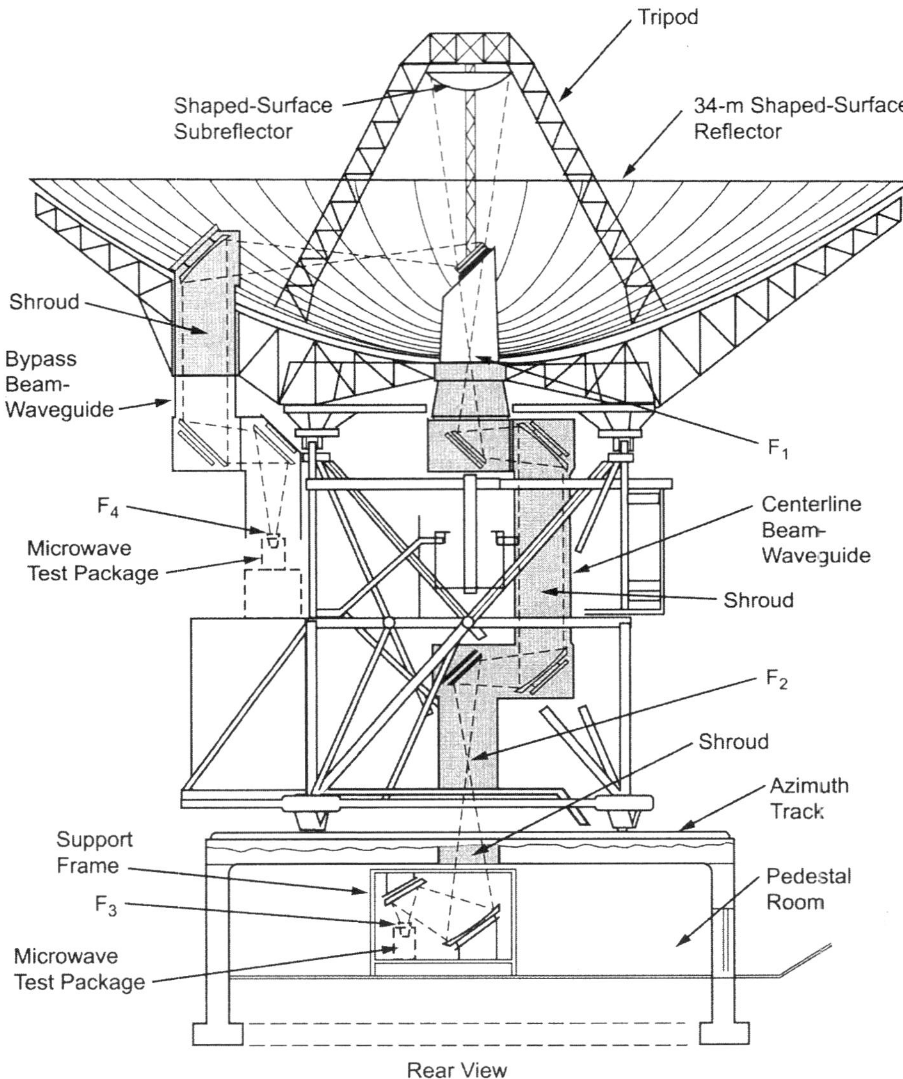


**Fig. 7-6. Beam-waveguide schematic.**

achieve acceptable spillover loss at the subreflector, M1, and M2. Normally, $F_1$ is in the neighborhood of the main reflector vertex with an 8- to 9-deg half-cone angle of illumination at the subreflector (compared to 17 deg for the normal Cassegrain feed of the 34-m HEF antenna).

The diameter of the subreflector, $D_s$, is determined according to the size of the main reflector. According to [9], a subreflector diameter of 1/10 of a main reflector is normally selected for good radiation efficiency of the antenna. For a 34-m antenna, the subreflector diameter of 135 in. (3.42 m) was chosen. The

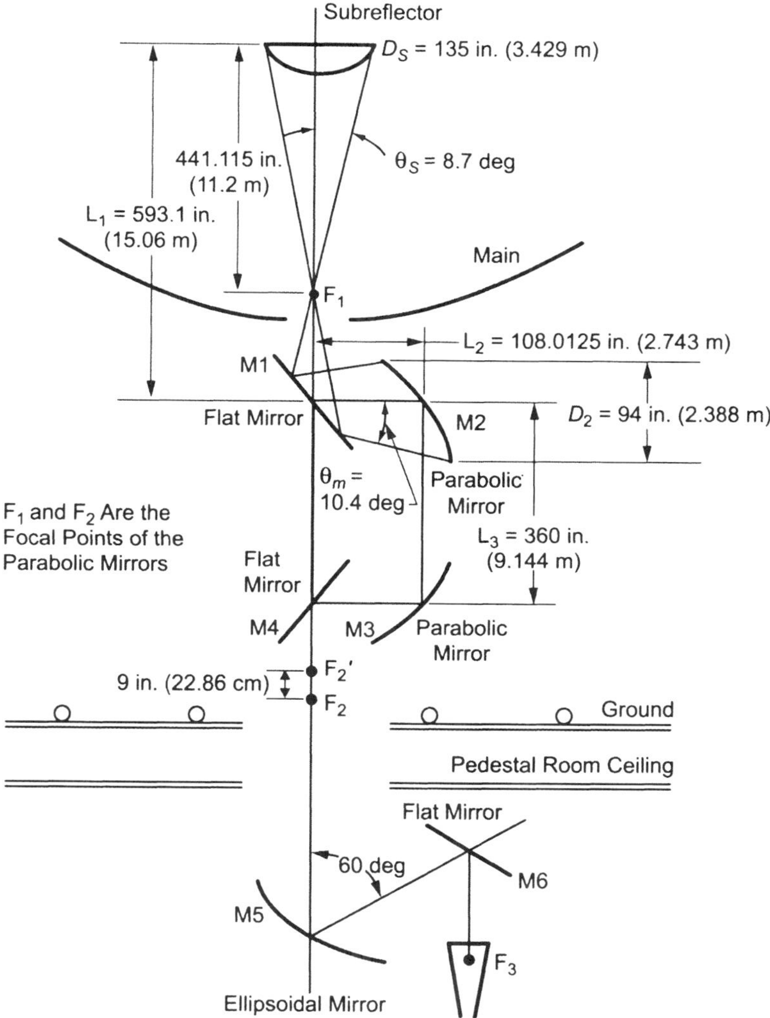


Fig. 7-7. Center-fed beam-waveguide layout. $F_1$ and $F_2$ are the focal points of the parabolic mirrors. $F_3$ and $F'_2$ are the focal points of the ellipsoidal mirror.

illumination at the subreflector $\theta_s$ is 8.7 deg. (compared to 17 deg for the normal Cassegrain feed of the 34-m HEF antenna).

For the same ratio as the HEF antenna and where $D_s$ = 135 in. (3.42 m), the distance $L_1$ = 593.1 in. (15.065 m) is obtained. Iterations are needed for a determination of $\theta_s$ and the location $F_1$. Known parameters are

$$D_s = 135 \text{ in. (3.42 m)}$$
$$L_1 = 593.1 \text{ in. (15.065 m)}$$

$D_2$ = 94 in. (2.388 m)

Variable parameters are

8.0 deg < $\theta_s$ < 9.0 deg

2.67 m < $L_2$ < 2.8 m

9.5 deg < $\theta_m$ < 11.0 deg.

The angle $\theta_m$ is the illumination angle at M2 with an edge taper of about –23 dB. The results of iterations of a GO ray geometry between $D_s$ and $D_2$ are

$\theta_s$ = 8.7 deg

$\theta_m$ = 10.4 deg

$L_1$ = 441.11 in. (11.204 m)

$L_2$ = 108.01 in. (2.743 m).

The focal length of M2 is equal to 260 in. (6.6 m). The exact dimensions are somewhat arbitrary but are constrained by limiting the M2 mirror projected diameter to 8 ft. (2.438 m); the larger the diameter, the higher the reflector above the elevation structure. The tube diameter was chosen such that the tube effects at S-band would be small [3]. It was subsequently necessary to design a feed system that provides a –18 to –20 dB taper at 8.7 deg (the illumination of the subreflector) and minimal spillover past 10.4 deg (the illumination of the BWG mirror).

Another important design parameter is the feed flare angle. Figure 7-8 shows the patterns and efficiencies (spillover times phase efficiency) for several different feed-horn flare angles. As can be seen, the patterns are not very sensitive to the flare angle. Since this was the case and the Jet Propulsion Laboratory (JPL) standard feed horn (designed for the DSN) has a flare angle of 6.25417 deg, it was decided to retain the standard flare angle in order to use existing feed horns and feed-horn designs.

Various feed-horn sizes with the JPL standard design angle of 6.25417 deg and frequency of 8.45 GHz were investigated. The goal was to find a horn with a –18-dB taper at $\theta$ = 8.7 deg (at subreflector distance, $r$ = 425 in. [10.8 m]) and a –23-dB taper at $\theta$ = 10.4 deg (at the parabolic mirror distance, $r$ = 260 in. [6.6 m]). The distances 6.6 m and 10.8 m are for a high-gain horn illuminating the BWG mirror M2 and the subreflector, respectively. The peak of the combined phase and spillover efficiencies ($\eta_{phase} \times \eta_{spill}$) should optimally be between 8.7 deg to 10.4 deg. The results from various trials show that a 19-in. (48-cm) aperture diameter meets these goals. Figures 7-9 and 7-10 show amplitude, phase, and efficiency plots of the 19-in. X-band aperture diameter at $r$ = 260 in. and 425 in., respectively. From Fig. 7-9, the edge taper at the rim of the subreflector ($\theta$ = 8.7 deg, $r$ = 425 in.) is equal to –18.7 dB, which is within

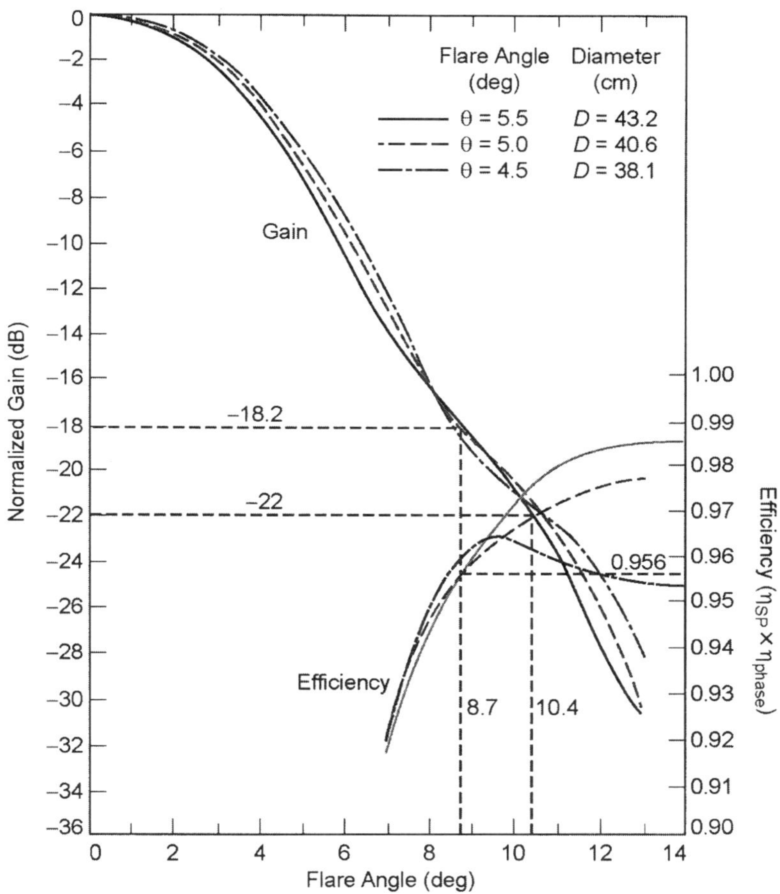


**Fig. 7-8. Gain and efficiency versus flare angle.**

the desired values of −18 dB to −20 dB. The combined phase and spillover efficiency is about 96.4 percent, where the maximum efficiency is about 97.8 percent at $\theta \sim 11.5$ deg. This is a typical design point for a DSN antenna, since to use the maximum efficiency point usually results in a slightly larger subreflector. The results for a 20.8-in. (53-cm) aperture were very similar to those for the 48-cm aperture, but the 20.8-in. aperture results in a feed horn 11 in. (28 cm) longer at X-band; hence, the smaller design was chosen.

From Fig. 7-10, the edge taper at the rim of a BWG mirror (at $r = 260$ in. [6.6 m]) is about −23.6 dB at $\theta = 10.4$ deg, with 96.5 percent efficiency. The maximum efficiency is equal to 96.7 percent at $\theta = 9.8$ deg, which is desirable because the value falls between 8.7 deg and 10.4 deg. It was concluded that the 48-cm X-band horn had radio frequency (RF) radiation characteristics that met

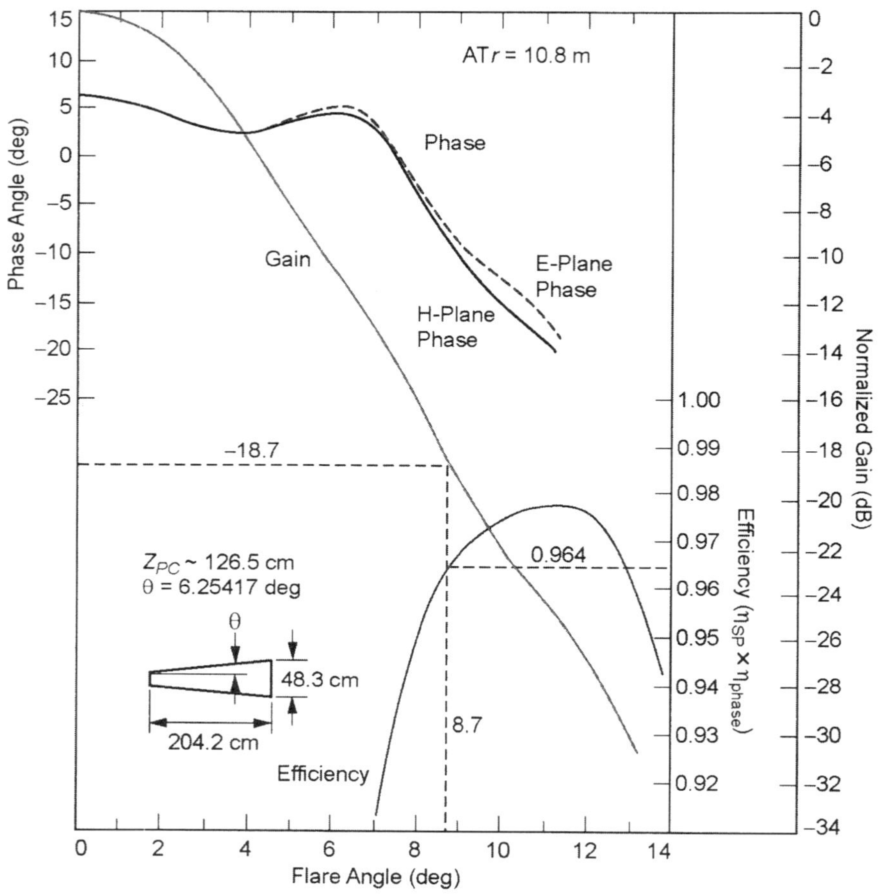


**Fig. 7-9. Horn radiation pattern; _R_ = 10.8 m.**

the requirements, and it was used in the design of BWG mirrors and the dual-shaped synthesis of the main and subreflector.

The choice of two identical paraboloidal sections for M2 and M3 has the following advantages:

- In the GO limit, a circularly symmetric input pattern still retains the original symmetric shape after reflection through both curved surfaces

- Since there is no caustic between the two curved mirrors, as there would be with ellipsoids, RF performance is not sensitive to the spacing ($L_3$) between the two mirrors, provided that the spillover loss remains small

- A highpass-type RF performance is obtained with very good X-band performance for 8-ft (2.438-m) mirrors (<0.1-dB loss for this path) and optimum performance for Ka-band

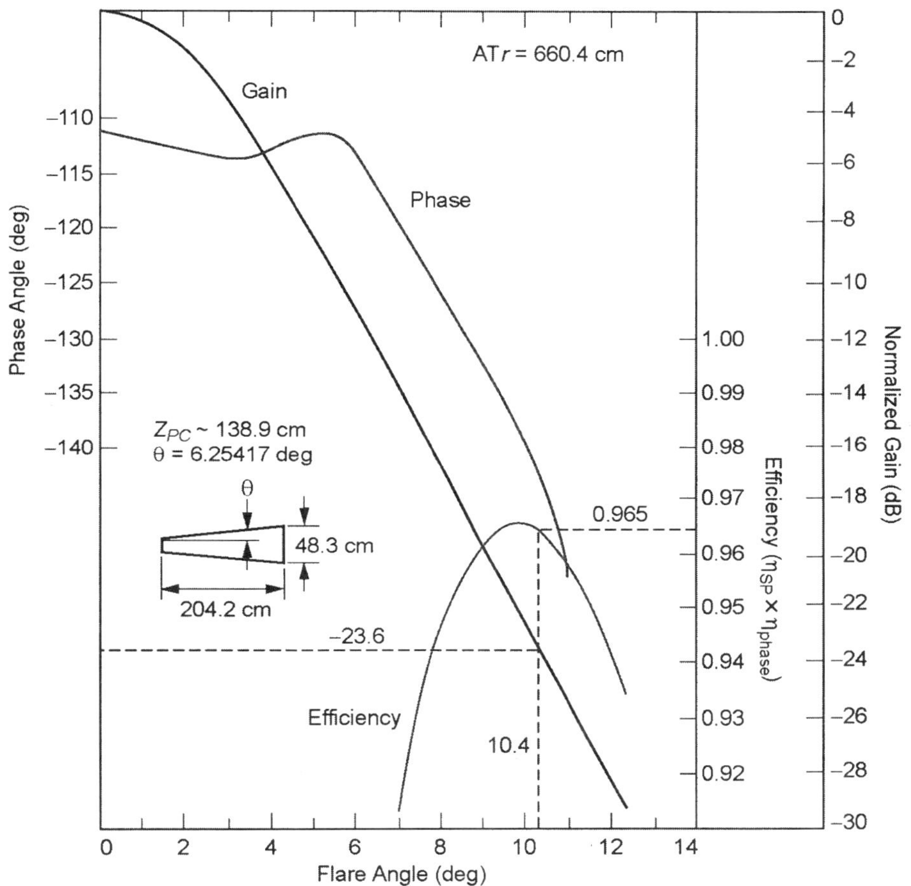


Fig. 7-10.  Horn radiation pattern; $R$ = 6.6 m.

- When paraboloidal surfaces are used in the design, it is possible to have four identical mirrors (two for center-fed BWGs and two for bypass BWGs)
- Identical mirrors are more economical.

The centerline (CL) BWG paraboloidal mirrors are positioned so that feed horns and instrumentation packages can be either in an alidade location (not presently planned for implementation) or the pedestal room. Spacing between the two paraboloids, $L_3$ = 360 in. (9.14 m), is chosen to allow enough head-room for vertical orientation of S-, X-, and Ka-band amplifier subassemblies. Also, S-band spillover loss at this distance is acceptably small. A flat plate, M4, reflects the RF beam downward, along the antenna azimuth axis, to the pedestal room, with focal point $F_2$ about 79 in. (2 m) above the azimuth floor and about 197 in. (5 m) above the pedestal room ceiling.

A significant decision was whether to locate the feeds on the alidade at focal point $F_2$ (requiring 29-dBi-gain feeds) or in the pedestal room, under the antenna, using focal point $F_3$. Despite an additional RF loss going from $F_2$ to $F_3$, the clear advantages of using the pedestal room (more available space, no cable wrap across the azimuth axis, smaller feeds required, etc.) led to its selection. The stable environment of the pedestal room was a major design determinant.

### 7.3.3   Pedestal Room Optics Design

Only X- and Ka-bands were planned for Phase I operation of DSS-13. However, the design was required to have capabilities for future S-/X-, X-/Ka-, C-, and Ku-band operations (S-band is 2 GHz, C-band is 4–6 GHz, and Ku-band is 13–15 GHz). Low-gain horns (~22 dBi) are desirable for all frequency bands. A basic layout for the RF design in the pedestal room is given in Fig. 7-11. Mirror M5 is an ellipsoidal surface used for magnifying gain (reducing beamwidth) from 22 to 29 dBi and switching among various feed horns by rotating M5 about the antenna azimuth. Mirror M6 is a flat plate used to reflect the RF beam from a vertically positioned feed horn to M5, with angle $\theta = 60$

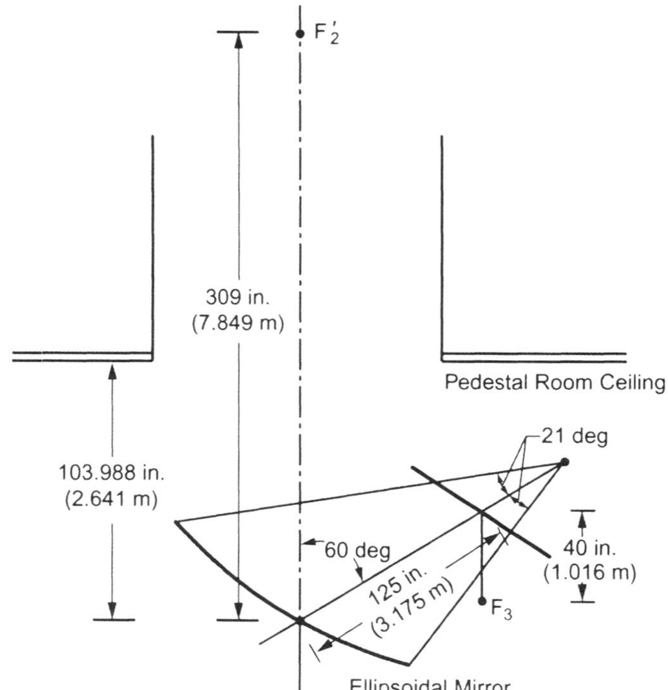


**Fig. 7-11.  Pedestal room geometry. $F_3$ and $F'_2$ are the foci of the ellipsoid.**

deg. The 60-deg angle is preferred because the existing JPL dichroic plate is designed with a 30-deg incident angle (equivalent to $\theta = 60$ deg). Therefore, the $\theta = 60$-deg angle will be convenient for simultaneous operation (S-/X- and X-/Ka-band). Even though a smaller angle of $\theta$ would yield a more symmetric beam pattern, angles smaller than 50 deg will have shadowing problems among M5, M6, and the feed horn. The curvature of M5 is determined by placing the near-field phase center of the 22-dBi X-band horn at a focal point of M5 ($F_3$) and calculating the field at M3 by using PO. Iteration continues by changing the surface curvature of M5 until the scattered field has an average edge taper at M3 of about $-23$ dB. The mirror M5 is adjusted vertically until the best-fit phase center of the scattered field of M5 overlays $F_2$. The curvature and position of M5 are designed at X-band, and there is no vertical adjustment of the mirror for other bands. There is a vertical offset of 22.9 cm (9 in.) between the output GO focal point of M5 and the input GO focal point of M3. The Ka-band horn (or other high-frequency horns) must be defocused and the gain increased slightly (from 22 to 23 dBi) to approximate the same edge taper and best-fit phase center as at X-band. The detailed RF design layouts in the pedestal room for X- and Ka-bands are shown in Figs. 7-12 and 7-13. There are small lateral translations of the feed horns to compensate for radiation pattern asymmetry due to the surface curvature of mirror M5.

### 7.3.4 Bypass Beam-Waveguide Design

A layout of the bypass BWG is shown in Fig. 7-14. All mirrors rotate in azimuth and all in the elevation plane except M10. To allow enough clearance between mirror M10 and the elevation bearing, the bypass BWG vertical tube is positioned about $\sim 403$ in. (10.2 m) from the antenna centerline. The flat plate, M7, is removed when the center-fed BWG mode is used. Mirrors M8 and M9 are paraboloidal surfaces positioned to satisfy Mizusawa's conditions. Mirrors M7, M8, and M9 are attached to, and move together with, the main reflector structure. A flat mirror, M10, is attached to an elevation bearing; it is not rotated with elevation rotation (but moves with azimuth rotation) in order to have a focal point $F_4$ always pointing straight downward to the alidade platform. By carefully adjusting $L_5$ and $L_6$ so that the distance from $F_1$ to the mirror M8 is equal to 260 in. (660.4 cm), the paraboloidal mirrors M8 and M9 are identical to the mirrors M2 and M3 in the center-fed BWG design. Thus, there are four identical curved mirrors in this double BWG feed system.

The value of $L_5$ used in this design is 290.645 in. (738.2 cm), which is the same as the spacing between mirrors M8 and M9. There is also enough clearance between an incident ray at the lower rim of M8 and the rim of the opening hole on the surface of the main reflector. Observe that the bypass performs slightly better than the center BWG, due to the absence of the ellipsoidal mag-

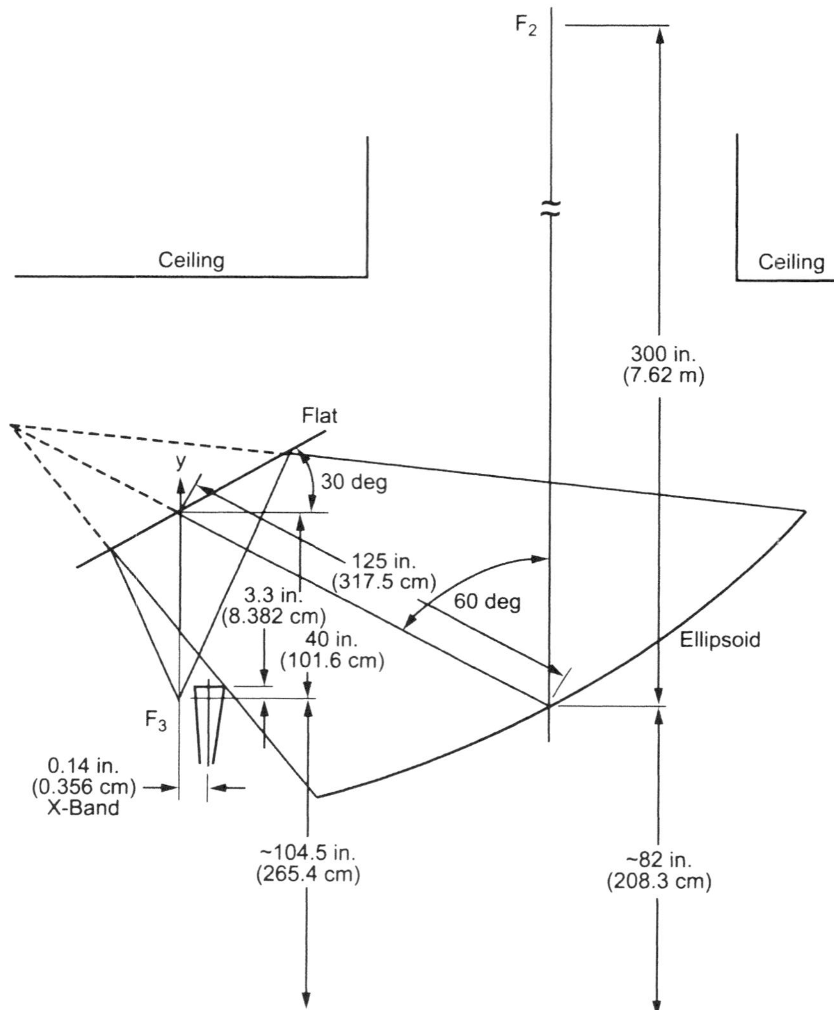



**Fig. 7-12. Pedestal feed system for X-band.**

nifier mirror and the shorter main path (290 in. versus 360 in. [736.6 cm versus 914.4 cm).

## 7.3.5   Theoretical Performance

The theoretical performance of the BWG system is determined by using various combinations of analytical software, as described in [6] and [10]. Figure 7-15 shows the measured pattern of the input of the X-band 22-dBi horn fed at $F_3$, the calculated output of the ellipsoid at $F_2$, and demonstrates the X-band gain-magnifying (beamwidth-reducing) property of the ellipsoid.

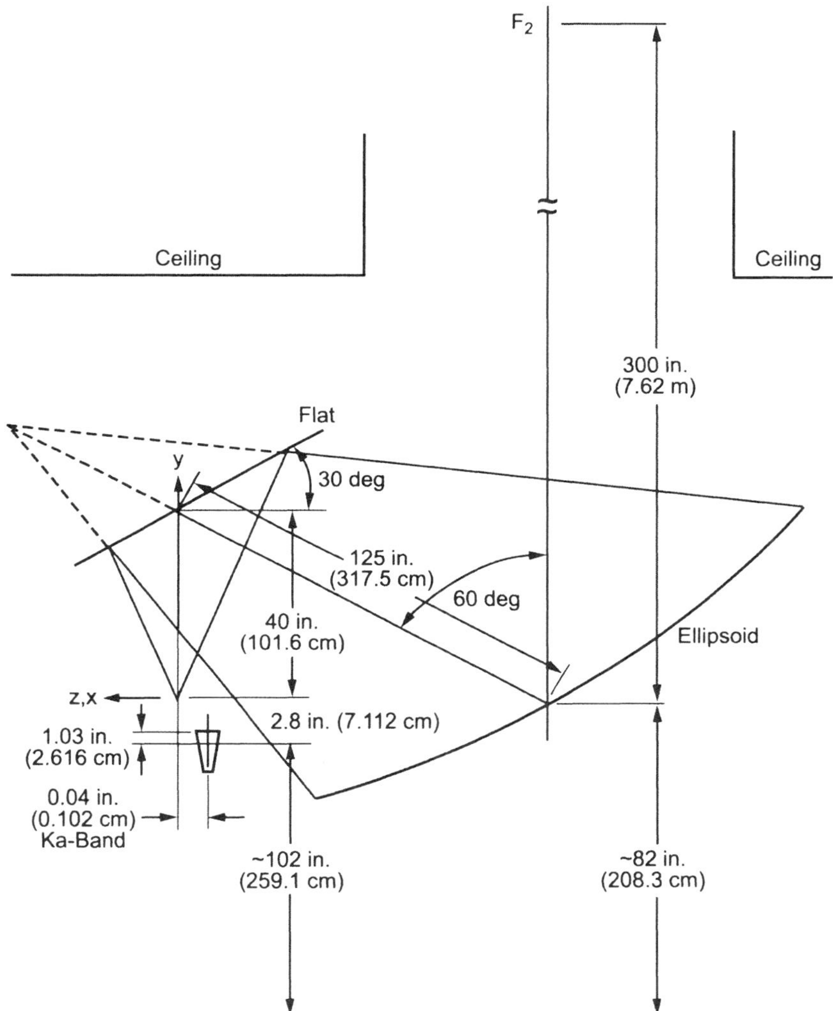


**Fig. 7-13. Pedestal feed system for Ka-band.**

Figure 7-16 shows the X-band output of the BWG at $F_1$ compared with both the calculated input at $F_2$ and the measured 29-dBi feed horn. Figure 7-17 shows a comparison of the E- and H-planes of the BWG output. The system is designed to image the 29-dBi horn at the input to the dual-reflector system. Figure 7-18 shows the input and output of the BWG at Ka-band and illustrates the nearly perfect imaging properties of the paraboloid pair. Figure 7-19 is a comparison of the 29-dBi feed horn located at $F_1$ and the BWG feed horn located at $F_3$ for the dual-reflector system at X-band.

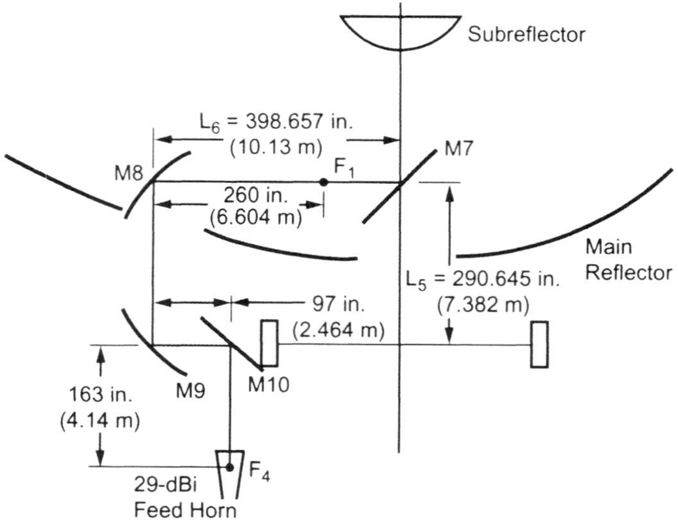



**Fig. 7-14. Details of the dimensions of the bypass beam-waveguide geometry.**

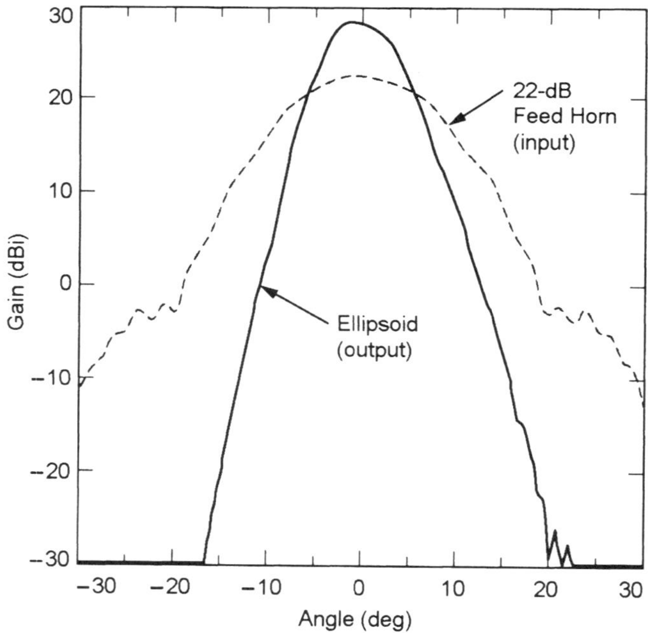



**Fig. 7-15. Beam-magnifier ellipse.**

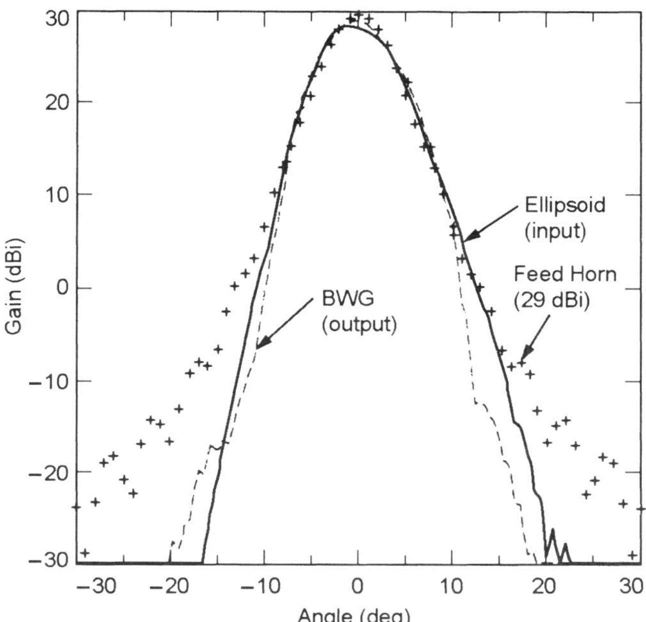



Fig. 7-16.  Center-fed output beam waveguide.

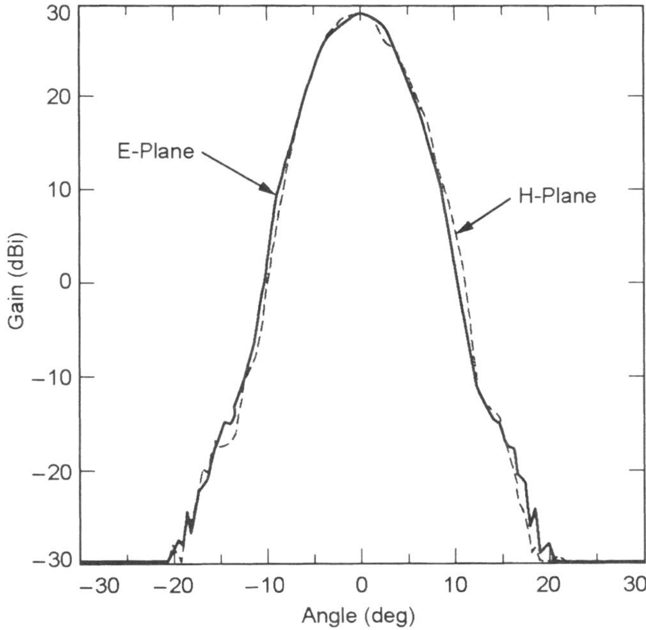



Fig. 7-17.  X-band E- and H-plane
beam-waveguide output

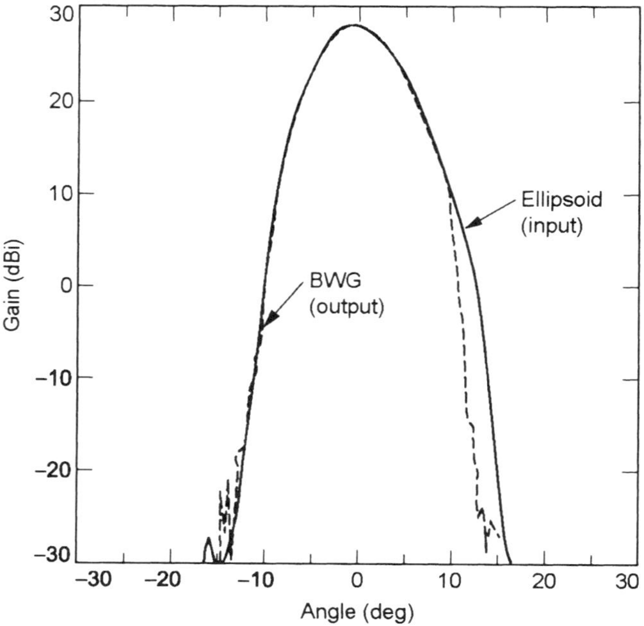



Fig. 7-18. Ka-band center-fed beam-waveguide
output.

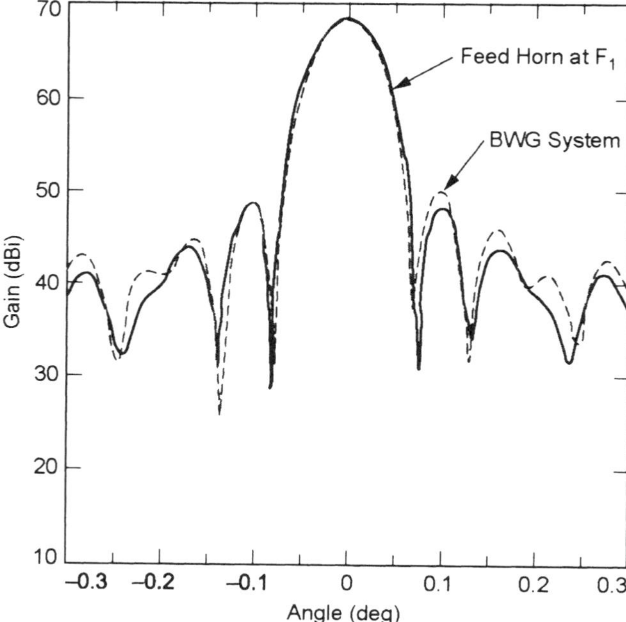



Fig. 7-19. Comparison of horn versus beam-
waveguide feeding the dual-reflector system.

Table 7-1 lists the BWG losses at X- and Ka-band for both BWG systems and shows the reference HEF performance. The loss due to spillover was calculated with the assumption that the mirrors are in free space and that the energy not impinging on the mirrors is lost.

**Table 7-1. Beam-waveguide performance. (Losses due to surface rms, BWG mirror misalignments, subreflector support blockage, and feed system ohmic loss are not included.)**

| Frequency (GHz) | Gain (dBi) (100% Efficient) | DSS-15 HEF Cassegrain Gain (dBi) | DSS-13 Bypass BWG Gain (dBi) | DSS-13 Bypass BWG Portion Due to Spillover | DSS-13 Center-Fed BWG Gain (dBi) | DSS-13 Center-Fed BWG Paraboloid Spill Portion | DSS-13 Center-Fed BWG Ellipsoid Spill Portion |
|---|---|---|---|---|---|---|---|
| 8.45 (X-band) | 69.57 | 69.21 | 69.13 | −0.06 | 69.06 | −0.06 | −0.06 |
| 31.4 (Ka-band) | 80.98 | 80.62 | 80.55 | −0.06 | 80.42 | −0.03 | −0.03 |

### 7.3.6 Dual-Shaped Reflector Design

The requirement for dual-shaped reflectors is to design the contour of the main reflector surface to be within ±0.5 in. (±1.2 cm) of the contour of the HEF antenna. The reason for this is to be able to use the existing backup structure together with the new panels.

The X-band feed-horn pattern at $r = 425$ in. (10.8 m) (the mean distance to the subreflector) is used as an input pattern to the synthesis program developed by Galindo [11]. The basic input parameters were similar to those of the HEF antenna design. The maximum difference between the main reflector surfaces of DSS-13 and those of the HEF antenna is 1.1 cm, which is below the 0.5-in. (1.2-cm) requirement. The corresponding rms is only 0.2 mm, which means the two contours are very close to each other over most of the surface of the dish.

### 7.3.7 The Effect of Using the DSS-15 Main Reflector Panel Molds for Fabricating DSS-13 Panels

In 1988, when the DSS-13 BWG antenna project was still in the planning stages, it was decided that to reduce costs, the main reflector panels for the DSS-13 antenna should be made from the available molds that were used to make the main reflector panels for the DSS-15 antenna. For both antennas, the main reflector surface is made up of nine rings of panels (see Fig. 7-20), with all panels in a given ring identical in shape. The differences in the shape of the panels for the two antennas were minor, and it was believed they would not significantly affect the required performance of the new R&D antenna.

Looking at each of the nine panel rings individually and assuming that the panels on the DSS-13 antenna were made accurately from the DSS-15 manufacturing contours, the panels were mathematically best-fitted to the DSS-13 design contour. The axial errors between these two contours were calculated for each of the nine panel rings by subtracting the reference DSS-13 required shape from the DSS-15 panel contour.

The errors in the first seven panel rings for DSS-13 are minor and cause no significant performance loss at 32 GHz. The errors in panels 8 and 9 are much more significant and contribute noticeably to RF performance degradation at 32 GHz. (It should be noted that the outer half of panel 9 is a noise shield and should not be viewed as contributing to RF gain performance.)

Figures 7-21 and 7-22 show the holographically derived surface errors for rings 8 and 9, respectively, overlaid on the predicted mechanical surface errors for these panels. These plots show good agreement between the manufacturing contour and the holography measurements.

The current antenna surface has an error of 0.28 mm for the inner seven panel rings, 0.60 mm for rings 8 and 9 together, and 0.40 mm for all of the antenna, excluding the noise shield.

The errors in the DSS-13 BWG antenna surface in the outer two panel rings have been measured accurately and are noticeable in 32-GHz performance.

**Fig. 7-20. The 34-m antenna main reflector panel rings.**

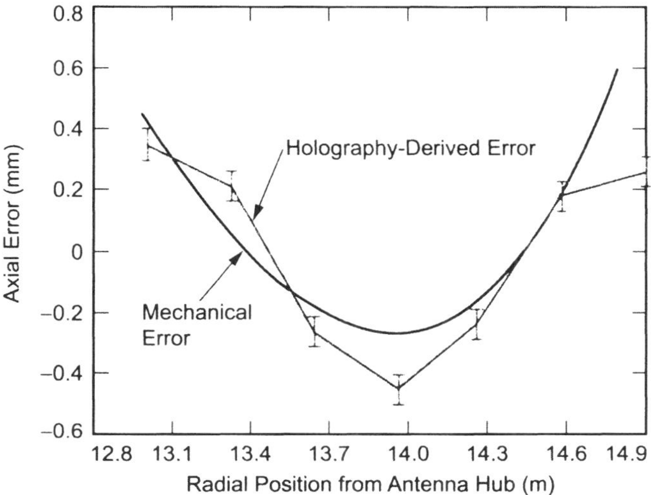



**Fig. 7-21. Mechanical error in panel ring 8, with holographically derived average phase-error function.**

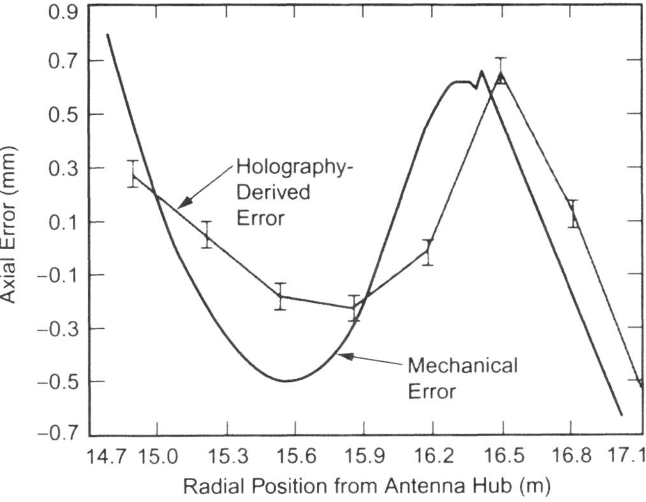



**Fig. 7-22. Mechanical error in panel ring 8, with holographically derived average phase-error function.**

Estimates have been made as to the possible gain improvement that would be obtained if these panels were replaced by panels that had surface errors on the order of those seen on the inner seven panel rings. The outer two rings incorporate 37 percent of the aperture area, not including the noise shield. If this area of the antenna were to have its rms error decreased from 0.60 mm to 0.28 mm, the increase in antenna gain would be 0.6 dB at 32 GHz. The present Ka-band area efficiency of 52 percent at $F_1$ would increase to 60 percent.


## 7.4  Phase I Measured Results

The testing method used on the BWG antenna is unique in that a direct experimental measurement can be made of the degradation contributed by the BWG mirror system. The methodology is to use a portable test package that can be transported to focal point locations $F_1$, $F_2$, and $F_3$. The phase center of the feed horn on the portable test package is made to coincide with the desired focal points, for example, $F_1$. Measurements are made of system temperatures and antenna efficiencies at $F_1$. The differences between system temperatures at $F_1$ and those on the ground give a measure of the additional contributions due to tripod scattering, main reflector spillover, and leakage. The test package is then taken to one of the other focal points ($F_2$ or $F_3$) and measurements are again made of system temperatures and efficiencies. The differences in measurements give a direct measure of the degradations caused by the BWG system mirror and surrounding shrouds. To the author's knowledge, this is the first use of a portable test package to test the integrity of a BWG antenna.

### 7.4.1  The X- and Ka-Band Test Packages

Figure 7-23 shows the system block diagram of the X-band test package [12]. Depicted are the usual Cassegrain front-end components such as a 22-dBi feed horn, polarizer, round-to-rectangular waveguide transition, waveguide switch, cryogenically cooled low-noise amplifier (LNA), and downconverter. The LNA is a high-electron-mobility-transistor (HEMT) assembly described in [13]. Noise-temperature calibrations are performed with the incorporation of a remotely controlled noise-diode assembly and a digital-readout thermometer embedded in an ambient-load-reference termination. For the X-band test package, the microwave signal is downconverted to 350 MHz and sent via coaxial cable to a total-power radiometer (TPR) system.

In order to test the antenna at $F_1$ and $F_3$, the test package is required to be convertible from 29-dBi to 22-dBi feed-horn configurations. This is accomplished through the removal of horn extensions of the same taper going from aperture diameters of about 19 to 7.08 in. (48 to 18 cm).

Figure 7-24 is a photograph of the fabricated and assembled X-band test package in its 22-dBi feed-horn configuration for testing the system on the ground. The test package is about 94.5 in. (2.4 m) high in the 22-dBi horn configuration.

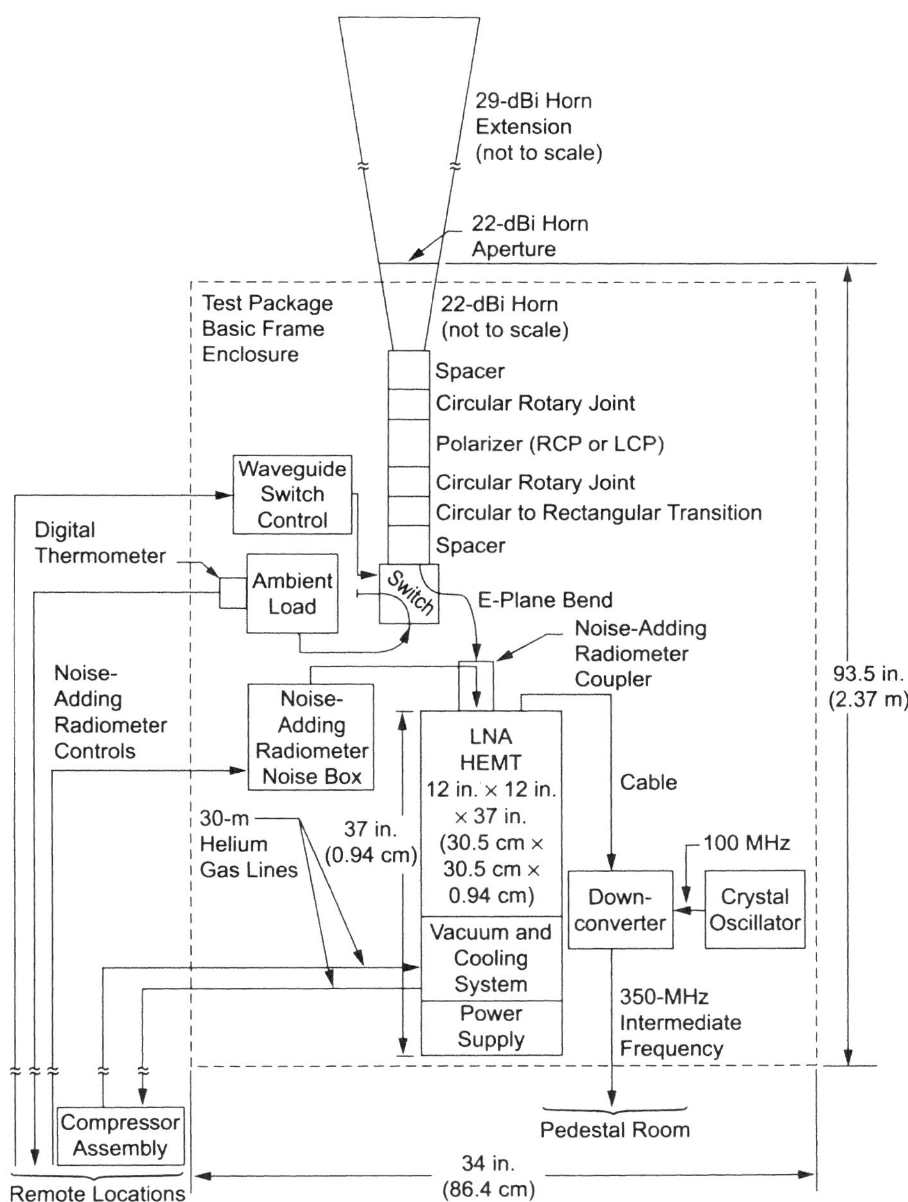


**Fig. 7-23. Block diagram of the X-band test package system.**

The Ka-band test package design is very similar to that of the X-band package and is described in [14].

**Fig. 7-24. X-band test package
in the 22-dBi feed-horn configuration for testing.**

## 7.4.2 Noise Temperature

Figure 7-25 shows the portable X-band front-end test package with the
29-dBi horn installed on the antenna at $F_1$ [15]. A tape measure indicated that
the desired and actual horn phase-center locations agreed to within 0.25 to
0.12 in. (6.3 to 3.1 mm), with measurements accurate to ±0.06 in. (±1.5 mm).
After completion of the noise-temperature and antenna-efficiency measurements at $F_1$, the X-band test package was removed and reconfigured to a
22-dBi feed-horn configuration, and installed at $F_3$ (Fig. 7-26). The mounting
table shown in Fig. 7-26 is a universal mount that can support any of the test
packages and provide three-axis adjustment of the test-package location. The
summary of the X-band zenith system temperatures at DSS-13, from June 10,
1990, to February 2, 1991, is shown in Table 7-2. The differential zenith system
temperatures for the various test configurations are shown in Table 7-3.
Observe that the degradation in noise temperature for the BWG antenna using
the 22-dBi feed horn was 8.9 K. This is significantly higher than originally
expected.

Fig. 7-25. X-band 29-dBi feed-horn test package and mounting assembly installed at $F_1$.

Fig. 7-26. X-band 22-dBi feed-horn test package and mounting table installed at $F_3$.

**Table 7-2. Summary of X-band zenith operating-system temperatures at DSS-13 from June 10, 1990, to February 2, 1991.**

| Configuration | Observation Dates | Grand Average[a] $T_{op}$ (K) | Peak Deviations from Grand Average (K) |
|---|---|---|---|
| Ground | 06/10/90 | 22.7 | +0.3 |
| | 01/21/91 | | |
| | 01/26/91 | | |
| $F_1$ | 10/04/90 | 25.9 | Not available |
| $F_3$ | 11/06/90 | 34.2 | +0.1/–0.1 |
| | 11/09/90 | | |

| | | | |
|---|---|---|---|
| After mirrors and ellipsoid were realigned on December 18, 1990 | | | |

| | | | |
|---|---|---|---|
| $F_3$ | 01/31/91 | 34.8 | +0.1/–0.1 |
| | 02/02/91 | | |

[a] See Table 5 in [15] for the average $T_{op}$ (system-noise temperature) for each observation period. These values formed the basis for obtaining the grand average for a particular test configuration.

**Table 7-3. Differential zenith operating-system temperatures for various test configurations at 8.45 GHz.**

| Configurations Differenced[a] | Delta $T_{op}$ (K) |
|---|---|
| $F_1$ – ground | 3.2 |
| $F_3$ – $F_1$ | 8.3 |

| | |
|---|---|
| After mirrors and ellipsoid were realigned on December 18, 1990 | |

| | |
|---|---|
| $F_3$ – $F_1$ | 8.9 |

[a] See Table 7-2 for ground, $F_1$, and $F_3$ values.

The Ka-band 29-dBi feed-horn test package [14] was installed on the antenna at the Cassegrain focal point $F_1$, and measurements were made of noise temperature and efficiency. After completing the measurements at $F_1$, the Ka-band test package was removed and then modified to a 23-dBi horn configuration and installed at $F_3$ on the universal mounting assembly [16]. The noise-temperature data is summarized in Tables 7-4 and 7-5. The degradation caused by the BWG system at Ka-band was 6.8 K.

**Table 7-4. Summary of Ka-band zenith operating-noise temperatures at DSS-13, from October 12, 1990, through January 31, 1991.**

| Configuration | Observation Dates | Grand Average[a] $T_{op}$ (K) | Peak Deviations from Grand Average (K) |
|---|---|---|---|
| Ground | 10/12/90, 11/09/90, 01/19/91, 01/31/91 | 84.7[b] | +1.6/−1.7 |
| $F_1$ | 10/13/90, 10/14/90, 01/11/91 | 91.8 | +0.4/−0.6 |
| $F_2$ | 01/16/91, 01/17/91 | 97.0[c] | +0.4/−0.4 |
| $F_3$ | 11/10/90, 12/18/90 | 102.4 | +0.1/−0.0 |
| After mirrors and ellipsoid were realigned on December 18, 1990 | | | |
| $F_3$ | 01/23/91, 01/25/91, 01/30/91 | 98.6[d] | +0.1/−0.1 |

[a] See Table 3 in [16] for the average $T_{op}$ for each observation period. These values formed the basis for obtaining the grand average for a particular test configuration.

[b] Ground values were reported in Part I of [14]. The measured ground value of 84.7 K agrees closely with the predicted value of 84.5 K under standard conditions.

[c] No calibrations longer than 10 min were done at $F_2$ with the antenna left at zenith.

[d] This number cannot be compared with the above $F_2$ value. It is probable that the new $F_2$ value was also lower after the mirror realignment, but a measurement was not made.

**Table 7-5. Differential zenith operating-noise temperatures for various test configurations at 32 GHz. (See Table 7-4 for $F_1$, $F_2$, and $F_3$ values. Do not compare the values for $F_3 - F_1$ after December 18, 1990, with the value for $F_2 - F_1$ because the value at $F_2$ might have become lower, but was not remeasured.**

| Configurations Differenced | Delta $T_{op}$ (K) |
|---|---|
| $F_1$ − ground | 7.1 |
| $F_2 - F_1$ | 5.2 |
| $F_3 - F_1$ | 10.6 |
| After mirrors and ellipsoid realigned on December 18, 1990 | |
| $F_3 - F_1$ | 6.8 |

### 7.4.3    Efficiency Calibration at 8.45 and 32 GHz

From July 1990 through January 1991, the 34-m-diameter BWG antenna at the NASA Goldstone deep space communications complex in California's Mojave Desert was tested as part of its postconstruction performance evaluation [17].

Efficiency and pointing performance were characterized at 8.45 and 32 GHz (X- and Ka-bands, respectively) at both the Cassegrain ($F_1$) and BWG ($F_3$) focal points. The $F_1$ focal point is located close to the vertex of the main reflector while the $F_3$ focal point is located about 35 m away, in a subterranean pedestal room.

Through the use of the X- and Ka-band portable test packages, located at both $F_1$ and $F_3$, a direct experimental measurement was made of antenna efficiency and gain degradation caused by the BWG mirror system.

The measurement of antenna efficiency involves boresighting the antenna, measuring the increase in noise temperature due to a radio source of known flux density, and correcting for atmospheric attenuation and radio source size. The technique is well described in both Chapter 1 of this monograph and [17] and will not be repeated here.

The predicted gains at X-band were 68.54 dBi at $F_1$ (the focus of the dual reflector system) and 68.29 dBi for $F_3$ (the focal point of the BWG feed system). The measured efficiency and gain at 8.45 GHz were reported to be 0.754 and 68.34 dBi for $F_1$ and 0.724 and 68.17 dBi for $F_3$ [18]. Since the antenna had a significant main reflector surface distortion as a function of elevation angles, the Ka-band predictions are applicable only for a main reflector adjustment or rigging angle near 45-deg elevation. The predicted $F_1$ total efficiency and overall gain were 0.527 and 78.36 dBi, respectively. The corresponding measured efficiency and gain were 0.523 and 78.33 dBi. At $F_3$, the predicted total antenna efficiency and overall gain at 32 GHz are, respectively, 0.452 and 77.70 dBi, as compared to the measured efficiency and gain values of 0.449 and 77.66 dBi. In general, the agreement between predicted and measured Ka-band efficiencies and gains was very good.

A summary of the X- and Ka-band measured points (adjusted for proper flux values) is shown in Fig. 7-27. The variation of Ka-band gain as a function of elevation angle is quite apparent.

The complete Phase I testing of the BWG antenna is summarized in [19].

### 7.4.4    Optimizing the *G/T* Ratio of the Beam-Waveguide Antenna

During Phase I testing of the DSS-13 BWG antenna, it was discovered that there was higher system-noise temperature than expected. The high noise temperature was caused by the spillover losses of the BWG mirrors having a greater effect than previously thought. It was experimentally determined that

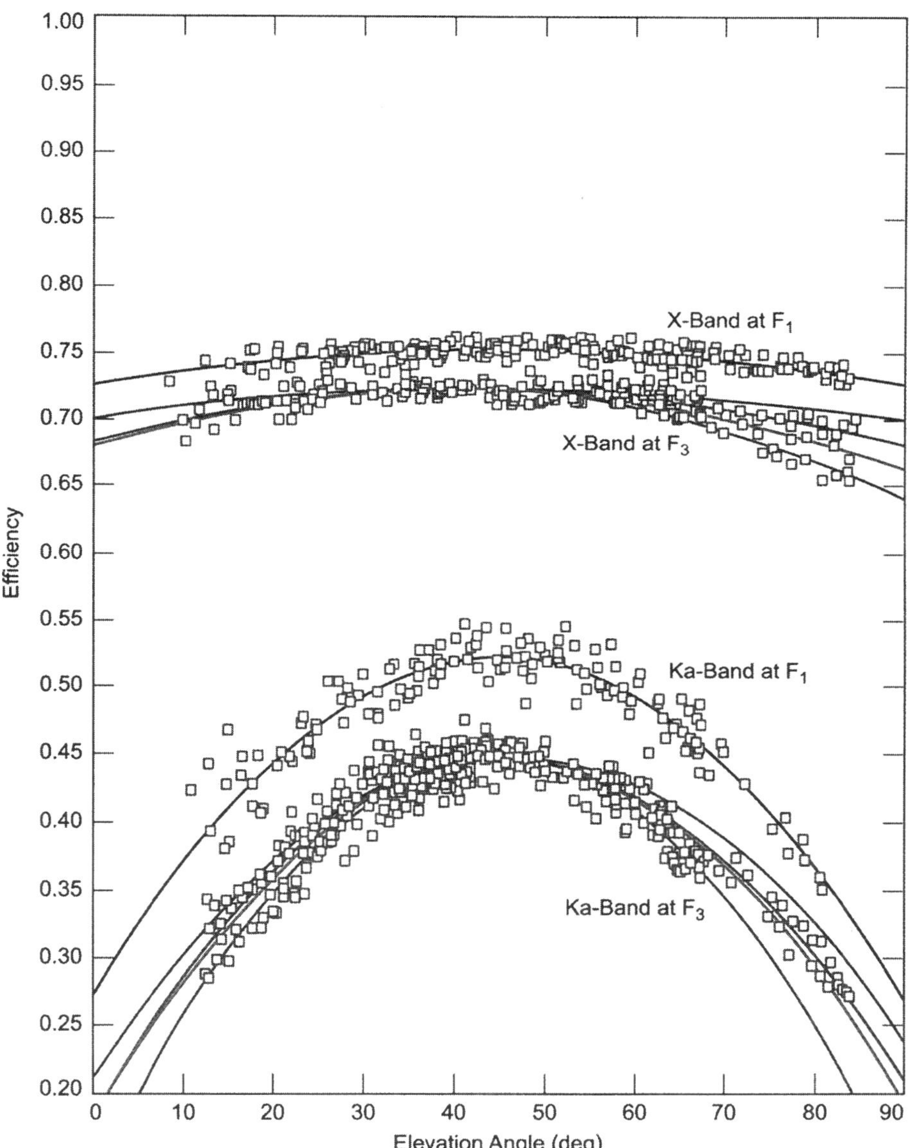


Fig. 7-27. DSS-13 X- and Ka-band efficiencies at $F_1$ and $F_3$ focal points, without atmosphere.

higher-gain feed horns would improve the *G/T* ratio of the antenna for two reasons: (a) there would be a lower spillover loss in the BWG mirrors and, hence, less noise temperature, and (b) when using higher-gain feed horns, the antenna gain would not decrease significantly.

With three different feed-horn patterns as inputs, the PO software was used to analyze the theoretical performance of the DSS-13 BWG antenna [20]. The three patterns used were those of corrugated horns with far-field gains of 22.5, 24.2, and 26.1 dB. Each horn pattern was placed at different positions along the z-axis from $F_3$, the focal point of the basement ellipsoid. From the PO analysis, the spillover of the BWG mirrors and the gain of the DSS-13 BWG antenna could be obtained and, in this manner, various $G/T$ values could be calculated. All of these calculations were done at 8.45 GHz.

The theoretical feed-horn patterns were modeled using the corrugated horn program described in Chapter 1 of this monograph. The far-field horn patterns were then converted into a set of spherical-wave-expansion (SWE) coefficients and then input into the PO software to obtain the gain of the antenna and spill-over of the BWG mirrors. The noise temperature was approximated by ignoring the effects of the BWG tubes and converting spillover energy into noise temperature, using an appropriate temperature factor for the particular mirrors. The technique is further described in [21]. The calculations are summarized in Fig. 7-28.

Figure 7-28(a) shows the efficiency of the DSS-13 BWG antenna for various feed horns as a function of the horn aperture position measured from $F_3$. The efficiency plotted in Fig. 7-28(a) includes the BWG spillover losses and the other losses associated with the antenna. For example, the PO software predicts a gain of 68.933 dB for the 22.5-dB corrugated horn with its aperture located $2\lambda$ from $F_3$. This gain corresponds to an efficiency of 86.36 percent at $f = 8.45$ GHz (69.57 dB is equivalent to 100 percent). The 86.36 percent efficiency would then be multiplied by the efficiency that represents the other losses (84.6 percent) to give a total efficiency of 73 percent [see Fig. 7-28(a)]. This result agrees well with the reported measured value of 72.4 percent.

Figure 7-28(b) shows the total noise temperature of the DSS-13 BWG antenna due to the ultralow-noise amplifier (ULNA) and the feeds as a function of horn aperture position with respect to $F_3$. The ULNA [22] was an X-band ultralow-noise maser specifically designed to operate with the BWG antenna. The total noise temperature was obtained by adding the baseline temperature of 11.05 K of the ULNA to the contributions of the six BWG mirrors, the subreflector and the 34-m main reflector. As expected, the noise temperature decreased as higher-gain horns were used, because they caused less spillover in the BWG mirrors than the lower-gain horns.

For example, for the 22.5-dB corrugated feed horn with its aperture located $2\lambda$ (7.1 cm) from $F_3$, the PO software predicts a spillover noise temperature of 9.41 K. This value is obtained by subtracting the amplifier noise temperature (11.05 K) from the value in Fig. 7-28(b) (20.46 K). This value was calculated by taking the spillovers computed by the PO software and modifying values, using the appropriate temperature factors. The measured BWG noise temperature

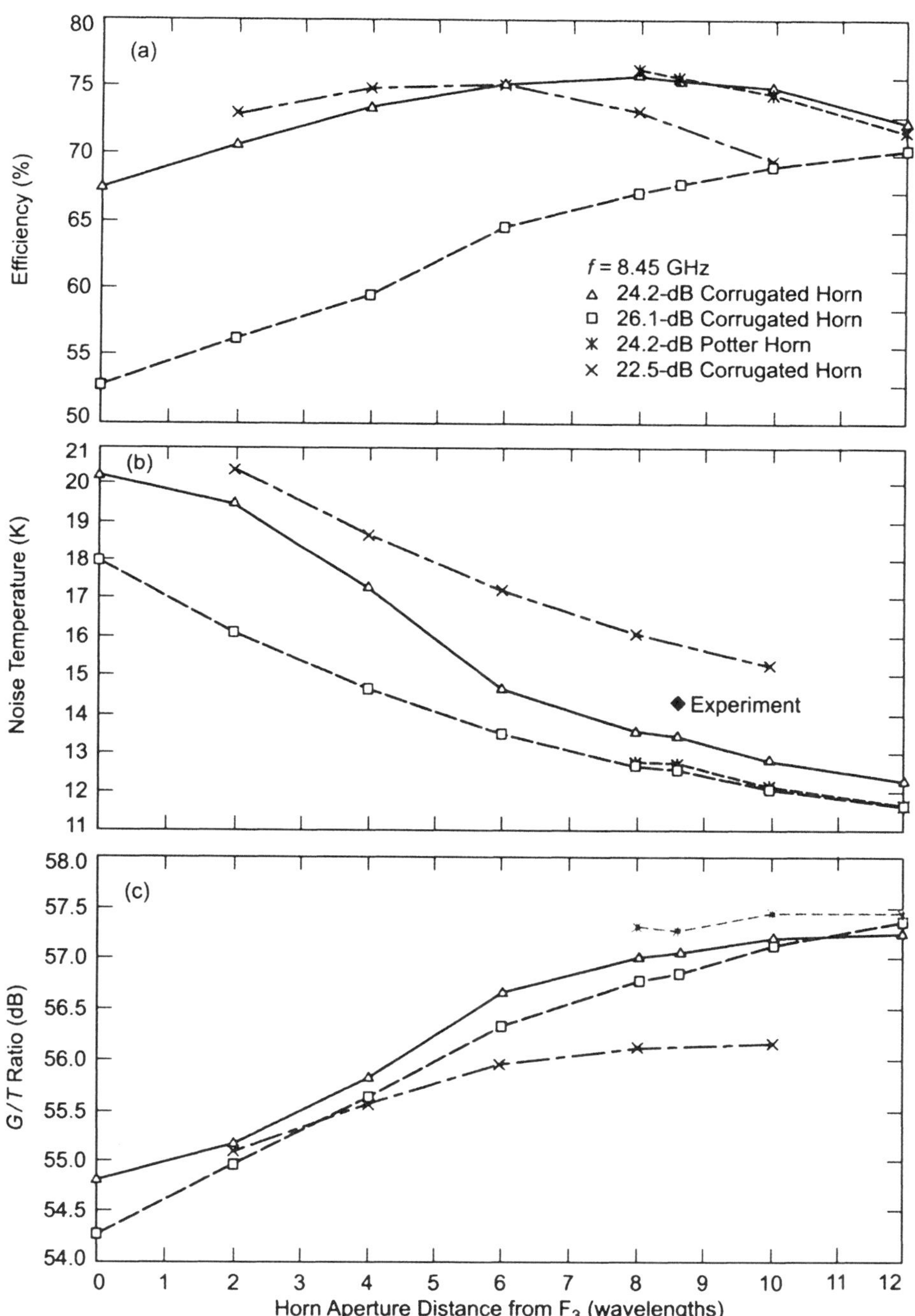


Fig. 7-28. DSS-13: (a) efficiency versus feed-horn aperture distance from $F_3$, (b) system temperature versus horn-aperture distance from $F_3$, and (c) $G/T$ ratio versus horn-aperture distance from $F_3$.

using this horn was 8.9 K when the aperture was placed about 3.3 in. (8.4 cm) from $F_3$.

Figure 7-28(c) shows the $G/T$ values of the DSS-13 BWG antenna for the horns as a function of their aperture position with respect to $F_3$. The $G/T$ ratio is calculated by converting the efficiencies of Fig. 7-28(a) to decibels and then subtracting the total noise temperature (in decibels).

Observe that a feed-horn gain in the 24–26-dBi range gives the optimum $G/T$. It was desirable to verify the improved performance experimentally and, because of cost constraints, it was decided to use a smooth-wall Potter horn instead of a corrugated horn. A 24.2-dB Potter horn was chosen because it had been determined to have nearly the same spillover as the 26.1-dB corrugated horn.

The new 24.2-dB Potter feed horn was then run through the PO software as the other horns had been. The results are shown in Fig. 7-28. Only four cases were run for this horn because results for the corrugated horns predicted that the optimum $G/T$ ratio would be achieved when the horn aperture-to-$F_3$ displacement was about $10\lambda$. Notice that the predicted $G/T$ ratio for the 24.2-dB Potter horn is better than for all the horns. This is because while the antenna efficiency is at the same level as for the 24.2-dB corrugated horn [see Fig. 7-28(a)], the noise temperature is about 0.7 K lower [see Fig. 7-28(b)].

The new 24.2-dB Potter horn was built, and radiation patterns were measured. Then, in November 1991, the system-noise temperature of the DSS-13 BWG antenna was measured using the 24.2-dB Potter horn with the ULNA. The horn's aperture was placed $8.61\lambda$ ($f = 8.45$ GHz) from $F_3$. A value of $N_i = 14.4$ K was observed. The measured point is shown in Fig. 7-28(b). The predicted noise temperature for the 24.2-dB Potter horn was 12.78 K.

### 7.4.5    Beam-Waveguide Antenna Performance in the Bypass Mode

The 34-m BWG antenna was designed with two BWG receiver paths: a centerline feed system, which terminates in the pedestal room at the $F_3$ focal point; and a bypass feed system, which terminates on the alidade structure at the $F_4$ focal point (Fig. 7-6). The centerline feed system has already been evaluated for zenith operating-noise temperature and antenna-area efficiency performance at X-band (8.45 GHz) and Ka-band (32 GHz) at the two focal points, $F_1$ and $F_3$ [15–17]. The bypass BWG feed system has also been evaluated at both X-band and Ka-band [23]. In order to maintain consistency, the same test packages that were used to evaluate the centerline feed BWG system were used to test the bypass BWG feed system.

**7.4.5.1  X-Band Measurements.** The X-band test package with its 29-dBi feed horn was used to make comparative zenith operating-noise temperature measurements at two locations: on the ground and on the antenna at the $F_4$ focal point. Table 7-6 shows the compiled results of the X-band $T_{op}$ (total system-noise temperature) measurements as well as the weather conditions present during the observation period. Table 7-6 also shows the corrections that were made to the operating-noise temperature measurements in order to normalize them to the weather conditions present during an average Goldstone day.

Normalizing the $T_{op}$ measurements to a fixed weather condition, it is easier to compare these measurements with each other as well as with those made at any other time of the year. As a means of comparison between these results for $F_4$ and the results obtained previously for $F_1$ and $F_3$, Table 7-7 contains a summary of some differential zenith operating-noise temperatures. All of the $T_{op}$ values presented in Table 7-7 have been normalized to standard DSS-13 atmospheric conditions at 8.45 GHz.

The antenna's efficiency performance was determined by continuously tracking a stellar radio source and measuring the peak received noise power in relation to the background atmospheric noise over a full 10- to 12-h track. By knowing the maximum signal strength of the radio source, the aperture efficiency with respect to the elevation angle of the antenna can be determined.

Figure 7-29 shows the combined antenna efficiency measurements at X-band versus elevation angle for the $F_4$ focal point, with the effects of atmospheric attenuation removed. An average antenna efficiency with a peak of 71.4 percent at an elevation angle of 38.6 deg is also shown in this figure.

**7.4.5.2  Ka-Band Measurements.** As with X-band, the Ka-band test package with its 29-dBi feed horn was used to make comparative zenith $T_{op}$ measurements on the ground and at the $F_4$ focal point. Continuous measurements at each of these locations were made over two-day periods, with operating-noise-temperature measurements taken every 30 min and ambient weather conditions measured every 10 min. As with the X-band measurements described above, corrections for gain changes are already accounted for in the operating noise-temperature measurements, and no corrections were made for linearity, which would have an effect of less than ±2 percent.

Table 7-6 shows the compiled results of the Ka-band $T_{op}$ measurements, as well as the weather conditions present during the observation period. The normalized $T_{op}$ values, which were calculated, are also included in this table.

Table 7-7 contains a comparison between these results for $F_4$ and the results obtained previously for $F_1$ and $F_3$. This comparison may be slightly erroneous due to the 6-month lapse between measurements. All of the $T_{op}$ values presented in Table 7-7 have been normalized to standard DSS-13 atmospheric conditions at 32 GHz.

**Table 7-6. Measured zenith operating-noise temperatures corrected for weather and waveguide-loss changes for X-band and Ka-band.**

| Configuration | Observation Period (UT) | Average Measured $T_{op}$ (K) | Average Weather During Observation | Computed $T_{atm}$[a] (K) | Computed $L_{atm}$[b] | Physical Waveguide Temperature (°C) | $T_{wg}$[c] (K) | Normalized $T_{op}$ (K) |
|---|---|---|---|---|---|---|---|---|
| X-band on the ground | 03/12/91 1600 03/13/91 1730 | 22.29 | 893.6 mbar 8.9 °C 37.3% RH[d] | 2.25 | 1.0085 (0.037 dB) | 13.34 | 4.58 | 22.32 |
| X-band at $F_4$ | 04/27/91 0500 | 28.84 | 891.3 mbar 17.6 °C 26.4% RH | 2.27 | 1.0086 (0.037 dB) | 13.96 | 4.59 | 28.84 |
| Ka-band on the ground | 04/23/91 1200 04/24/91 2330 | 83.09 | 890.8 mbar 9.96 °C 69.9% RH | 10.94 | 1.0418 (0.178 dB) | 5.05 | 16.77 | 80.33 |
| Ka-band at $F_4$ | 05/11/91 0030 05/12/91 2230 | 93.51 | 894.1 mbar 12.6 °C 34.4% RH | 8.22 | 1.0312 (0.133 dB) | 7.65 | 16.77 | 80.33 |

[a] $T_{atm}$ = atmospheric-noise temperature.
[b] $L_{atm}$ = atmospheric loss
[c] $T_{wg}$ = waveguide-noise temperature.
[d] RH = relative humidity.

**Table 7-7. Differential zenith operating-noise temperatures at X-band.**

| Configurations Differenced | X-Band Delta $T_{op}$ (K) | Ka-Band Delta $T_{op}$ (K) |
| --- | --- | --- |
| $F_1$ – ground | 3.2[a] | 7.1[b] |
| $F_2$ – $F_1$ | Not measured | 5.2[b] |
| $F_3$ – $F_1$ | 8.9[a] | 6.8[b] |
| $F_4$ – $F_1$ | 3.3 | 5.7 |
| $F_2$ – ground | Not measured | 12.3 |
| $F_3$ – ground | 12.1 | 13.9 |
| $F_4$ – ground | 6.5 | 12.8 |

[a] These values obtained from [15].
[b] These values obtained from [16].

During Ka-band testing, 8 full days of antenna efficiency measurements were conducted using four radio sources. Figure 7-30 shows the combined antenna efficiency measurements versus elevation angle for Ka-band at the $F_4$ focal point, with the effects of atmospheric attenuation removed. An average antenna efficiency having a peak of 45.0 percent at an elevation angle of 41.4 deg is also shown. The standard deviation of these data points is 5.0 percent.

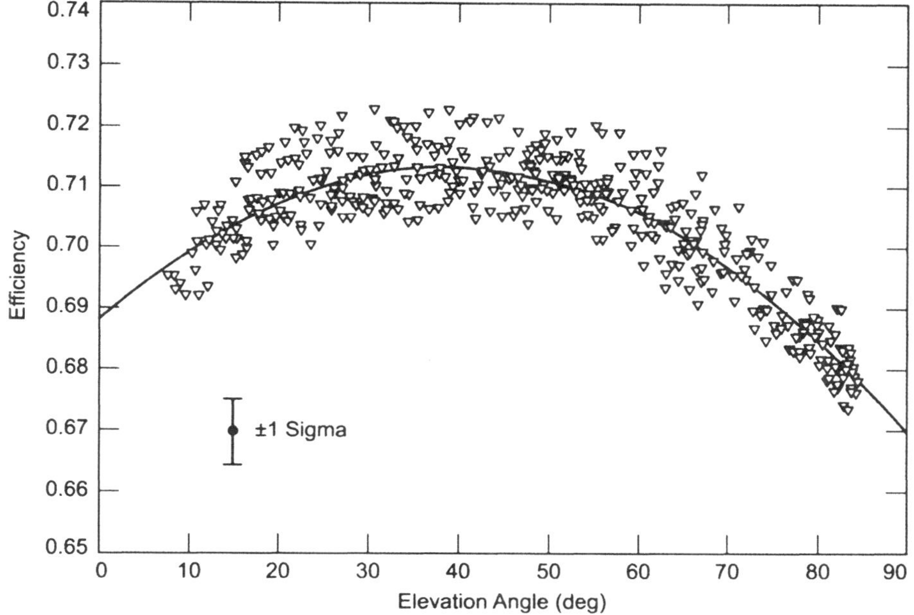


**Fig. 7-29.  X-band efficiency plot for bypass mode.**

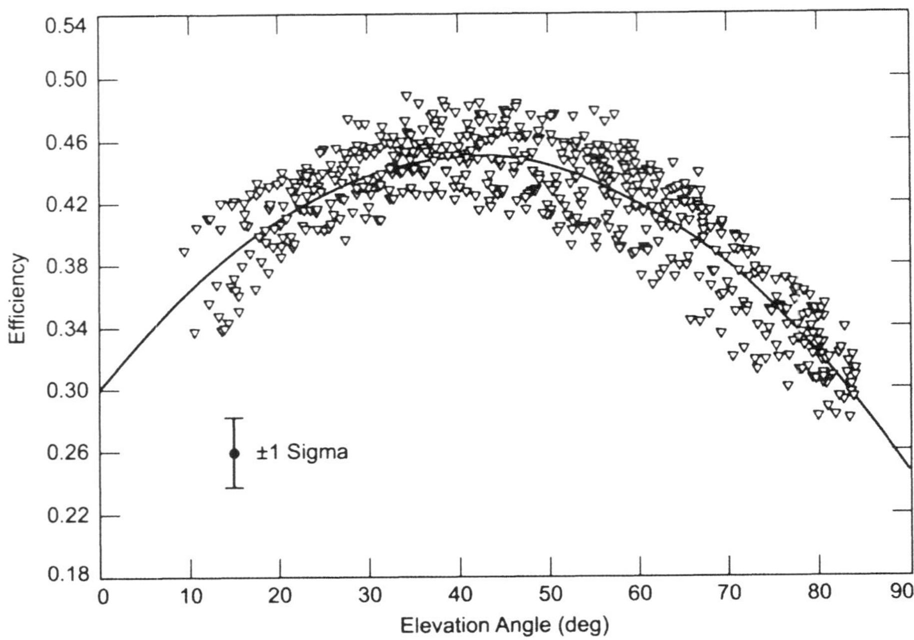

**Fig. 7-30. Ka-band efficiency for bypass mode.**

Figure 7-31 contains the combined results of antenna-efficiency measurements at both X-band and Ka-band for the three focal points, $F_1$, $F_3$, and $F_4$. The results shown in this figure, like those in Figs. 7-29 and 7-30, have had the effects of atmospheric attenuation removed.

## 7.5  Removal of the Bypass Beam Waveguide

Efficiency measurements showed reasonable X-band performance approximately equivalent to that of the DSN predecessor 34-m HEF antennas. However, Ka-band performance (Figs. 7-27, 7-30, and 7-31) at elevations away from the 45-deg rigging angle was worse than that anticipated from the finite-element method (FEM) model predictions for the effects of gravity loading at these elevations.

Figure 7-32 shows curves for the root-mean-square (rms) surface-path-length-error variation with the antenna elevation angle [24]. The highest valued curve is derived from the Ka-band measurements by converting gain loss to rms, using the Ruze equation [25]. One of the other two curves is derived from the JPL iterative design and antenna structures (IDEAS) [26] FEM software program that was used in the original design. The remaining curve comes from a later NASTRAN [27] FEM program analysis model that incorporated model-

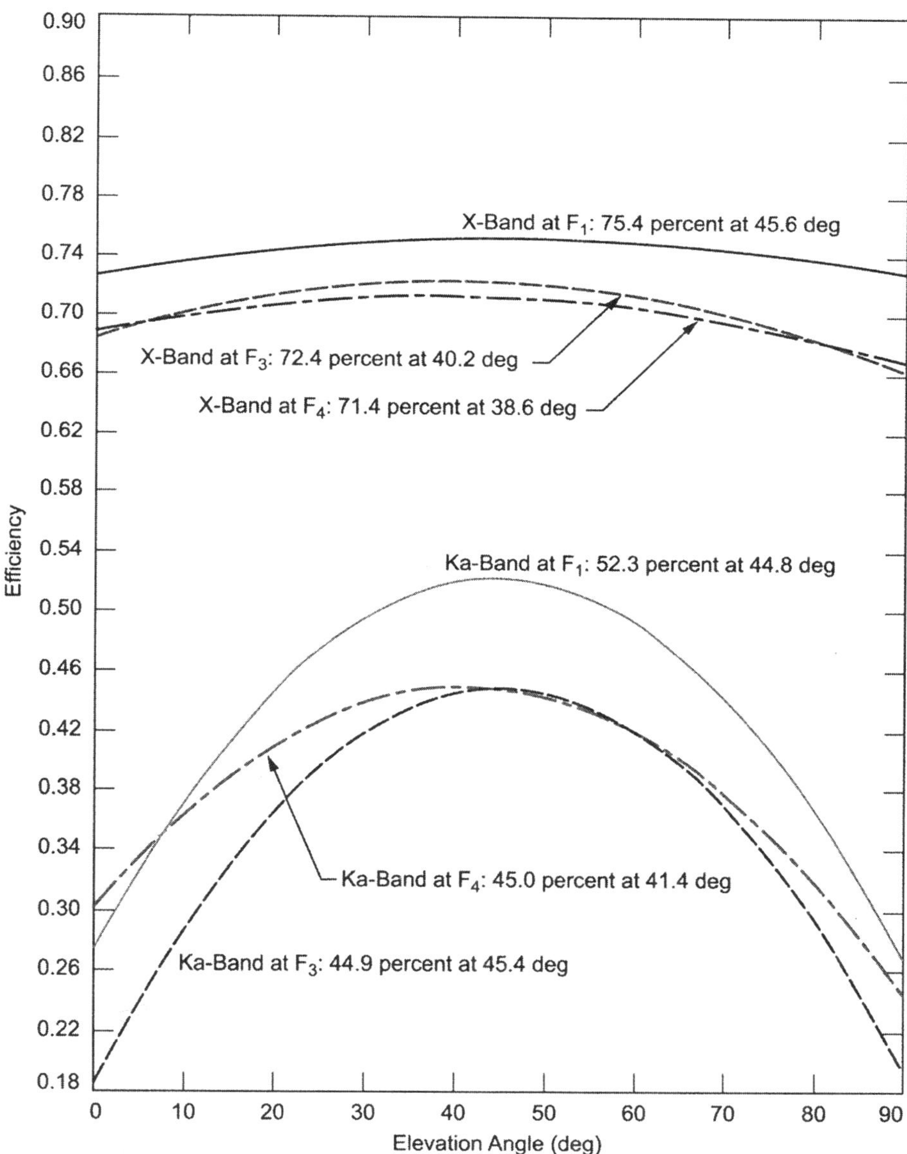


**Fig. 7-31. Comparative results for all the efficiency measurements on the DSS-13 antenna.**

ing refinements and as-built modifications. The errors found from measurements significantly exceed those predicted by the models. The measurement-derived curve amounts to a gain degradation of about 2.6 dB at the extreme elevations, while the NASTRAN model produces only from about 0.6- to 1.1-dB losses. The IDEAS model produces even smaller losses.

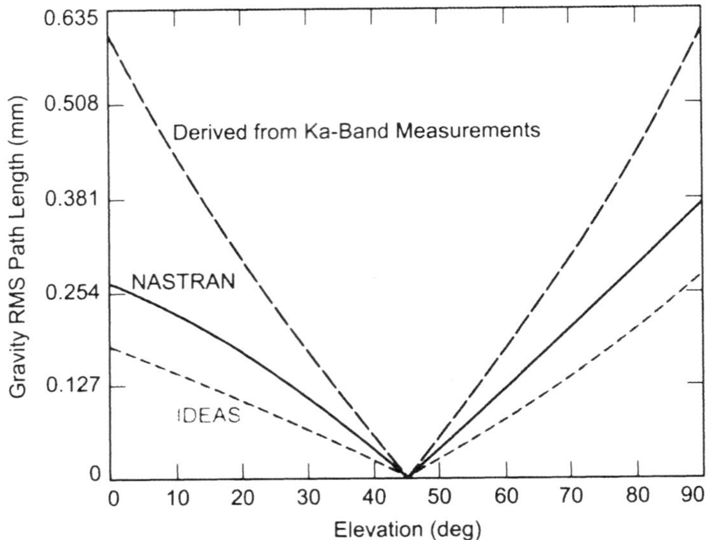


**Fig. 7-32. DSS-13 path-length errors from measurements and models.**

Neither of the analytical models accounted for forces from two BWG shroud structures interacting with the backup structure. As shown in Fig. 7-6, one shroud is at the center of the antenna, and the other is a bypass shroud offset from the center. These shrouds are steel shells approximately 8 ft (2.43 m) in diameter with 0.19-in. (4.8-mm)-thick walls that are connected to both the backup structure through bolts and clip angle plate brackets and to the alidade through large-diameter bearings. The models, which were constructed in advance of the eventual detailed design drawings, did not anticipate that shroud loads would interact with the structure. Nevertheless, depending upon how they are connected, loads transmitted by the shroud shells from the alidade into the backup structure can be significant.

A further analytical study examined the potential influence of the shroud-backup interaction forces, and the model with simulated shrouds was capable of approaching the measurements. A major field-measurement strain gauge program was therefore undertaken to measure the actual shroud-backup structure interaction forces.

The primary measurement program consisted of strain gauge measurements and theodolite surface measurements. Supplementary measurements included laboratory load testing of shroud brackets and dial-indicator measurements of relative shroud–antenna motion.

The strain gauge instrumentation and numerical integration provided the magnitudes of the shroud forces. The bypass-shroud forces caused a major disturbance to the surface shape. Analytical FEM models confirmed the sensitivity of the surface to shroud forces of these magnitudes. Figure 7-33 shows the theodolite measurement points and Fig. 7-34 the best-fit surface errors derived from the theodolite measurements for the difference between the 45-deg rigging angle and the 6-deg elevation angle. The corresponding plot from the FEM model is shown in Fig. 7-35. The effect of the shroud can be clearly seen in the measured data.

The recommendations of the study were either to reduce the interaction of the bypass shroud with the backup structure or else remove the shroud completely. In October 1991, it was proposed to remove the bypass shroud, which no longer was considered to be a microwave requirement, and to free the structure from the associated interaction forces. There appeared to be general agreement that the structure performance would improve through the elimination of the bypass-shroud forces, which acted unsymmetrically on the structure and, therefore, were inconsistent with the desirable objective of homologous struc-

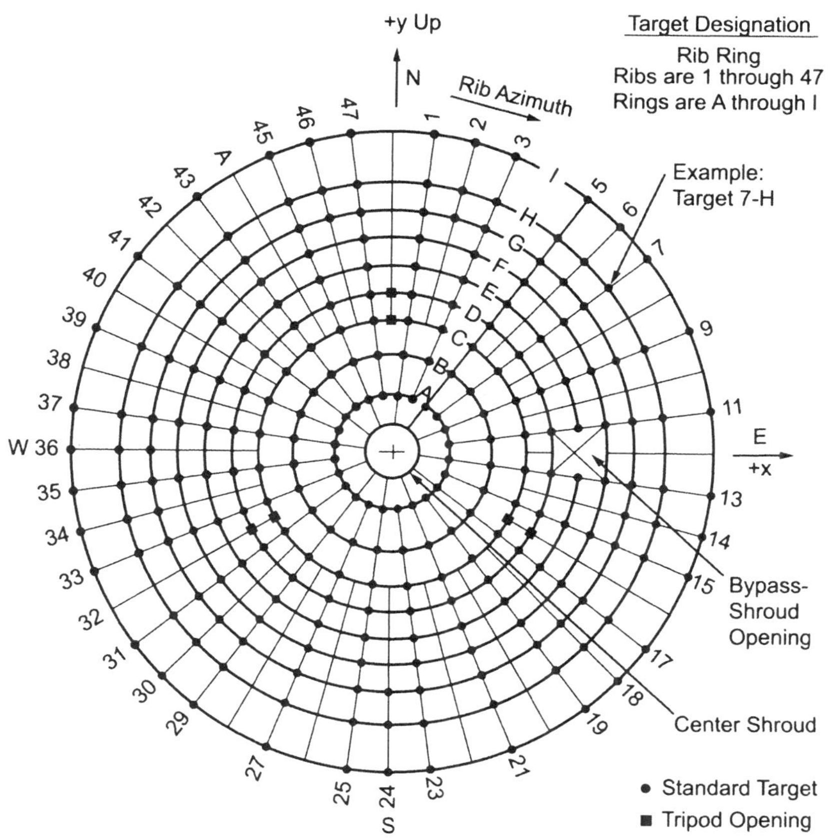


**Fig. 7-33. Theodolite measurement points.**

ture response. Theodolite and holography measurements clearly established unsymmetrical and nonhomologous surface-distortion patterns.

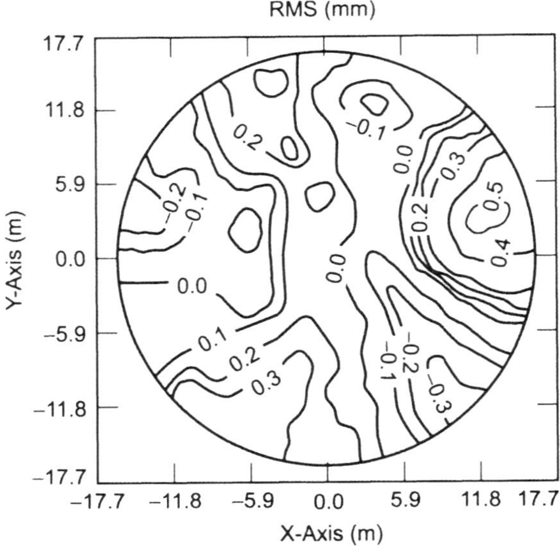



Fig. 7-34. Best-fit surface errors, elevation 6–45 deg, from theodolite measurements.

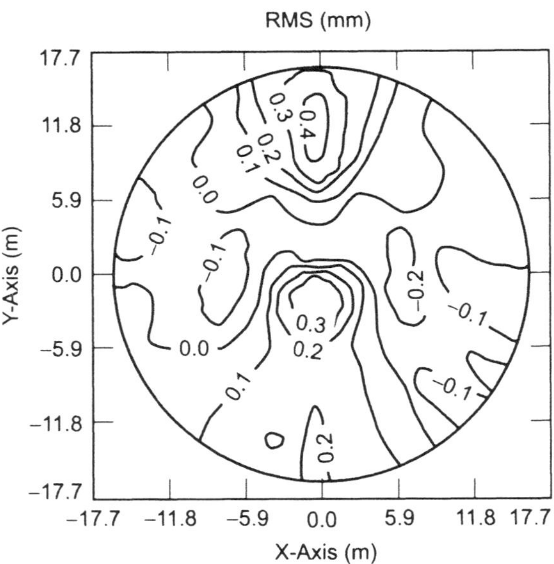



Fig. 7-35. Best-fit surface errors, elevation 6–45 deg, from FEM model.

Ka-band efficiency as a function of elevation angle at $F_1$ before and after removal of the bypass shroud is shown in Fig. 7-36. Table 7-8 shows the theodolite and holography measurements (rms values) and FEM model predictions. Unfortunately, efficiency roll-off as a function of elevation angle is not substantially different from before removal of the bypass. Thus, even though FEM models the structure with the bypass removed, it still lacks the fidelity to accu-

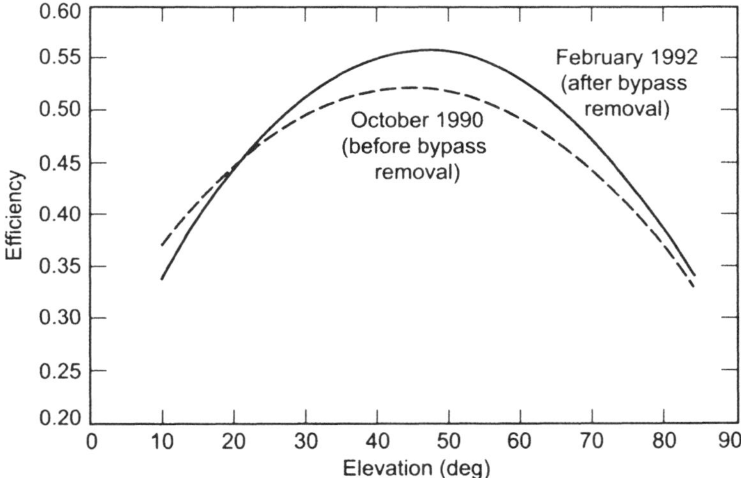


**Fig. 7-36. DSS-13 Ka-band efficiency at $F_1$ before and after bypass removal.**

**Table 7-8. DSS-13 theodolite measurements and model predictions for gravity loading (best-fit rms half-path-length errors, in mm). Approximate 1-sigma theodolite measurement noise at elevation = 6 deg, noise = 0.41; elevation = 45 deg, noise = 0.23; elevation = 90 deg, noise = 0.18.**

| Case | Theodolite Measurement | Model Prediction | Holography | Derived from Gain |
|---|---|---|---|---|
| Elevations 6 deg from 45 deg | | | | |
|    With bypass | 0.36 | 0.18 | 0.38 | 0.46 |
|    Without bypass | 0.46 | 0.23 | 0.53 | 0.48 |
| Elevations 90 deg from 45 deg | | | | |
|    With bypass | 0.61 | 0.38 | Not measured | Not measured |
|    Without bypass | 0.51 | 0.36 | | |

rately predict Ka-band performance. Much of the difficulty is traced to the structure joints in that the model does not accurately reflect the actual joint design implemented. A significant effort was made to bring the actual joint design and FEM modeling more in line, and the antenna drawings were updated with the new design. The operational antenna (see Chapter 8 of this monograph) was built with the updated joint design, and performs substantially better than DSS-13. A photograph of the antenna after the bypass removal is shown in Fig. 7-37.

## 7.6  Multifrequency Operation

Most planetary spacecraft use multiple frequencies for both radio science and reliability reasons. Simultaneous dual frequency is provided in the ground system either through the use of a dichroic plate or a multifrequency feed. The initial dual-frequency implementation at DSS-13 utilized dichroic plates. Later implementations of multiple frequencies in BWG systems utilized multiple-frequency feeds.

### 7.6.1  X-/Ka-Band System

The design of the X-/Ka-band system is shown in Figs. 7-38 and 7-39. The optics design uses an additional refocusing elliptical mirror at Ka-band to pro-

**Fig. 7-37. DSS-13 after bypass beam-waveguide removal.**

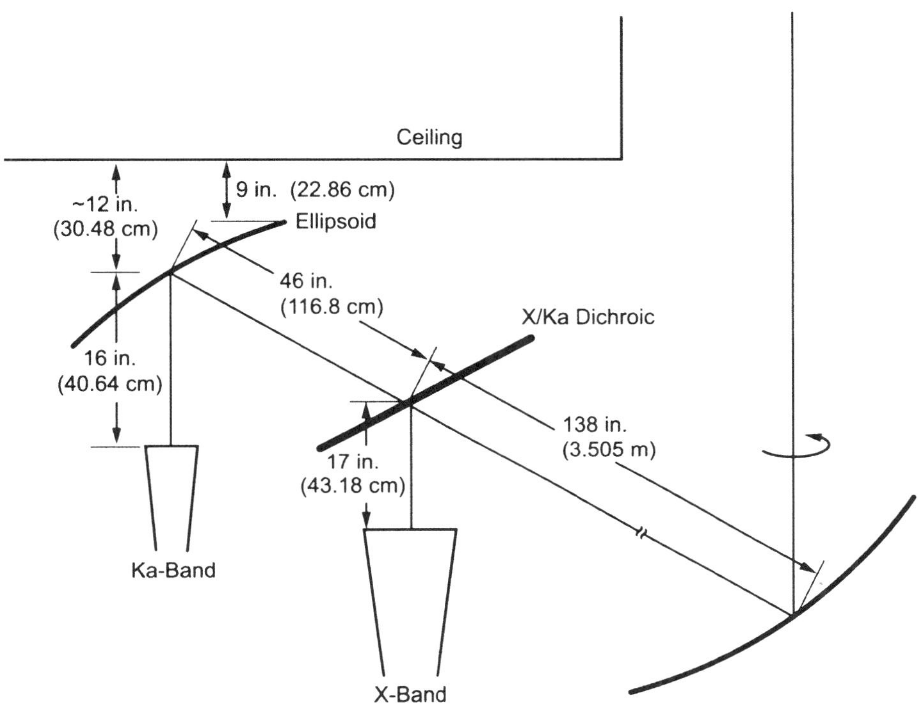


**Fig. 7-38. Optics design configuration of the X- and Ka-band feed system for the DSS-13 Phase II.**

vide sufficient room to fit both receive systems by placing the Ka-band focal point further from the M5 elliptical mirror than the X-band focal point. This is accomplished by putting the Ka-band feed horn at one focus of the ellipse and the other focal point at the input to the M5 magnifying ellipse. In addition, to reduce the spillover loss, the gain of the X- and Ka-band feed horns is increased from the Phase I values of about 22.5 dB, to 25.0 dB and 26.1 dB, respectively. The horn section required to be added on to the X- (as well as Ka-) band (Phase I) horn is identical to the first extended horn section from the JPL standard 22-dB to 29-dB horn (see Fig. 7-40). The ellipse geometry is shown in Fig. 7-41 and the geometry of the dichroic plate in Fig. 7-42. A 20-kW X-band transmitter at 7.7 GHz was also included in the X-band as part of the X-/Ka-band system using a waveguide diplexer. Tests were successfully conducted in the early 1990s with X-band transmitting while simultaneously receiving X- and Ka-bands. There were no noise bursts or intermodulation product signals detected. About the same time, tests were also conducted with the Mars Observer spacecraft to demonstrate the first-ever Ka-band

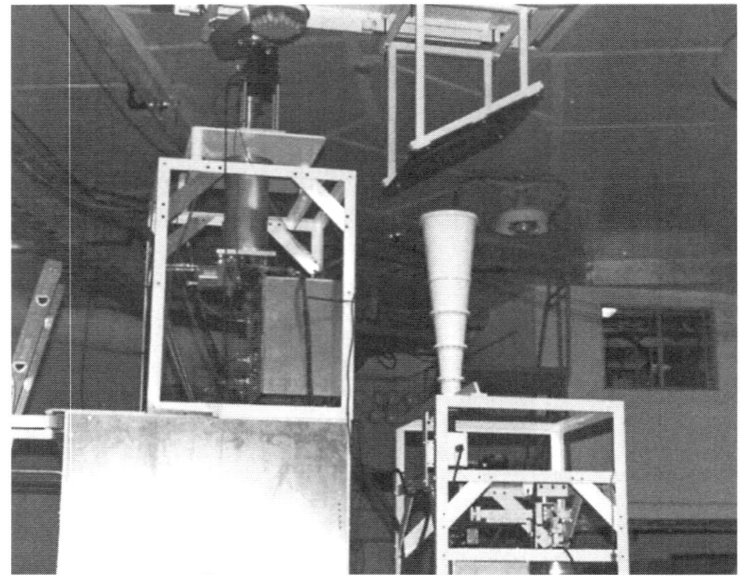

Fig. 7-39.  DSS-13 X- and Ka-band feed system.

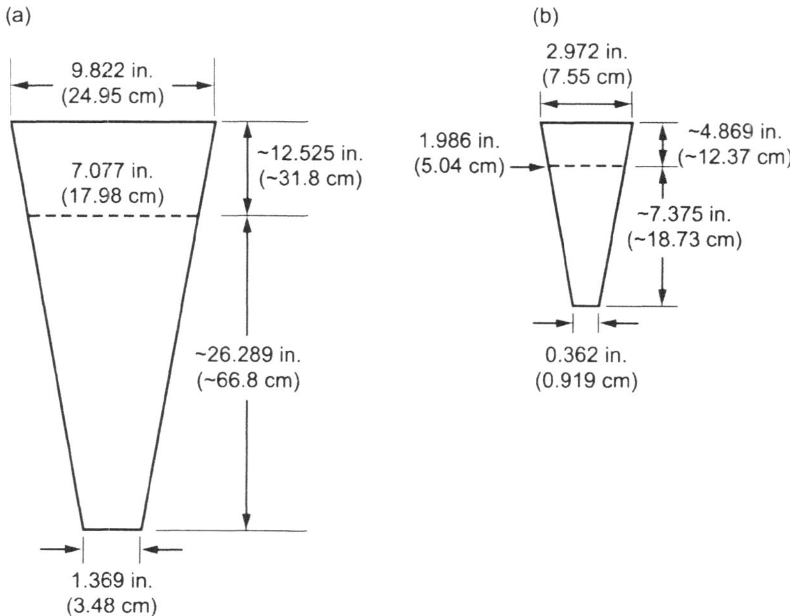


Fig. 7-40.  Geometry of the X- and Ka-band feed horns for DSS-13 Phase II.
All diameters are inside diameters.

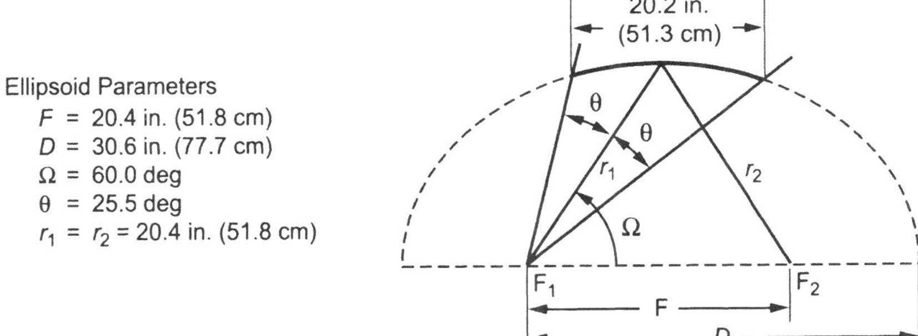



Ellipsoid Parameters
$F$ = 20.4 in. (51.8 cm)
$D$ = 30.6 in. (77.7 cm)
$\Omega$ = 60.0 deg
$\theta$ = 25.5 deg
$r_1 = r_2$ = 20.4 in. (51.8 cm)

**Fig. 7-41. Detail geometry of the ellipsoid above the Ka-band feed horn as shown in Fig. 7-1.**

(33.67-GHz) telemetry reception with a deep-space planetary spacecraft [28,29]. The measured X- and Ka-band efficiencies are shown in Fig. 7-43.

Since the loss of Mars Observer (1993) and the launch of Mars Global Surveyor (1996) spacecraft, tests at Ka-band have continued to assess the quality of the Ka-band link [30,31].

The X-/Ka-band system was subsequently modified to include Ka-band transmit. This was accomplished by adding a Ka-band transmit feed and using a five-layer dichroic plate [32] to separate the transmit and receive Ka-band frequencies. A sketch of the design is shown in Fig. 7-44. Simultaneous Ka-band transmit with Ka- and X-band reception was successfully demonstrated.

### 7.6.2　S-Band Design

As stated earlier, the DSS-13 BWG system was initially designed (Phase I) for operation at 8.45 GHz (X-band) and 32 GHz (Ka-band) and utilized a GO design technique. In Phase II, 2.3 GHz (S-band) was to be added. At S-band, the mirror diameter is less than 20 wavelengths, and it is clearly outside the original GO design criteria.

If a standard 22-dB S-band feed horn were to be placed at the input focus of the ellipse, the BWG loss would be greater than 1.5 dB, due primarily to the fact that, for low frequencies, the diffraction phase centers are far from the GO mirror focus, resulting in a substantial spillover and defocusing loss. This defocusing is especially a problem for the magnifier ellipse, where the S-band phase center at the output of the ellipse is 3 m away from the GO focus.

A potential solution was to redesign the feed horn to provide an optimum solution for S-band. The question was how to determine the appropriate gain and location for this feed.

A straightforward design by analysis would have proved cumbersome because of the large number of scattering surfaces required for the computa-

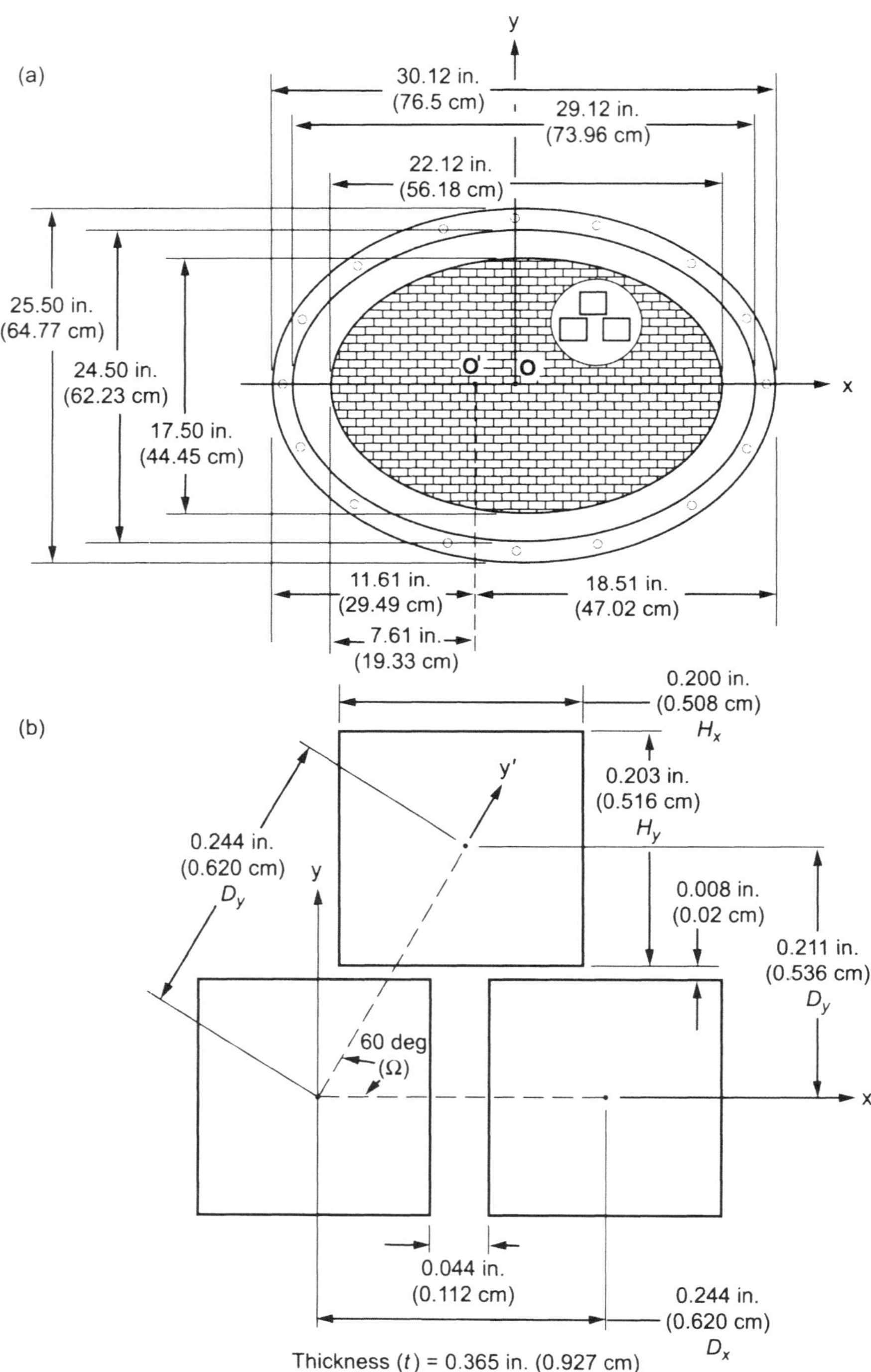


Fig. 7-42. X-/Ka-band dichroic geometry for DSS-13 Phase II: (a) layout of dichroic plate and (b) detailed dimensions of holes.

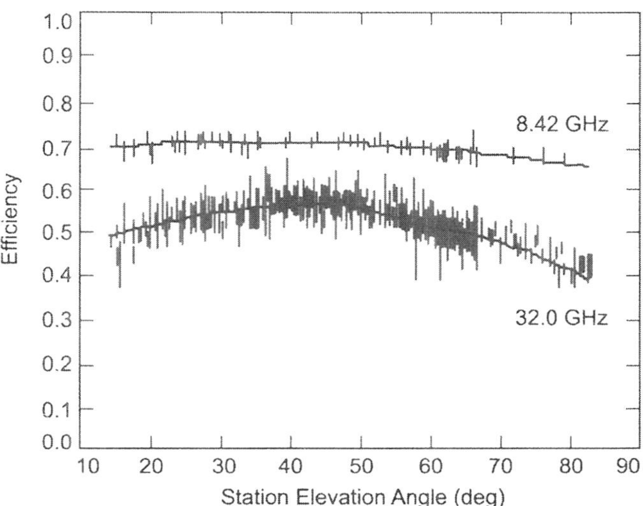



**Fig. 7-43. Antenna efficiency versus elevation angle of the DSS-13 beam-waveguide antenna at X-band (8.4 GHz) and Ka-band (32.0 GHz) along with best-fit curves. Redrawn from [31].**

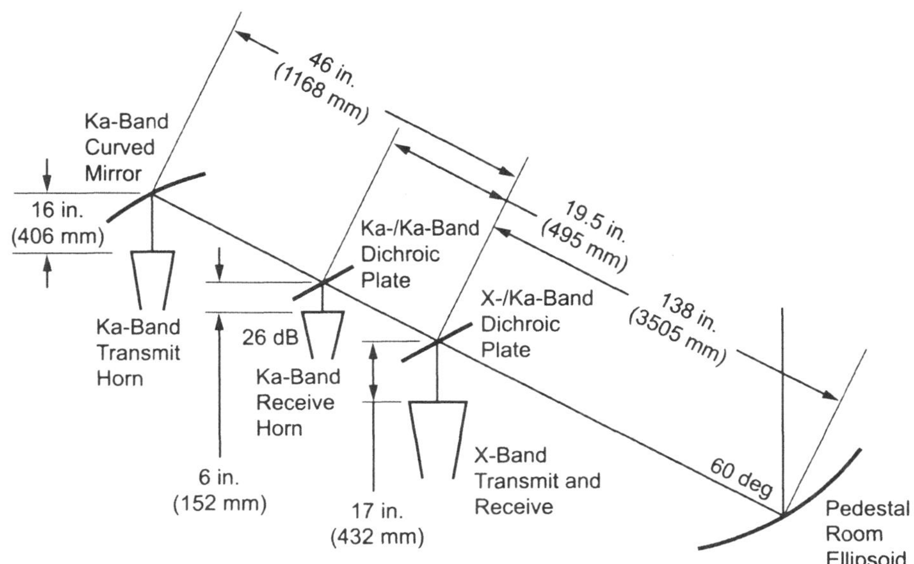



**Fig. 7-44. Ka-band transmit mode optical layout at DSS-13.**

tion. Rather, a unique application was made of the conjugate phase-matching technique to obtain the desired solution [33–35].

A plane wave was used to illuminate the main reflector, and the fields from the currents induced on the subreflector were propagated through the BWG to a plane centered on the input focal point of the M5 ellipse. By taking the complex-conjugate of the currents induced on the focal plane and applying the radiation integral, the far-field pattern of a feed horn that would theoretically maximize antenna gain was obtained.

There was no a priori guarantee that the pattern produced by this method could be produced by a single corrugated feed horn. However, the pattern was nearly circularly symmetric, and the theoretical horn pattern was able to be matched fairly well by an appropriately sized circular corrugated horn.

Figure 7-45 shows the near-field E-plane patterns of the theoretical horn pattern and a 19-dBi circular corrugated horn. The agreement in amplitude and phase is quite good out to $\theta = 21$ deg, the angle subtended by the beam-magnifier ellipse.

The corrugated horn performance was only 0.2 dB lower than that of the optimum theoretical horn, but about 1.4 dB above that which would be obtained using a standard 22-dB horn. A system employing the corrugated horn was built and tested and installed in the 34-m BWG antenna as part of a simultaneous S-/X-band receiving system. Figure 7-46 shows the geometry, and Fig. 7-47 a photograph of the actual S-/X-band system.

Table 7-9 lists the PO analysis results of the antenna at S-band. In this table, the spillover of the antenna mirrors, the antenna efficiency, and system-noise temperature are listed for the 19-dBi corrugated feed horn and the theoretical horn pattern predicted by the focal-plane method.

Also, for comparison purposes, the calculated performance of a 22-dBi corrugated horn is presented from [36]. The S-band feed is part of a simultaneous

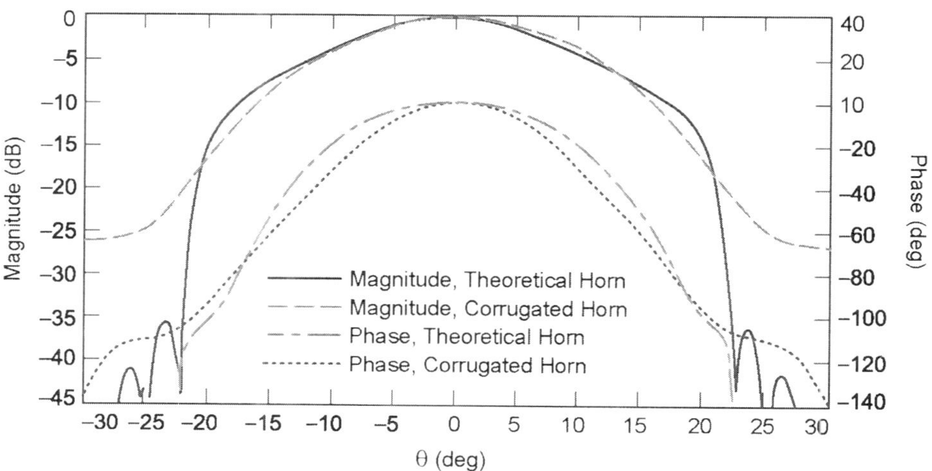



**Fig. 7-45. E-plane near-field patterns (input to the ellipse).**

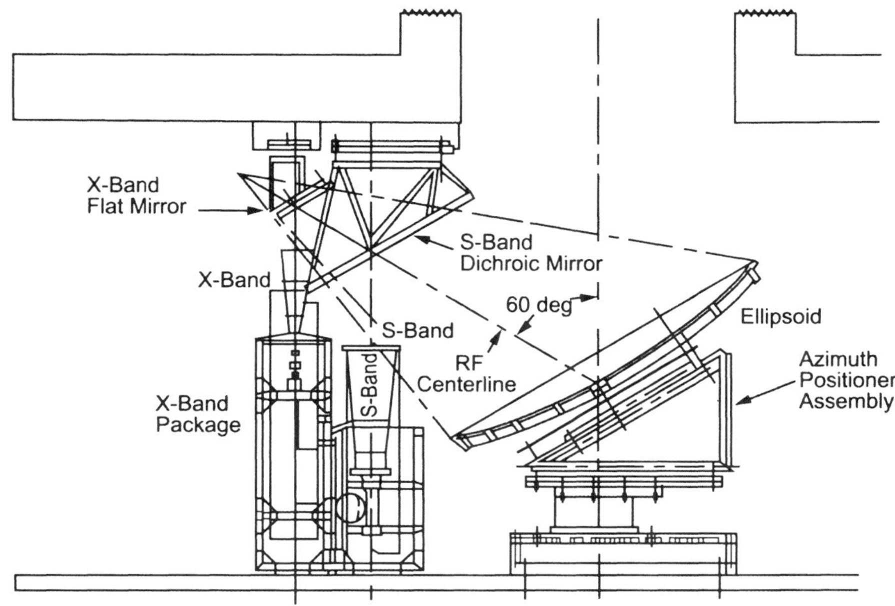



Fig. 7-46. Feed system installed in the beam-waveguide antenna.

Fig. 7-47. DSS-13 S-/X-band
feed-system geometry.

S-/X-band receive system implemented on the new BWG antenna. The general configuration of the feed system, the detail design, and measured feed system performance are described in [37].

Table 7-9 shows the 34-m BWG antenna predicted S-band efficiency was 68 percent and the measured efficiency was 67.5 percent, demonstrating the successful design and implementation. For comparison, the predicted X-band efficiency (at the rigging angle of 45 deg) was 72.7 percent, and the measured efficiency, including the dichroic plate, was 70.1 percent.


## 7.7  Beam-Waveguide Versatility

After Phase I and at the beginning of Phase II, the basement ellipsoid was placed on a rotating platform to allow easy switching between a number of feed stations assembled in the pedestal room. This versatility of the BWG antenna can be seen in Fig. 7-48. At the time of the photograph, there were five feed stations in the R&D BWG antenna. One was utilized for S-/X-band simultaneous receive operation and one for X-/Ka-band receive and X-band transmit. There was a vernier beam-steering mirror position, which enables conscan operation at Ka-band without scanning the main reflector. There was a seven-element array feed system at this position. The other two positions are for general R&D use and have housed at various times a Ku-band (11.7–12.2-GHz) holography system, the X-band ULNA, and a 49-GHz radio science receiver

**Fig. 7-48.  Pedestal room at DSS-13.**

**Table 7-9. S-band physical optics calculations.**

| Optical Element | Spillover (percent) | | |
| --- | --- | --- | --- |
| | 22-dBi Corrugated Horn | 19-dBi Corrugated Horn | Theoretical Horn |
| M6 | 0.00 | 0.41 | 0.00 |
| M5 | 2.05 | 2.46 | 0.24 |
| M4 | 1.57 | 0.70 | 1.19 |
| M3 | 5.91 | 0.73 | 0.86 |
| M2 | 5.55 | 0.96 | 1.29 |
| M1 | 1.26 | 0.26 | 0.46 |
| Main reflector | 0.00 | 0.94 | 3.61 |
| Subreflector | 0.00 | 1.14 | 1.94 |
| Total efficiency (percent) | 48.415 | 68.27 | 69.502 |
| Total efficiency (dB) | 55.102 | 56.594 | 56.672 |

system. Since this photograph was taken, two more positions for R&D use have been added.

# References

[1] J. W. Layland, R. L. Horttor, R. C., Clauss, J. H Wilcher, R. J. Wallace, and D. J. Mudgway, "Ka-band Study-1988," *Telecommunications and Data Acquisition Progress Report 42-96*, vol. October–December1988, http://tmo.jpl.nasa.gov/progress_report/issues.html Accessed August 2001.

[2] A. G. Cha, W. A. Imbriale, and J. Withington, "New Analysis of Beam-waveguide Antennas Considering the Presence of the Metallic Tube and its Experimental Verification," *1990 Antennas and Propagation Symposium Digest*, Dallas, Texas, pp. 1506–1509, May 1990.

[3] A. G. Cha and W. A. Imbriale, "A New Analysis of Beam Waveguide Antennas Considering the Presence of the Metal Enclosure," *IEEE Transactions on Antennas and Propagation*, vol. 40, no. 9, pp. 1041–1046, September 1992.

[4] J. R. Withington, W. A. Imbriale, and P Withington, "The JPL Beam-waveguide Test Facility," *1991 Antennas and Propagation Symposium*, Ontario, Canada, pp. 1194–1197, June 1991.

[5]  T. Veruttipong, J. R. Withington, V. Galindo-Israel, W. A. Imbriale, and D. Bathker, "Design Considerations for Beam Waveguide in the NASA Deep Space Network," *IEEE Transactions on Antennas and Propagation*, vol. AP-36, pp. 1779–1787, December 1988.

[6]  T. Veruttipong, W. Imbriale, and D. Bathker, "Design and Performance Analysis of the New NASA Beam Waveguide Antenna," National Radio Science Meeting, Boulder, Colorado, January 3–5, 1990.

[7]  T. Veruttipong, W. Imbriale, and D. Bathker, "Design and Performance Analysis of the DSS-13 Beam Waveguide Antenna," *Telecommunications and Data Acquisition Progress Report 42-101*, vol. January–March 1990, http://tmo.jpl.nasa.gov/progress_report/issues.html   Accessed   August 2001.

[8]  M. Mizusawa and T. Kitsuregawa, "A Beamwaveguide Feed Having a Symmetric Beam for Cassegrain Antennas," *IEEE Transactions on Antennas and Propagation*, vol. AP-21, pp. 844–846, November 1973.

[9]  M. Mizusawa, "Effect of Scattering Pattern of Subreflector on Radiation Characteristics of Shaped-Reflector Cassegrain Antenna," *Transactions of the International Electronics and Communications Exhibition*, 52-B, Japan, pp. 78–85, February 1969.

[10] V. Galindo-Israel, T. Veruttipong, and W. Imbriale, "GTD, Physical Optics, and Jacobi-Bessel Diffraction Analysis of Beam Waveguide Ellipsoids," *IEEE AP-S International Symposium Digest*, vol. 2, Philadelphia, Pennsylvania, pp. 643–646, June 1986.

[11] V. Galindo-Israel, "Circularly Symmetric Dual-Shaped Reflector Antenna Synthesis with Interpolation Software-User Manual," (internal document) Jet Propulsion Laboratory, Pasadena, California, January 1988.

[12] T. Y. Otoshi, S. R. Stewart, and M. M. Franco, "A Portable X-Band Front-End Test Package for Beam-Waveguide Antenna Performance Evaluation—Part I: Design and Ground Tests," *Telecommunications and Data Acquisition Progress Report 42-103*, vol. July–September 1990, http://tmo.jpl.nasa.gov/progress_report/issues.html Accessed August 2001.

[13] L. Tanida, "An 8.4-GHz Cryogenically Cooled HEMT Amplifier for DSS 13," *Telecommunications and Data Acquisition Progress Report 42-94*, vol. April–June 1988, http://tmo.jpl.nasa.gov/progress_report/ issues.html   Accessed August 2001.

[14] T. Y. Otoshi, S. R. Stewart, and M. M. Franco, "A Portable Ka-Band Front-End Test Package for Beam Waveguide Antenna Performance Evaluation—Part I: Design and Ground Tests," *Telecommunications and Data Acquisition Progress Report 42-106*, vol. April–June 1991, http://tmo.jpl.nasa.gov/progress_report/issues.html Accessed August 2001.

[15] T. Y. Otoshi, S. R. Stewart, and M. M. Franco, "A Portable X-Band Front-End Test Package for Beam Waveguide Antenna Performance Evaluation—Part II: Tests on the Antenna," *Telecommunications and Data Acquisition Progress Report 42-105*, vol. January–March 1991, http://tmo.jpl.nasa.gov/progress_report/issues.html Accessed August 2001.

[16] T. Y. Otoshi, S. R. Stewart, and M. M. Franco, "A Portable Ka-Band Front-End Test Package for Beam Waveguide Antenna Performance Evaluation – Part II: Tests on the Antenna," *Telecommunications and Data Acquisition Progress Report 42-106*, vol. April–June 1991, http://tmo.jpl.nasa.gov/progress_report/issues.html Accessed August 2001.

[17] S. D. Slobin, T. Y. Otoshi, M. J. Britcliffe, L. S. Alvarez, S. R. Stewart, and M. M. Franco, "Efficiency Calibration of the DSS 13 34-Meter Diameter Beam Waveguide Antenna at 8.45 and 32 GHz," *Telecommunications and Data Acquisition Progress Report 42-106*, vol. April–June 1991, http://tmo.jpl.nasa.gov/progress_report/issues.html Accessed August 2001.

[18] S. D. Slobin, T. Y. Otoshi, M. J. Britcliffe, L. S. Alvarez, S. R. Stewart, and M. M. Franco, "Efficiency Measurement Techniques for Calibration of a Prototype 34-m Diameter Beam-waveguide Antenna at 8.45 and 32 GHz," *IEEE Transactions on Microwave Theory and Techniques*, vol. 40, no. 6, June 1992.

[19] M. J. Britcliffe, editor, "DSS-13 Beam Waveguide Antenna Project: Phase I Final Report," JPL D-8451 (internal document), Jet Propulsion Laboratory, Pasadena, California, pp. 5–12, March 1991.

[20] M. S. Esquivel, "Optimizing the G/T Ratio of the DSS-13 34-meter Beam Waveguide Antenna," *Telecommunications and Data Acquisition Progress Report 42-109*, vol. January–March 1992, http://tmo.jpl.nasa.gov/progress_report/issues.html Accessed August 2001.

[21] W. Imbriale, W. Veruttipong, T. Otoshi, and M. Franco, "Determining Noise Temperatures in Beam Waveguide Systems," *Telecommunications and Data Acquisition Progress Report 42-116*, vol. October–December 1993, http://tmo.jpl.nasa.gov/progress_report/issues.html Accessed August 2001.

[22] G. W. Glass, G. G. Ortiz, and D. L. Johnson, "X-band Ultralow-Nise Maser Amplifier Performance," *Telecommunications and Data Acquisition Progress Report 42-116*, vol. October–December 1993, http://tmo.jpl.nasa.gov/progress_report/issues.html Accessed August 2001.

[23] S. Stewart, "DSS-13 Beam-Waveguide Antenna Performance in the Bypass Mode," *Telecommunications and Data Acquisition Progress Report 42-108*, vol. October–December 1991, http://tmo.jpl.nasa.gov/progress_report/issues.html Accessed August 2001.

[24] R. Levy, "DSS-13 Antenna Structure Measurements and Evaluation," JPL D-8947 (internal document), Jet Propulsion Laboratory, Pasadena, California, October 1, 1991.

[25] J. Ruze, "Antenna Tolerance Theory-A Review," *Proceedings of the IEEE*, vol. 54, no. 4, pp. 633–640, April 1966.

[26] R. Levy and D. Strain, "JPL-IDEAS, Iterative Design and Antenna Structures," COSMIC, NPO 17783, University of Georgia, Athens, Georgia, October 1988.

[27] *The NASTRAN Theoretical Manual*, COSMIC, NASA SP 221(06) University of Georgia, Athens, Georgia, January 1981.

[28] D. D. Morabito and L. Skjerve, "Analysis of Tipping-Curve-Measurements Performed at the DSS-13 Beam-Waveguide Antenna at 32.0 and 8.45 GHz," *Telecommunications and Data Acquisition Progress Report 42-122*, vol. April–June 1995, http://tmo.jpl.nasa.gov/progress_report/issues.html Accessed August 2001.

[29] D. Morabito, "The Efficiency Characterization of the DSS-13 34-Meter Beam-Waveguide Antenna at Ka-Band (32.0 and 33.7 GHz) and X-band (8.4 GHz)," *Telecommunications and Data Acquisition Progress Report 42-125*, vol. January–March 1996, http://tmo.jpl.nasa.gov/progress_report/issues.html Accessed August 2001.

[30] D. Morabito, S. Butman, and S. Shambayati, "The Mars Global Surveyor Ka-Band Link Experiment (MGS/KaBLE II)," *Telecommunications and Data Acquisition Progress Report 42-137*, vol. January–March 1999, http://tmo.jpl.nasa.gov/progress_report/issues.html Accessed August 2001.

[31] D. Morabito, "Characterization of the 34-Meter Beam-Waveguide Antenna at Ka-band (32.0 GHz) and X-band (8.4 GHz)," *IEEE Antennas and Propagation Magazine*, vol. 41, no. 4, pp. 23–34, August 1999.

[32] J. C. Chen, P. H. Stanton, and H. F. Reilly, Jr., "A Prototype Ka-/Ka-Band Dichroic Plate With Stepped Rectangular Apertures," *Telecommunications and Data Acquisition Progress Report 42-124*, vol. October–December 1995, http://tmo.jpl.nasa.gov/progress_report/issues.html Accessed August 2001.

[33] W. A. Imbriale, "Solutions to Low-Frequency Problems with Geometrically Designed BWG Systems," National Radio Science Meeting, Boulder, Colorado, p. 60, January 7–10, 1992.

[34] W. A. Imbriale and M. S. Esquivel, "A Novel Design Technique for Beam-waveguide Antennas," 1996 *IEEE Aerospace Applications Conference*, Aspen, Colorado, vol. 1, pp. 111–119, February 3–10, 1996.

[35] W. A. Imbriale, M. S. Esquivel, and F. Manshadi, "Novel Solutions to Low-Frequency Problems with Geometrically Designed Beam-Waveguide Systems," *IEEE Transactions on Antennas and Propagation*, vol. 46, no. 12, pp. 1790–1796, December 1998.

[36] T. Cwik and J. C. Chen, "DSS-13 Phase II Pedestal Room Microwave Layout," *Telecommunications and Data Acquisition Progress Report 42–106*, vol. April–June 1991, http://tmo.jpl.nasa.gov/progress_report/issues.html Accessed August 2001.

[37] F. Manshadi, "The S/X-Band Microwave Feed System for NASA's First Beam-waveguide Antenna," *1996 IEEE Aerospace Applications Conference*, Aspen, Colorado, February 3–10, 1996.

# Chapter 8
# The 34-Meter Beam-Waveguide Operational Antennas

With the successful completion of the Deep Space Station 13 (DSS-13) research and development (R&D) beam-waveguide (BWG) antenna, the decision was made to introduce 34-m BWG antennas into the operational network. The initial plan was to build three 34-m antennas at each of the Deep Space Network (DSN) complexes (Australia, Spain, and the U.S.). Arraying the three 34-m BWG antennas with the already existing 34-m high-efficiency (HEF) antenna would provide a $G/T$ capability similar to that of the 70-m antenna and could, therefore, serve as a backup to the DSN's aging 70-m antennas. Due to funding constraints, however, three antennas (DSS-24, DSS-25, and DSS-26) were built at the Goldstone, California, complex, but only one antenna at each of the overseas complexes. Currently, a second BWG antenna is under construction in Spain.

## 8.1  Beam-Waveguide Design

The initial design was for simultaneous S-band (2.270–2.3 GHz receive and 2.110–2.120 GHz transmit) and X-band (8.4–8.5 GHz receive and 7.145–7.190 GHz transmit) operation, although the back structure, main reflector panels, subreflector, and BWG mirrors were to be designed and manufactured accurately enough to permit Ka-band (31.8–32.2 GHz receive and 34.2–43.7 GHz transmit) operation. At the time that the antenna was being designed, only Phase I (X- and Ka-band operation) of the R&D antenna had been completed, and the low-frequency problem had been identified but not yet solved using the focal plane method [1]. Consequently, another design technique was used.

While geometrical optics (GO) was useful for designing high-frequency or electrically large mirrors (>50 wavelengths in diameter with 20-dB edge taper), this BWG needed to be operated at low frequencies, where the mirrors are as small as 20 wavelengths in diameter. Due to diffraction effects, the characteristics of a field propagated between small BWG mirrors (<30 wavelengths in diameter) are substantially different from the GO solution [2]. For these cases, the Gaussian beam technique can be utilized.

Gaussian-beam modes are an approximate solution of the wave equation describing a beam of radiation that is unguided but effectively confined near an optical axis. The zero-order mode is normally used in the design. A major advantage of the Gaussian technique is the simplicity of the Gaussian formula, which is easy to implement and requires negligible computation time.

Because of the short computation time, a Gaussian solution can be incorporated with an optimization routine to provide a convenient method to search design parameters for a specified frequency range, mirror sizes and locations, and horn parameters.

Although Gaussian-beam analysis is fast and simple, it is less accurate than the physical optics (PO) solution for smaller mirrors (<30 wavelengths in diameter). Further comparisons between Gaussian-beam techniques and PO analysis can be found in [3] and [4]. However, by designing with Gaussian beam analysis and checking and adjusting using PO analysis, an accurate and efficient tool can be fashioned. Veruttipong [5] developed such a tool for designing the operational 34-m BWG antenna for the DSN. The goal was to provide performance over the range of 2–32 GHz.

The design is similar to that of the DSS-13 antenna in that it uses three curved mirrors (one in the basement room and two rotating in azimuth) and a 34-m dual-shaped reflector antenna (Fig. 8-1). Multiple frequency operation is provided by the use of dichroic mirrors or multiple frequency feeds. The desire is to have the radius of curvature (i.e., phase center location) and −18-dB beam diameter of the Gaussian beam at the subreflector be the same at all frequencies. The size and locations of the mirrors are relatively fixed because of the basic structure geometry, so the pertinent variables are the feed-horn diameters, feed-horn positions, and mirror curvatures. Approximating the mirrors by a thin lens formula and utilizing Gaussian-mode analysis to iterate the various design parameters, a design was achieved that met the initial design constraint of identical patterns at the subreflector.

The resulting design is shown in Fig. 8-2. The feed-horn designs and mirror curvatures are well documented in [6]. A picture of the three operational antennas at Goldstone is shown in Fig. 8-3.

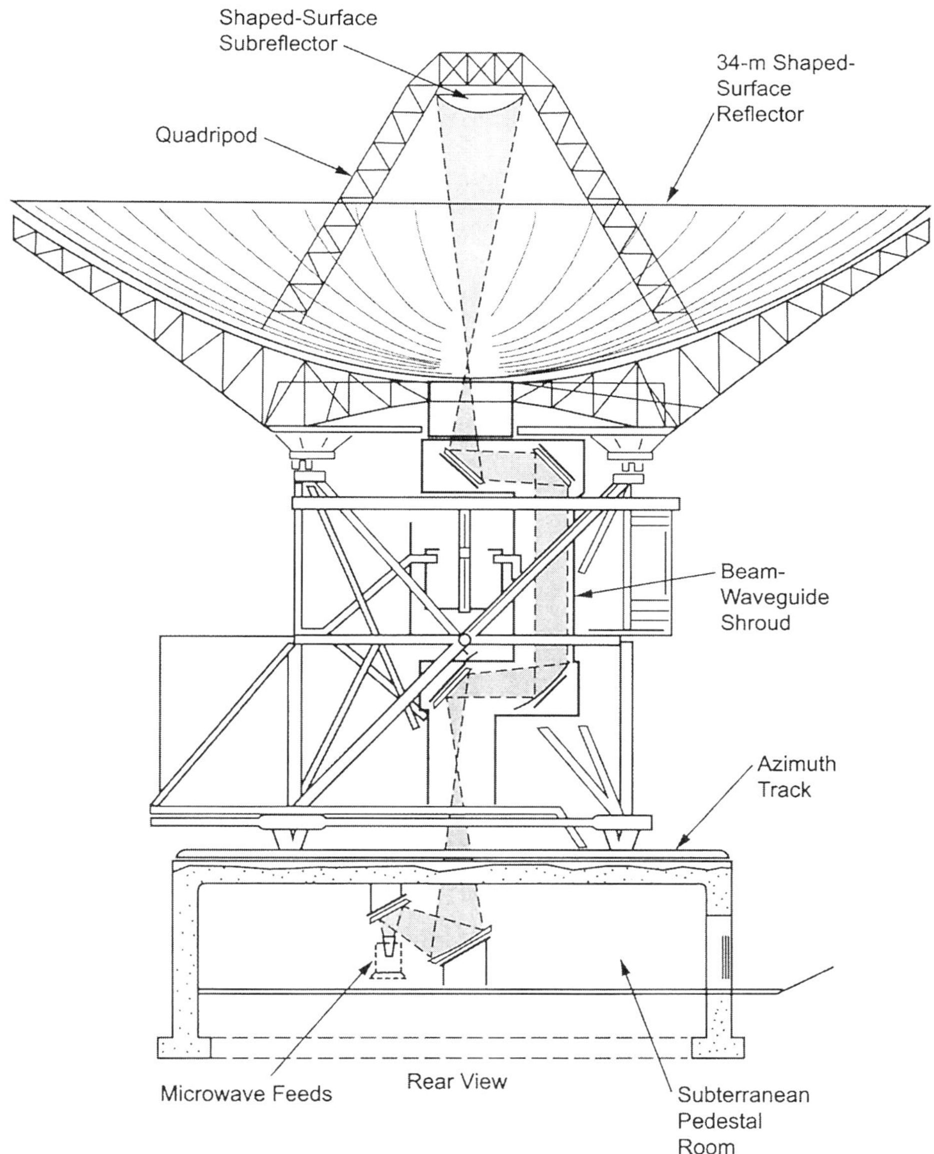


## 8.2  Initial Testing

The radio frequency (RF) performance measurements taken during the start-up of DSS-24 after completion of the antenna construction contract with TIW (now Tripoint Global Communications) is described. The measurements

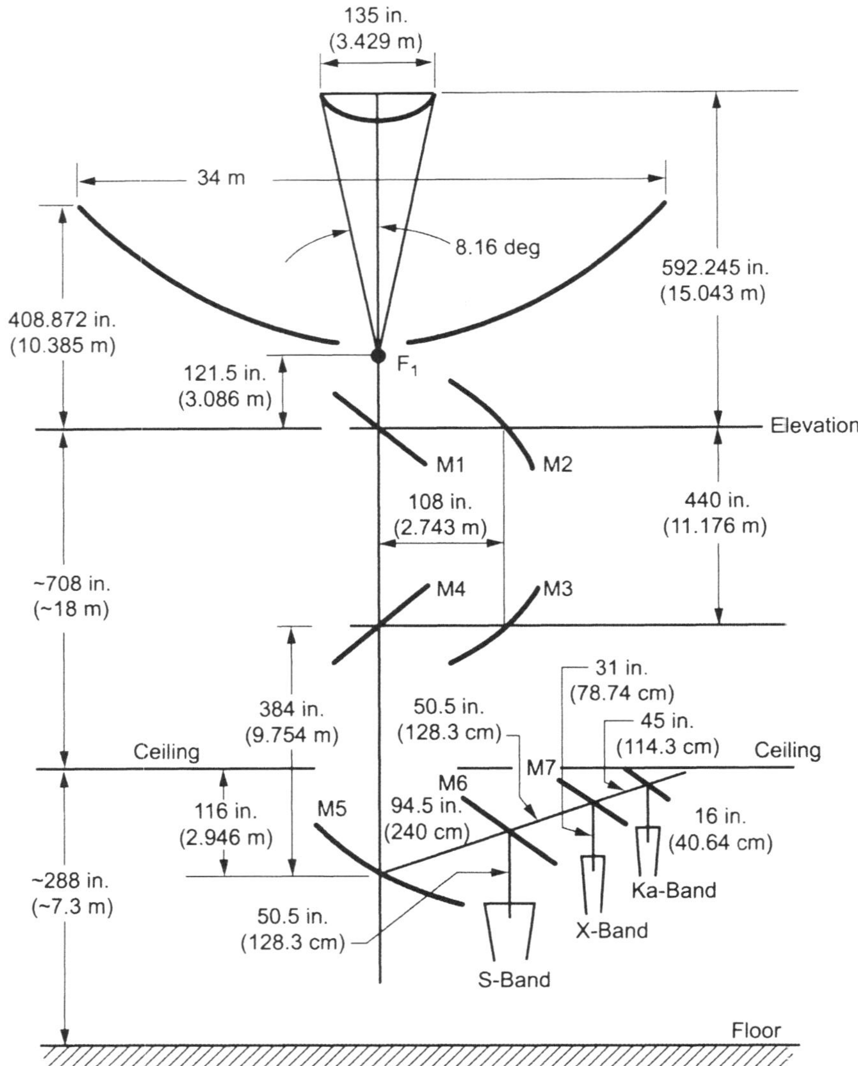



**Fig. 8-2. Detailed dimensions of the BWG configuration for DSS-24.**

were taken during an intensive 8-week program from May 6 to July 11, 1994, according to a test plan described in [7]. During this period, the antenna mechanical subsystem, the microwave subsystem, and portions of the frequency and timing subsystem were tested as a complete system.

The measurements included microwave holography to verify the surface accuracy of the main reflector and to optimize the position of the subreflector. Aperture efficiency, noise temperature, and pointing accuracy at Ka- and X-bands were measured using the portable test package technique in both the

Fig. 8-3. The three operational BWG antennas at
Goldstone, California.

Cassegrain ($F_1$) configuration and in the normal BWG ($F_3$) configuration to accurately determine the difference in performance between the two configurations. Additional noise-temperature measurements were taken with the test packages on the ground to measure the actual noise-temperature contribution of the antenna.

X-band and S-band aperture efficiency, noise temperature, and pointing accuracy measurements were also made with the operational feeds installed in the pedestal. The operational feeds were tested in the low-noise, listen-only mode and the S-/X-band mode. During the testing, a systematic error correction model for the antenna-pointing computer (APC) was developed.

The details of the testing are well documented in [8]; therefore, only a brief summary will follow.

## 8.2.1 Microwave Holography Measurements

The first operational 34-m BWG was named DSS-24 and constructed at Goldstone, California. As part of the construction contract, the antenna panels were approximately set using theodolite measurements, and the subreflector position settings were analytically determined using a theoretical mechanical model. Then the Jet Propulsion Laboratory (JPL) Microwave Antenna Holography System (MAHST) [9] was utilized to provide accurate panel-setting and subreflector position information. Table 8-1 shows the performance improvement achieved by using the data obtained from the MAHST system to reset the panels and update the subreflector position tables. Using MAHST enabled set-

**Table 8-1. Performance improvement by microwave
holography at 45-deg elevation.**

| Frequency | Panel Setting (dB) | Subreflector (dB) | Total (dB) |
|---|---|---|---|
| 8.45 GHz (X-band) | 0.10 | 0.25 | 0.35 |
| 32 GHz (Ka-band) | 1.27 | 3.60 | 4.87 |

ting the main reflector panels of DSS-24 to 0.25 mm root-mean-square (rms),
making DSS-24 the highest-precision antenna in the National Aeronautics and
Space Administration (NASA)/JPL DSN. The precision of the DSS-24 antenna
(diameter/rms) is $1.36 \times 10^5$, and the gain limit is 95 GHz.

### 8.2.2 Efficiency Measurements

In a similar manner to that used for DSS-13, efficiency measurements were
made at the $F_1$ and $F_3$ focal points, using the R&D test packages at X-band
(8.45 GHz) [10] and Ka-band (32 GHz) [11].

Figure 8-4 shows the X-band efficiencies at the $F_1$ and $F_3$ focal points as a
function of elevation angle. The peak X-band efficiency at $F_1$ was determined
to be 75.25 percent at an elevation angle of 42.5 deg, with a one standard deviation uncertainty of 2.04 percent. The peak X-band efficiency at $F_3$ was calculated to be 72.67 percent at an elevation angle of 49 deg with a computed
uncertainty on the peak efficiency of 1.81 percent.

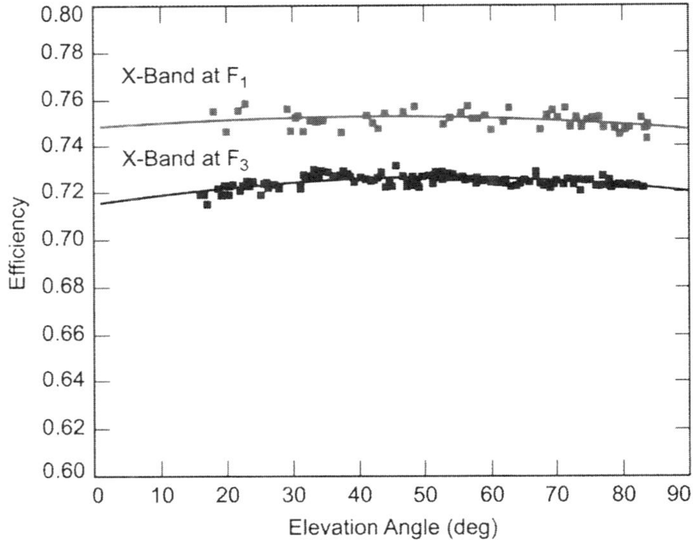


**Fig. 8-4. X-band (8.45-GHz) efficiencies at $F_1$ and $F_3$
focal points, without atmosphere.**

Figure 8-5 shows the Ka-band efficiency measured at $F_1$ and Fig. 8-6 the efficiency measured at $F_3$. The Ka-band peak efficiency at $F_1$ was calculated to be 60.60 percent at an elevation angle of 44.5 deg, with an uncertainty of 2.18 percent. The Ka-band peak efficiency at $F_3$ was determined to be 57.02 percent at an elevation angle of 42.0 deg, with an uncertainty of 2.51 percent. The gravity-induced roll-off of the $F_3$ efficiency curve shown in Fig. 8-6 is slightly flatter than that measured at $F_1$, shown in Fig. 8-5. This is expected since the final subreflector offset curves determined at $F_3$ with the automated subreflector optimization scheme provided better focusing than those used at $F_1$. The $F_3$ gravity loss profile, computed from the best-fit efficiency curve, is shown in Fig. 8-7. It is assumed that there is zero loss at the peak gain elevation angle of 42 deg. Observe that this is significantly better than that of DSS-13 (also shown on Fig. 8-7) and indicates that the redesign of the structure to provide a better match to the finite-element method (FEM) model was successful.

Efficiency measurements were made of the X-band operational feed with the S-/X-band dichroic plate in place. See Fig. 8-8 for a picture of the S-/X-band system and Fig. 8-9 for the efficiency measurements as a function of elevation angle. The peak efficiency was computed to be 71.10 percent, with an estimated uncertainty of 2.31 percent.

S-band efficiency measurements were also made on the operational feed system. Figure 8-10 shows the measurements as a function of elevation angle. The azimuth angles for the data points range from 60 to 295 deg, with southern (180-deg azimuth) transit elevation angles of 67.5 deg for star 3C274 and

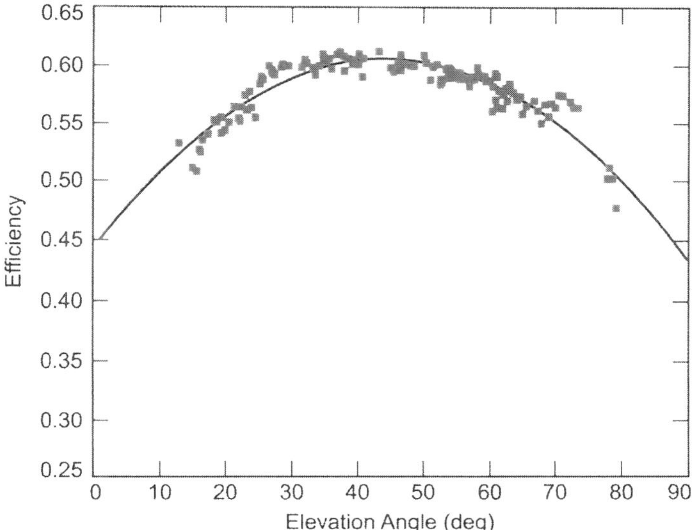

**Fig. 8-5.  Ka-band (32-GHz) efficiency at $F_1$ focal point, without atmosphere.**

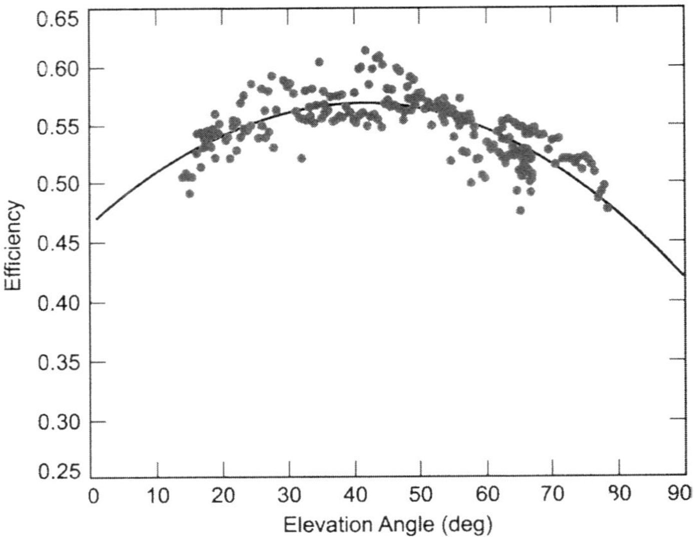

**Fig. 8-6.  Ka-band (32-GHz) efficiency at $F_3$ focal point, without atmosphere.**

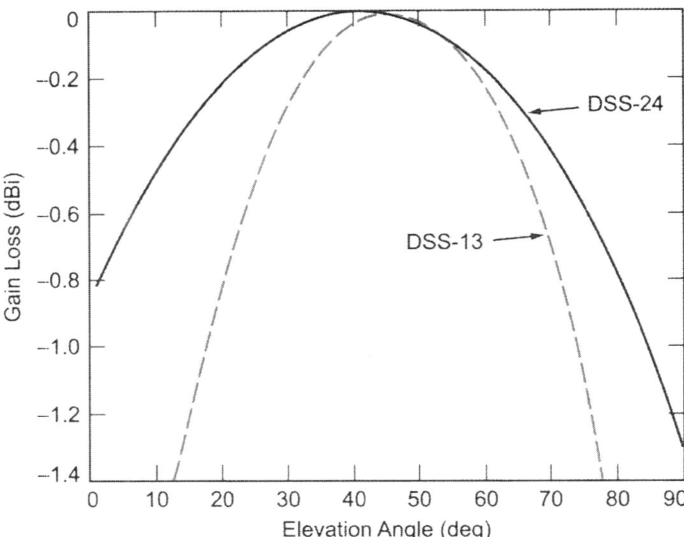


**Fig. 8-7.  Ka-band gravity loss for the $F_3$ focal point.**

83 deg for star 3C123. The peak S-band efficiency was computed to be 71.50 percent, measured for both sources at the 180-deg azimuth angle. The uncertainty on the peak value is 1.82 percent, which corresponds to a peak S-band gain of 56.79 dBi $\pm$ 0.11 dB.

Fig. 8-8. The S-/X-band feed system at DSS-24.

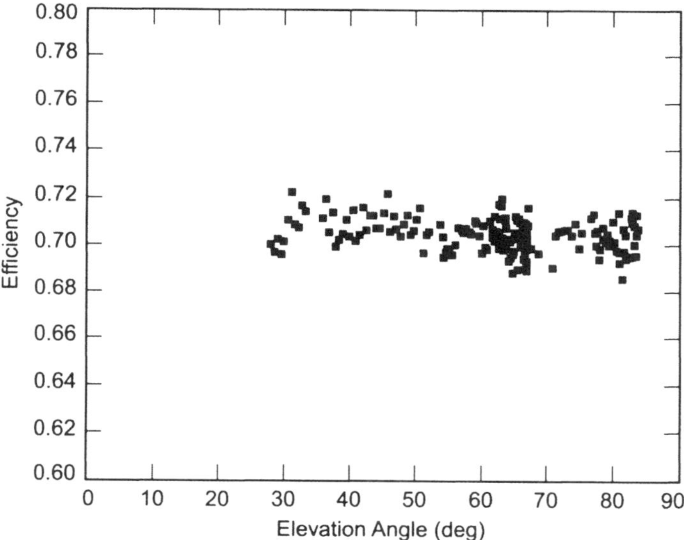

Fig. 8-9. X-band (8.45 GHz) efficiency measurements
on the operational feed horn at $F_3$ focal point,
without atmosphere.

The variation of antenna efficiency as a function of azimuth angle is also contained in Fig. 8-10. The peak operational gain at the 180-deg azimuth, was predicted by a previous RF analysis [12] for the S-band feed at the 270-deg pedestal room position. The predicted antenna efficiency variation between the azimuth range of 60 to 295 deg, computed for a perfectly aligned BWG mirror system, was 2 percent. As measured at DSS-24 and shown in Fig. 8-10, the variation is 4 percent over the same angle range.

The efficiency data is summarized in Table 8-2 for both the R&D and operational feed packages.

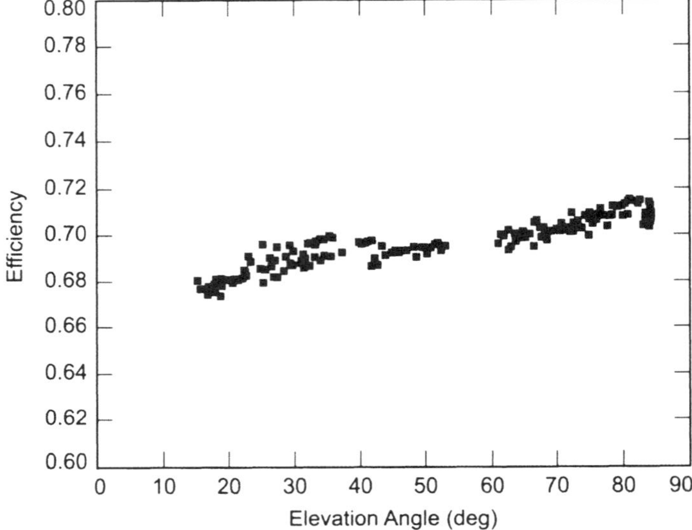

Fig. 8-10.  S-band (2.295-GHz) efficiency measurements on the operational feed horn at $F_3$ focal point, without atmosphere.

Table 8-2.  Summary of antenna peak gains and efficiencies at the S-, X-, and Ka-band frequencies.

| Frequency and Focal Point | Gain (dBi) | 1-Sigma Gain Error (dB) | Efficiency (percent) | 1-Sigma Efficiency Error (percent) |
|---|---|---|---|---|
| 8.45 GHz at $F_1$ | 68.33 | 0.12 | 75.25 | 2.04 |
| 8.45 GHz at $F_3$ | 68.19 | 0.11 | 72.67 | 1.81 |
| 32 GHz at $F_1$ | 78.97 | 0.16 | 60.60 | 2.18 |
| 32 GHz at $F_3$ | 78.70 | 0.20 | 57.02 | 2.51 |
| 8.45 GHz OPS[a] feed at $F_3$ | 68.09 | 0.14 | 71.10 | 2.31 |
| 2.295 GHz OPS feed at $F_3$ | 56.79[b] | 0.11 | 71.50 | 1.82 |

[a]Operational.
[b]Measured when at azimuth angle = 180 deg.

### 8.2.3    Noise-Temperature Results

Noise-temperature measurements were also made with both the R&D test packages and the operational S-/X-band feed system. Table 8-3 shows the measured system-noise-temperature data of these test packages off and on the DSS-24 antenna. If measurements were to be made with the mirrors taped (aluminum tape covering the gap between the central alignment plug and the mirror), the values would be 1–2 K lower than those shown in Table 8-3. Table 8-4 shows the differential noise contribution of the DSS-24 antenna with the data from the DSS-13 antenna for comparison. These values are derived by taking the difference of the system noise temperatures with the test package on the ground, at $F_1$ and $F_3$.

Table 8-5 shows the measured noise-temperature data of the operational S-/X-band feed systems in different polarization and feed configurations (that is, diplexed or listen-only, S/X or X only). The operational S-/X-band feed systems cannot be tested at $F_1$, but differential values for the $F_3$–ground case are

**Table 8-3. Antenna noise-temperature data (K) with R&D test packages. Measurements were made before the taping of M5 and M6. (X-band LNA input temperature = 12 K; Ka-band LNA input temperature = approximately 60 K; Ka-band waveguide feed components temperature contribution = approximately 18 K.)**

| R&D Feed | $F_1$ | $F_3$ | Ground |
|---|---|---|---|
| X-band | 26.6 | 30.6 | 21.3 |
| Ka-band | 93.1 | 101.3 | 86.8 |

**Table 8-4. Antenna noise-temperature contributions: (a) R&D X-band; estimated uncertainty = ± 0.2 K (actual $F_3$-related values are expected to be 1.4 K lower than shown in the table) and (b) R&D Ka-band; estimated uncertainty = ± 1 K. Measurements for both were made before mirror taping of M5 and M6.**

| Antenna | (a) R&D X-band Temperature (K) | | |
|---|---|---|---|
| | $F_1$ – Ground | $F_3$ – $F_1$ | $F_3$ – Ground |
| DSS-24 | 5.3 | 4 | 9.3 |
| DSS-13 | 3.2 | 2.4 | 5.6 |
| | (b) R&D Ka-band Temperature (K) | | |
| | $F_1$ – Ground | $F_3$ – $F_1$ | $F_3$ – Ground |
| DSS-24 | 6.3 | 8.2 | 14.5 |
| DSS-13 | 7.1 | 6.8 | 13.9 |

**Table 8-5. Noise temperature data using operational feed horns: (a) S-band (2.295 GHz) antenna and (b) X-band (8.420 GHz) antenna.**

|  | (a) S-Band[a] | | |
|---|---|---|---|
| Polarization | Configuration | $F_3$ Noise Temperature (K) | $F_3$ – Ground (K) |
| RCP | Low-noise | 30.8 | 15 |
| LCP | Low-noise | 30.6 | Not measured |
| RCP | Diplexed | 37.3 | 15 |
| LCP | Diplexed | 37.0 | Not measured |

[a] S-band LNA temperature = 5.5 K; follow-on temperature = 0.2 K.
S-band measurements performed before mirror taping of M5 and M6; actual noise temperature is expected to be 0–1 K lower than above.
Estimated uncertainty = ± 1 K.

|  | (b) X-Band[b] | | |
|---|---|---|---|
| Polarization | Configuration | $F_3$ Noise Temperature (K) | $F_3$ – Ground (K) |
| RCP | Low-noise | 25.5 | 10 |
| LCP | Low-noise | 25.2 | Not measured |
| RCP | S-/X-band | 27.0 | 12 |
| LCP | S-/X-band | 26.5 | Not measured |

[b] X-band LNA temperature = 3.2 K; follow-on temperature = 0.1 K.
Above values include mirror taping of M5 and M6, noise reduction of 1.4 K.
Estimated uncertainty = ± 1 K.
$F_3$ – ground values derived from testing performed with feed horn at JPL.

included. The ground measurements were made at JPL as part of the feed acceptance testing [13]. These measurements were corrected for the atmospheric difference between DSS-24 and JPL. A noise budget for the X-band operational feed horn is shown in Table 8-6. System-noise temperature of the operational S-/X- and Ka-band feed systems as a function of elevation angle is shown in Fig. 8-11 (all other noise measurements were made at zenith elevation angle).

## 8.2.4   The Shroud

One of the requirements for versatility on the BWG antenna was to be able to upgrade or repair electronics not currently in use in the pedestal room while the antenna was otherwise transmitting. This could be the case if there was a second feed system in the BWG similar to the multiple feed systems in the DSS-13 BWG. This might be accomplished if the beam path was enclosed in a

**Table 8-6. Antenna system-noise-temperature frequency budget using X-band operational feed. $T_{maser}$ = 3.5 K and $T_{follow-on}$ = 0.1 K.**

| Source | Contribution (K) | Difference from 890-261 [14] Predicted (K) |
|---|---|---|
| Cosmic background | 2.5 | 0.0 |
| Atmosphere | 2.5 | 0.0 |
| Main and subreflectors | 5.3 | 1.4 (3.9) |
| BWG | 4.0 | 1.6 (2.4) |
| S-/X-band dichroic plate | 1.5 | −1.8 (3.3) |
| Mirror tape | −1.5 | −1.5 |
| Feed/LNA | 12.4 | 0.0 |
| Total | 26.7 | −0.3 |

metal shield so that energy scattered off the edge of the BWG mirrors could be contained.

The selected design was a metal shroud enclosure in the form of an inverted pyramid; it was installed around the basement ellipse, as shown in Fig. 8-12. The apertures of the feed horn penetrate the shroud, and the radiated beam is entirely enclosed and separated from the rest of the pedestal room. A survey was made on the radiated power in the pedestal room outside the shroud, and it was determined that the levels were below the accepted 1 mW/cm$^2$ standard for

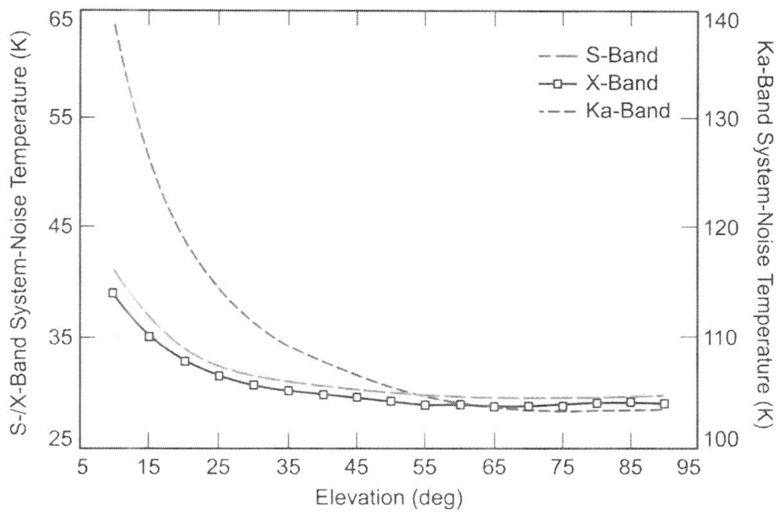


**Fig. 8-11. System-noise temperature as a function of elevation angle.**

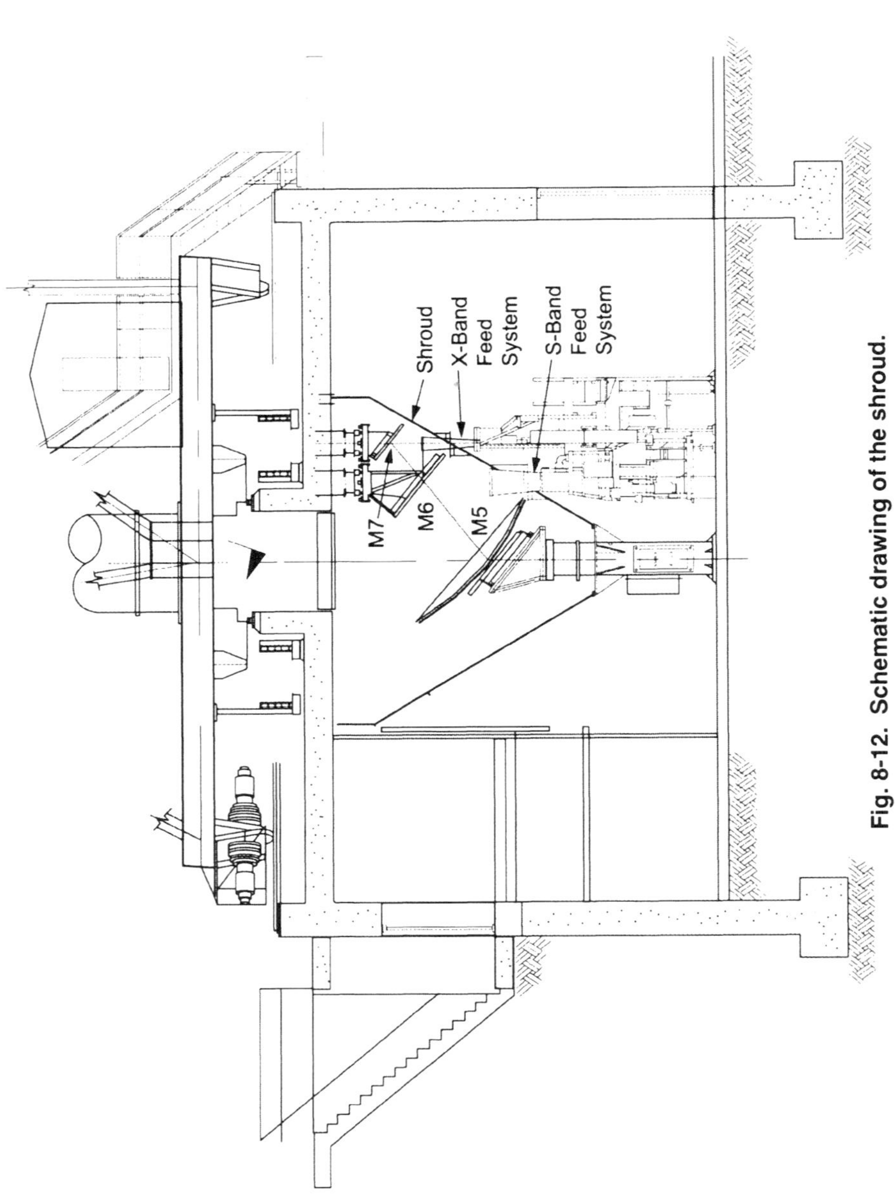


Fig. 8-12. Schematic drawing of the shroud.

safety. Hence, it is quite safe to be in the pedestal room when the antenna is transmitting. A photograph of the DSS-24 shroud is shown in Fig. 8-13.

## 8.3   Adding Ka-Band to the Operational 34-Meter Beam-Waveguide Antennas

There are two different implementations of Ka-band in the operational network: a Ka-band system installed at DSS-25 at Goldstone, California, in support of Cassini Project radio science and a second, yet-to-be-installed (as of 2002) system intended for all the other BWG systems. The Ka-band system at DSS-25 includes Ka-band transmit-and-receive capability and utilizes separate Ka-band feed horns for transmit-and-receive, with a dichroic plate to separate the frequencies. The system intended for the rest of the network is receive-only at Ka-band and includes a multifrequency X-/X-/Ka-band feed that also provides a feed diplexing system for X-band. In 1999, an engineering model of the front end was constructed and tested at DSS-26 (also at Goldstone, California); it validated the design.

### 8.3.1   The Cassini Radio Science Ka-Band Ground System

A schematic of the basic geometry for the X-/Ka-band system at DSS-25 is shown in Fig. 8-14. Additions to the BWG baseline design include transmit-and-receive Ka-band feed horns, a Ka-/Ka-band dichroic plate (M8), an X-/Ka-band dichroic plate (M7), and a hyperboloidal mirror (M9) [15].

**Fig. 8-13. The shroud at DSS-24.**

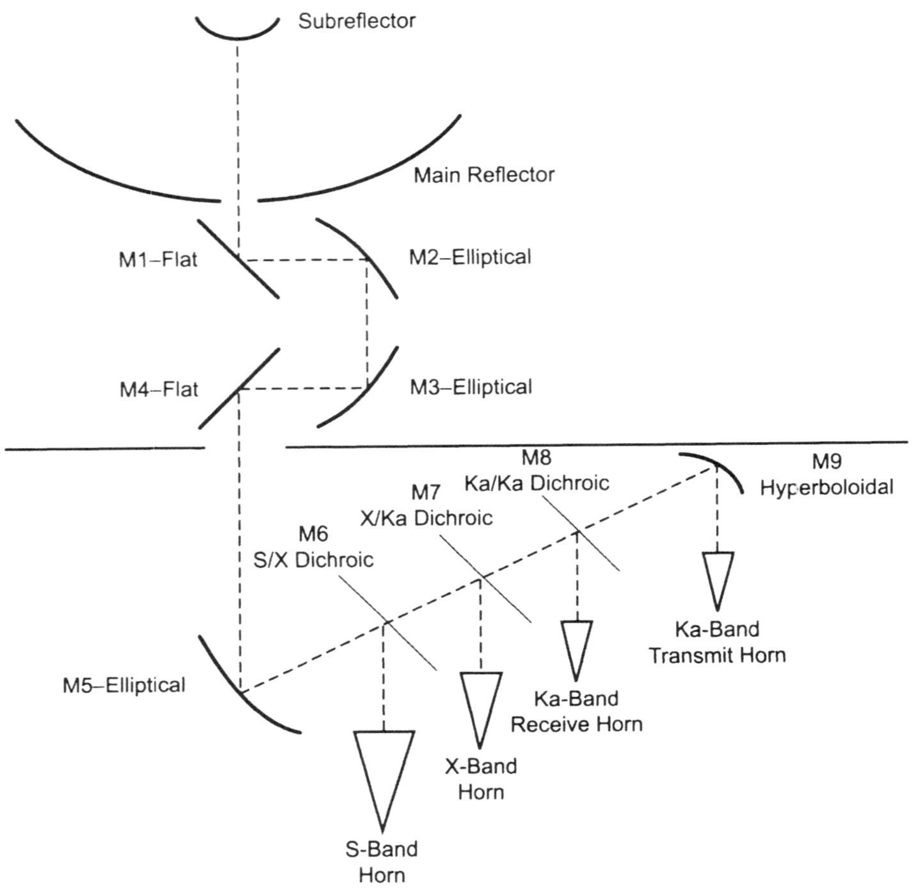



**Fig. 8-14. Basic geometry of DSS-25.**

**8.3.1.1 Optics Design.** For an initial sizing of the Ka-band receive horn, the focal-plane-analysis method was used [1]. The Cassegrain shaping pattern was used to excite the BWG from $F_1$ in a receive-mode analysis. The far-zone horn pattern that maximizes the antenna gain is the far-zone pattern produced by the complex conjugate of the PO currents on M5 (or the focal plane at $F_3$). A horn gain of 30.3 dBi produces the maximum gain; however, the significant antenna performance parameter is $G/T_{sys}$. The antenna performance ($G/T_{sys}$) was calculated for several gain feed horns (Table 8-7) around the maximum gain horn, and a horn design with 31.47-dB gain was chosen because it yielded the maximum $G/T_{sys}$ and is for all practical purposes a maximum-gain horn.

The DSS-25 basement ceiling is very close to the M5 focus, and using an M9 flat mirror does not allow enough space for the transmit horn, as approxi-

**Table 8-7. Dependence of DSS-25 Ka-band receive
performance on feed-horn gain.**

| Horn Gain (dBi) | BWG Spill (percent) | Subreflector Spill (percent) | Main Reflector Spill[a] (percent) | Noise Temperature (K) | Gain (dBi) | $\Delta(G/T_{sys})$[b] (dB) |
|---|---|---|---|---|---|---|
| 29.3 | 1.07 | 3.29 | 1.60 | 6.81 | 80.64 | −0.11 |
| 30.46 | 0.77 | 2.80 | 1.45 | 5.66 | 80.62 | +0.00 |
| 30.96 | 0.52 | 2.09 | 1.04 | 3.98 | 80.67 | +0.16 |
| 31.47 | 0.34 | 1.57 | 0.94 | 3.25 | 80.63 | +0.23 |
| 31.47[c] | 0.24 | 1.18 | 0.68 | 2.34 | 80.72 | +0.32 |

[a] Main reflector spillover values were computed using an insufficient number of integration steps.
[b] $\Delta(G/T_{sys})$ values were computed using S. Slobin's parameters as described by Veruttipong in [14] for LNA temperature, ohmic losses, atmospheric attenuation, etc.
[c] Feed refocused.

mately 38 cm is needed between the receive and transmit horn axes to accommodate the corresponding amplifiers. Also, the size of the X-band feed horn and its amplifier does not allow the Ka-band receive horn to be moved closer to M5 to create more space for the Ka-band transmit horn. A solution to the lack-of-space difficulty, without sacrificing antenna performance, is to use a lower-gain horn for transmit in conjunction with a gain-magnifying mirror for M9 (i.e., a concave hyperboloidal mirror). A lower-gain horn is physically smaller and has a phase center closer to its aperture, allowing more clearance between the horn aperture and M9. An M9 mirror that magnified the horn gain by approximately 3.5 dB was selected to accommodate the required horn and amplifiers. This yielded a horn gain of approximately 26.9 dB. The overall antenna gain was calculated with a variety of horn gains, and it was determined that the overall antenna gain is only weakly dependent on the horn gain. A horn of 26.64 dBi was selected for the design.

The M7 dichroic mirror utilizes the same design as that used for DSS-13. To reduce manufacturing costs, a 30-dB rim taper was adopted for the perforated regions of all dichroic mirrors. The Ka-band transmit mode determines the dimensions of the M7 perforated region and yields an elliptical area approximately 94.5 by 81.8 cm.

The M8 dichroic plate was designed to pass Ka-band uplink (34.2–34.7 GHz) and to reflect the downlink (31.8–32.3 GHz). The plate utilized stepped apertures [16] in order to separate the two-frequency band with only a 1:1.07 ratio. The geometry of the five-layer plate is shown in Fig. 8-15; the measured and calculated reflection coefficient versus frequency for circular polarization is shown in Fig. 8-16.

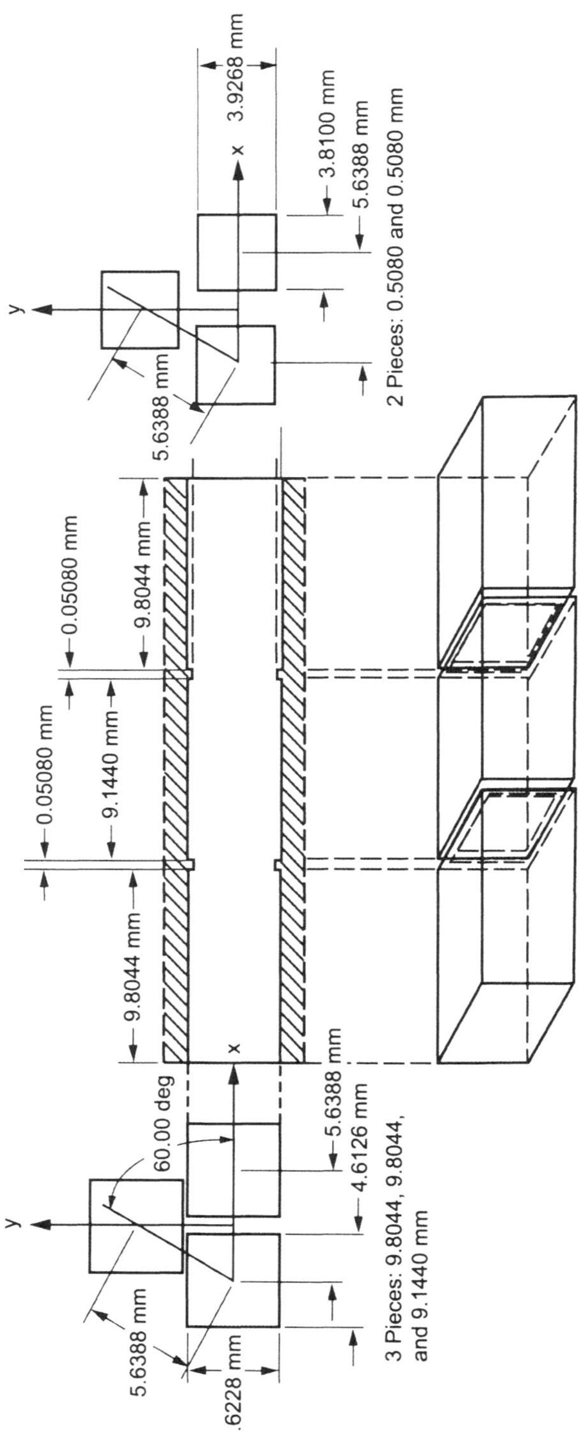



Fig. 8-15.  Layout of the holes and hole pattern for the five-layer Ka-band dichroic plate.

**8.3.1.2  Monopulse Pointing System.** Traditionally, the DSN has employed the conical scanning (CONSCAN) algorithm for pointing the antennas at 2.2–2.305 GHz (S-band) and 8.2–8.6 GHz (X-band). In CONSCAN, the pointing error is estimated by moving the antenna in a circle with respect to the best estimate of the spacecraft location. The received power detected on this circle is used to generate a new best estimate of the spacecraft location. Nominally, the circle chosen is such that the detected power is 0.1 dB less than the peak of the antenna beam. In the BWG antennas, the 0.1-dB beamwidths are 22 mdeg at S-band (2.295 GHz), 5.9 mdeg at X-band (8.45 GHz), and 1.5 mdeg at Ka-band (32 GHz). At S- and X-band frequencies, a typical jitter of 1 mdeg in the antenna pointing (under favorable environmental conditions, with no wind) does not result in significant fluctuations in detected power levels. At Ka-band, the same jitter causes significant fluctuations in the detected power levels; these fluctuations are not acceptable for radio science application. For this reason, it became necessary to adopt a new technique that allowed the antenna to be pointed directly at the spacecraft. An additional advantage is a 0.1-dB gain in signal by not having to point away from the spacecraft.

The selected pointing system uses a single feed horn and a monopulse tracking coupler, as shown in Fig. 8-17. This coupler has both a sum port ($TE_{11}$ mode) and a difference port ($TE_{21}$ mode). As described in [17], the $TE_{21}$ mode is excited by means of eight symmetrically placed Ka-band waveguides, which are, in turn, connected through a series of combiners to generate circular polarization. When radiated by the corrugated horn, the $TE_{21}$ mode has a theoretical

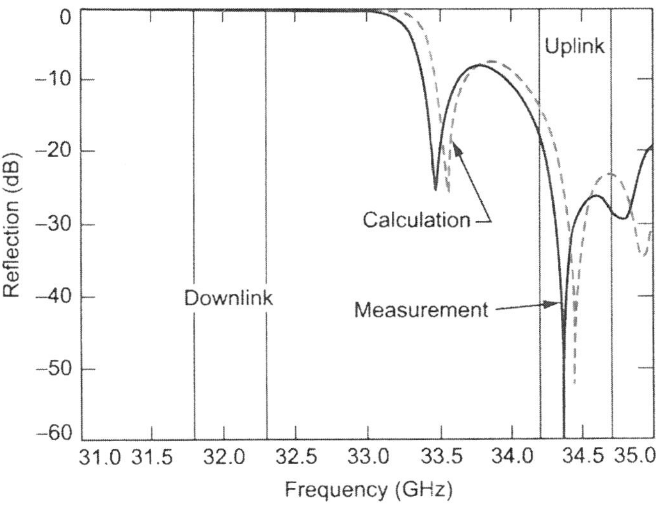


**Fig. 8-16.  Performance of the five-layer Ka-band
dichroic plate.**

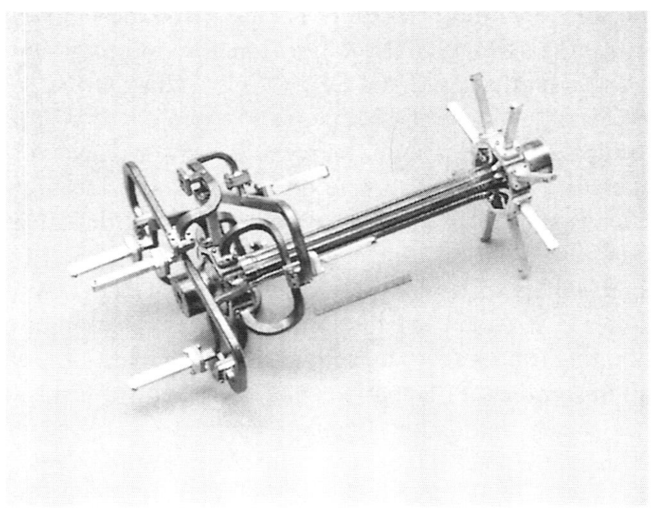

**Fig. 8-17. Ka-band monopulse coupler.**

null on axis. The magnitude of the power received in the $TE_{21}$ mode relative to the $TE_{11}$ mode (sum beam) is proportional to the $\theta$ pointing error (angle relative to the antenna boresite—see Fig. 8-18). The azimuthal error, $\phi$, is proportional to the phase difference between the sum and difference signals. Thus, by measuring the complex ratio of the sum and difference signals, pointing updates can be generated that will drive the difference signal and, consequently, the antenna pointing error to zero.

An additional consideration in a BWG antenna is that there is a rotation of the coordinate system from the feed system to the reflector system. The tracking coupler measures the $\phi$ angle in the feed coordinate system. The $\phi$ angle in the reflector system rotates linearly with both the azimuth and elevation motion of the antenna.

The performance of the monopulse pointing system of the 34-m BWG antennas is well documented in [18].

### 8.3.1.3 Measured Performance After Installation of Ka-Band.

The measured efficiencies before installation of Ka-band (single-frequency system) were R&D X-band 76.6 percent, operational X-band 76.5 percent, and R&D Ka-band 57.5 percent [19].

After M7 dichroic installation (X-/Ka-band feed system), the measured efficiencies were 75.5 percent for X-band and 57.9 percent for Ka-band.

The noise temperature of the maser amplifier system at X-band was measured with M7 as a flat plate, 23.8 K; and with M7 as the X-/Ka-band dichroic plate, 23.6 K.

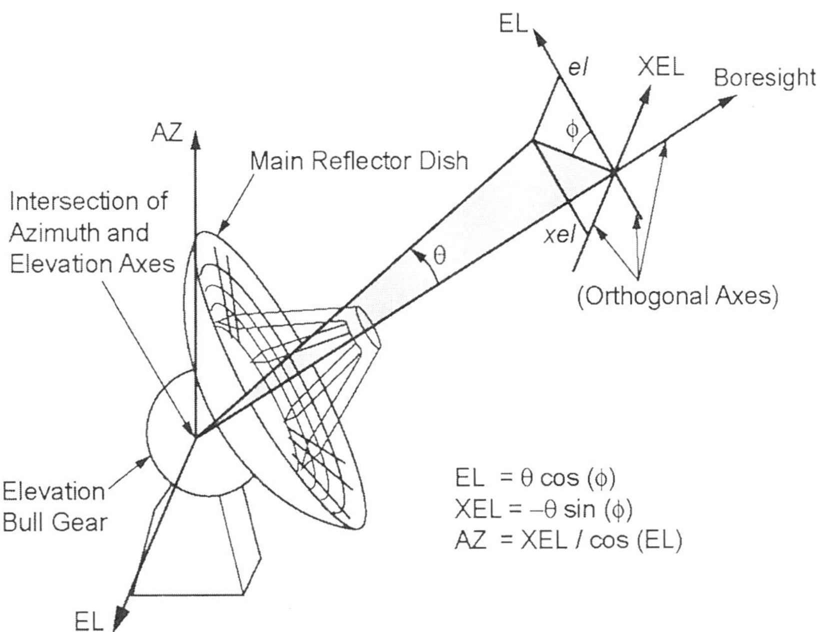


**Fig. 8-18. The relationship between the multiple coordinate systems within the monopulse implementation.**

The system-noise temperature of the Ka-band feed horn at DSS-25 was measured with the indium phosphide high-electron-mobility-transistor (HEMT) technology low-noise-amplifier (LNA) module [20]. The LNA module has an input noise temperature of 12.3 K and represents the debut of this technology in the DSN.

Measurement of the LNA package input noise temperature, $T_e$ (referenced at the feed-horn aperture), was accomplished using a Y factor method. An ambient load and a liquid-nitrogen-cooled load were used as reference power sources. A 32-GHz measurement made at JPL was 20.2 K; when repeated at DSS-25, it was 19.9 K.

Measurements were also made of the total system-noise temperature ($T_{op}$), both outside the antenna and in its final installed position inside the antenna. The average $T_{op}$ measured outside the antenna was 34.6 K at 32 GHz. Subtracting the measured $T_e$ of 20 K results in a sky-noise temperature of 14.6 K. The measured $T_{op}$ with the feed horn installed in its final operational position was 45.0 K. Using the measured air temperature and humidity, the sky-noise temperature was estimated to be 11.7 K. The sky noise measured by the water vapor radiometer at DSS-13 at 31 GHz was 10 K. Subtracting the measured $T_e$ of 20 K and the estimated sky-noise temperature of 10–12 K results in an antenna noise temperature of 13 to 15 K.

**8.3.1.4 Beam-Aberration Correction.** The aberration effect occurs when the uplink and downlink beams must be pointed in different directions for simultaneous uplink and downlink communications. It arises solely from the fact that a spacecraft is moving in the right-ascension/declination (RA/DEC) coordinate system as seen from a point (the antenna) on Earth. At any instant of time, the downlink beam must be pointed where the spacecraft was one round-trip light time (RTLT) ago, and the uplink beam must be pointed where the spacecraft will be in one RTLT. Kurt Liewer [21] was the first person to draw attention to the fact that aberration effects on the radio link between a ground station and a spacecraft must result in an offset angle between the optimal uplink and downlink directions. For Saturn and Earth, this results in a distinct aberration effect (the angular distance between the two RA/DEC positions), with a period of about 400 days. The maximum aberration effect due to planetary motion of the Earth/Saturn is about 15 mdeg. At S- and X-bands, the loss due to this separation of pointing directions is small because the beamwidths of the 34-m antenna are approximately 230 mdeg at S-band and 64 mdeg at X-band. At Ka-band, however, the loss can be as large as 10 dB for a 15-mdeg offset.

In order to obtain uplink and downlink communications in the presence of planetary aberration, it is required that the receive beam be pointed at the predicted spacecraft position (position 1 on Fig. 8-19) and the transmit beam be pointed at the spacecraft position according to the space aberration correction

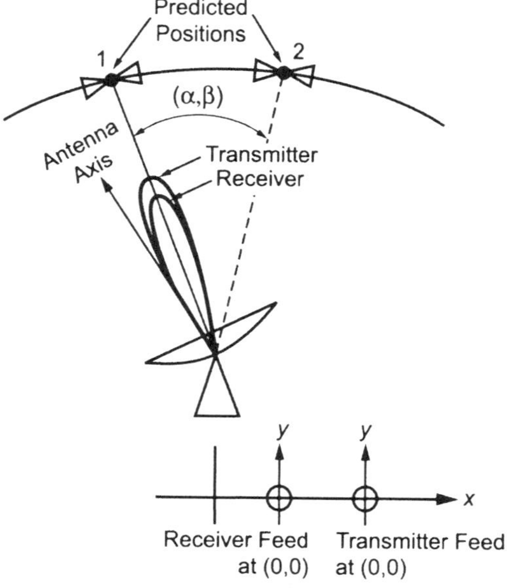


**Fig. 8-19. Without antenna-pointing correction
for beam aberration.**

(position 2). With no correction, both feed horns will be at their nominal positions, termed (0,0). The separation of the feed horns in space is due to the dichroic mirror that separates the transmit and receive frequencies. The relative position between the receive and transmit beams required for aberration correction is identical to the relative position between the two spacecraft positions, which is defined by $(\alpha,\beta)$ regardless of how we point the antenna. Repointing of one beam with respect to the other can be accomplished by either tilting the mirrors above the feed horn or moving the feeds. For minimum effect on the downlink performance, the option selected was to move the transmit feed horn and klystron. Thus, with correction, the transmit feed is moved a distance $(x,y)$, depending on the predicted correction $(\alpha,\beta)$, as shown in Fig. 8-20. The required angles for the Cassini mission are given in [22].

Due to the complex relationship between the $(x,y)$ position of the feed horn and the beam pointing direction, a conceptual demonstration of the beam aberration and point-ahead capability was performed at DSS-13. The demonstration was first done at X-band in the receive mode. The antenna was pointed at an X-band source, then moved off source by a known $(\alpha,\beta)$. The position of the source was recovered by the $(x,y)$ motion of the feed system calculated from the aberration correction model. The experiment was repeated at Ka-band, with the beam maintained on source to within the accuracy of the blind pointing model.

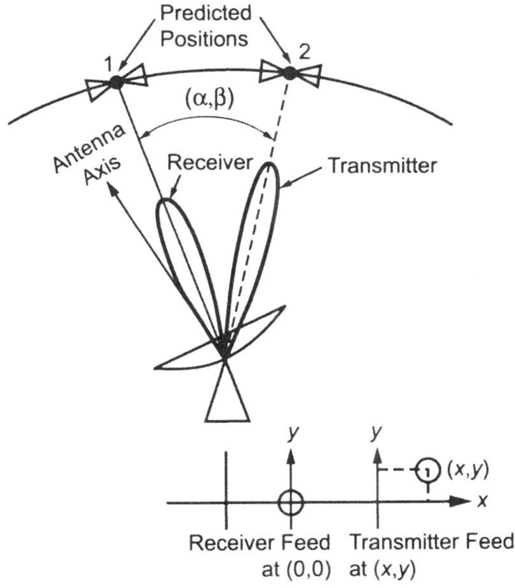


**Fig. 8-20. With antenna-pointing correction (by moving the feed horn) for beam aberration.**

In DSS-25, the Ka-band feed horn and transmitter were mounted on an "x,y table," and the connection to the shroud was made with a flexible cloth. In June 2001, the system was successfully used with both the Cassini Gravity Wave System tests and the Solar Conjunction System test. It will be used for both the Gravity Wave and Solar Conjunction Experiments.

## 8.3.2    Ka-Band Upgrades—Receive-Only System

The Cassini radio science system uses a frequency-selective surface (FSS) to separate the X-band and Ka-band signals [23]. This approach has the advantage of requiring relatively simple X- and Ka-band feed horns. Disadvantages include the real estate required for the two feed horns, the need for separate dewars for the X- and Ka-band LNAs, and the noise added due to scattering from the FSS used to diplex the signals.

To overcome these disadvantages, a single feed horn that accommodates all the required frequencies in one package was developed [24]. This feed horn provides for the Ka-band receive signal as well as diplexing the X-band transmit and receive signals. A monopulse tracking coupler is also utilized for Ka-band pointing. This feed system is compact and has low noise relative to separating the X- and Ka-band signals with an FSS. Also, a single, but larger dewar, is used to house all of the LNAs. An additional complication in the design of the feed horn is that the phase center of the feed horn at X-band is not coincident with the phase center at Ka-band, as the positions of the phase centers are prescribed by the BWG geometry.

**8.3.2.1 X-/X-/Ka-Band Feed.** Figure 8-21 shows a photograph of the X-/X-/Ka-band horn with monopulse coupler and linear polarization combiners attached to the X-band ports. The basic internal configuration of the X-/X-/Ka-band feed horn is shown in Fig. 8-22.

The feed horn is fed at Ka-band through a monopulse coupler, similar in design to that described in [17]. The sum mode at Ka-band, a circularly polarized $TE_{11}$ mode, is excited using a commercial orthomode junction and polarizer. The sum mode passes through the center of the monopulse coupler and enters the horn, as shown at the left in Fig. 8-22. The difference mode, a circularly polarized $TE_{21}$ mode, is excited through the monopulse coupler arms and enters the feed horn as well. These modes are transformed to the hybrid $HE_{11}$ and $HE_{21}$ modes in the Ka-band mode-converter section. The X-band downlink signal is extracted in the X-band downlink junction, which consists of four waveguides that exit the circular guide radially at 90-deg intervals. These waveguides excite a pair of orthogonal $TE_{11}$ modes in the feed horn. A straight corrugated section and slight flare follow. The X-band uplink signal is injected in the X-band uplink junction, which is similar in design to the downlink junction. Next the $TE_{11}$ modes at X-band are transformed into the $HE_{11}$ mode in the

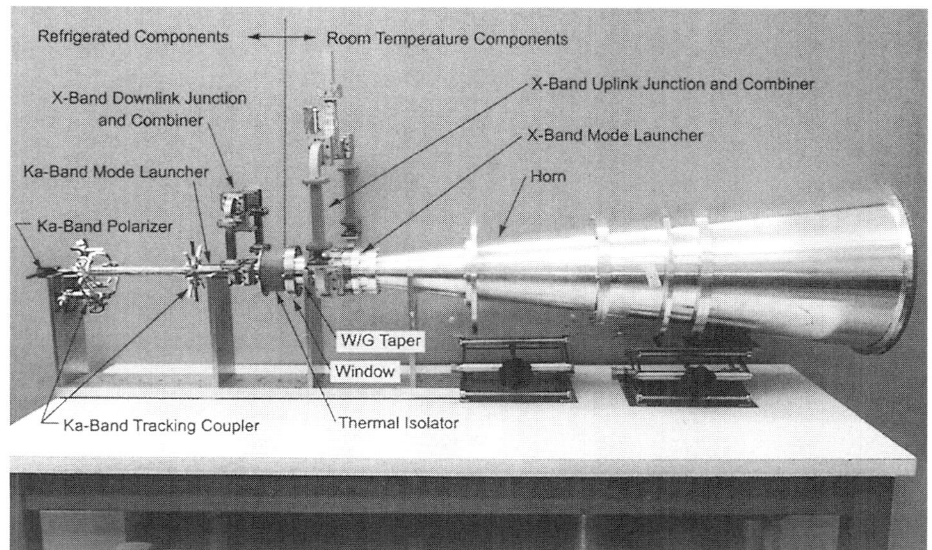


**Fig. 8-21. The X-/X-/Ka-band feed horn with monopulse coupler.**

X-band mode-converter section. Finally, the feed horn flares at a constant angle with a constant groove profile to the final aperture size. Stanton [24] describes the various feed-horn components in greater detail.

**8.3.2.2 BWG Geometry.** Only minimal changes in the existing S-/X-band BWG optics is required to incorporate the X-/X-/Ka-band feed horn. This feed horn is just a direct replacement for the existing X-band feed horn. A slight modification of the shroud interface is required to accommodate the different-size feed aperture. The feed system layout is shown in Fig. 8-23.

**8.3.2.3 Demonstration at DSS-26.** A breadboard X-/X-/Ka-band feed horn was fabricated and tested on DSS-26. The predicted efficiency at X-band was 72.5 percent, and the predicted noise temperature at zenith was 19.83 K. The predicted efficiency at Ka-band is 58.7 percent, with a noise temperature of 37.8 K at zenith [25]. The early tests indicated some problems with the initial feed-horn design, as the noise temperature was 4 K too high at Ka-band and it did not meet the bandwidth requirements at X-band. This led to several redesigns that are discussed in depth in [24]. The design modifications included (a) a new X-band mode launcher design that required only two corrugations instead of seven, (b) the external room temperature waveguides reduced 4.1 in. (10.4 cm) in length, (c) the window and thermal isolator integrated into one component, and (d) the X-band downlink junction modified to a dual-slot design to improve bandwidth and impedance match.

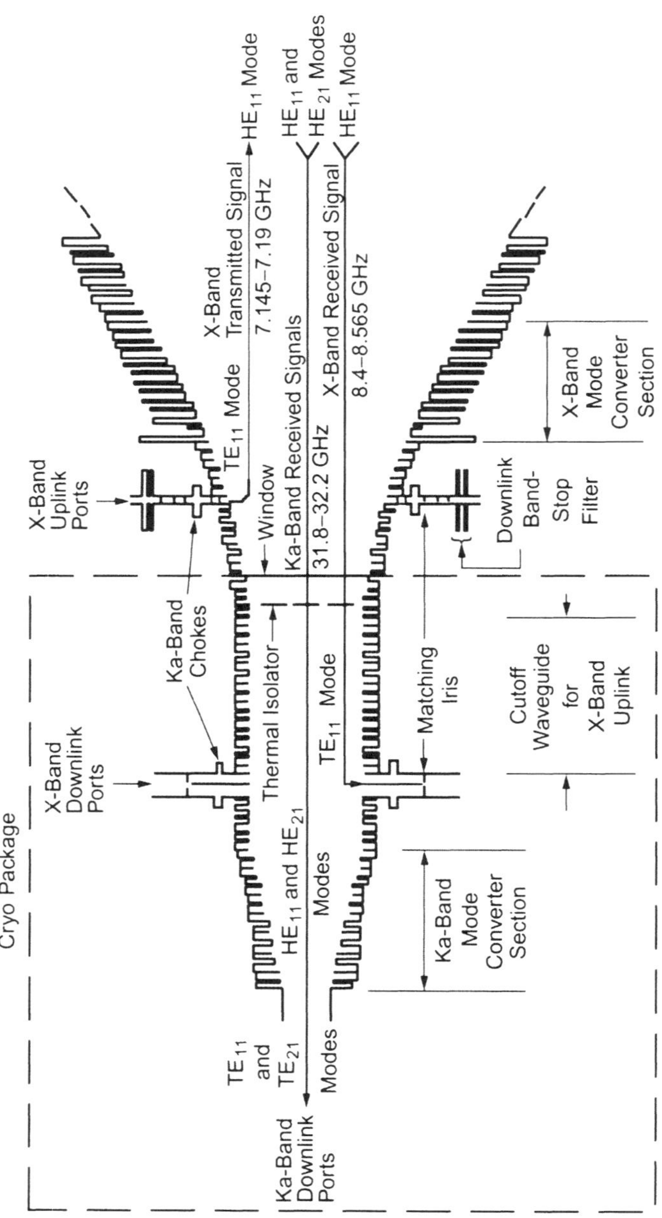


Fig. 8-22. The **X-/X-/Ka-band feed horn internal geometry.**

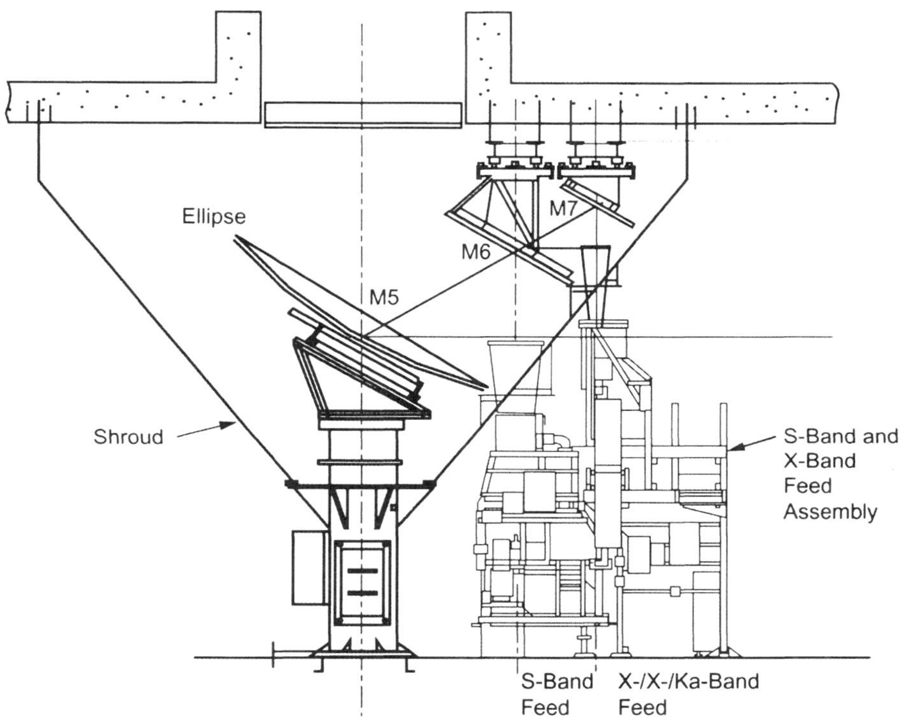



**Fig. 8-23. Feed-system layout at DSS-26.**

Efficiency measurements were made at DSS-26 with the modified X-/X-/Ka-band feed horn and are shown in Fig. 8-24. A problem was discovered with the subreflector position mechanism and subsequently corrected. After the subreflector fix, the efficiencies (72 percent at X-band and 58 percent at Ka-band) were comparable to those of DSS-25 and the predicted values of 72.5 percent and 58.7 percent, respectively.

Noise-temperature measurements were also made at DSS-26. The data is given in Table 8-8. There was an unaccounted-for antenna noise temperature at X-band. The problem was traced to an untaped center plug on M7. After taping, $T_{ant}$ dropped by 6.6 K at X-band and by 0.9 K at Ka-band. After taping, $T_{ant}$ was 6.4 K at X-band and 10.2 K at Ka-band.

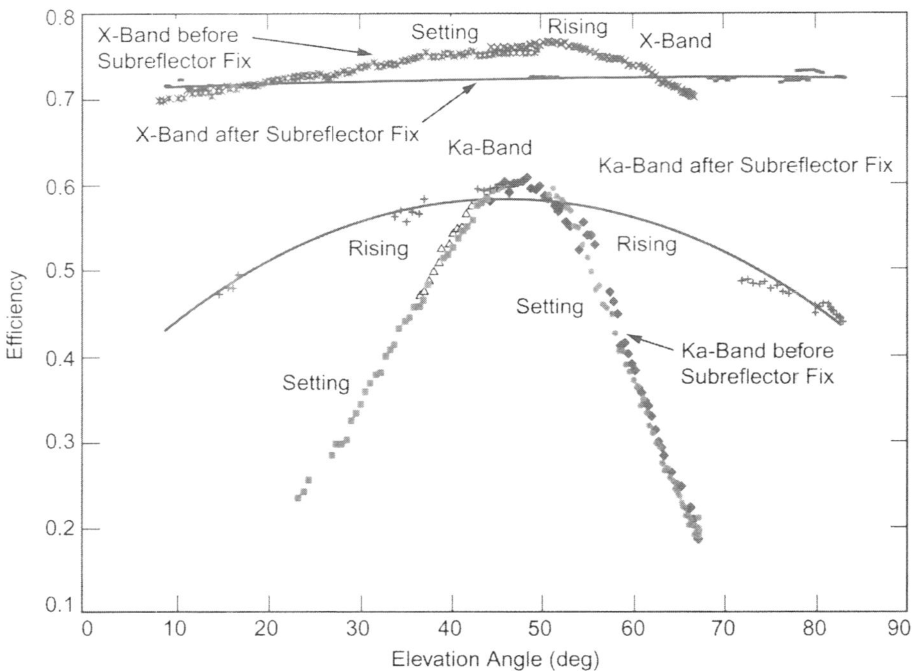


**Fig. 8-24.  DSS-26 operational X-/X-/Ka-band feed efficiencies.**

**Table 8-8.  Noise-temperature measurements of the X-/X-/Ka-band feed horn at DSS-26. At X-band, 5.8 K was unaccounted for. The problem was traced to an untaped center plug on M7.**

| | Noise Temperature (K) | | |
| --- | --- | --- | --- |
| Noise Component | X-Band (8.425 GHz) | Ka-Band (32 GHz) | |
| | Feed System | Sum Channel | Error Channel |
| $T_e$ (feed/LNA) | 16.4 | 16.8 | 25.4 |
| $T_{op}$ | 34.3 | 39.4 | 47.2 |
| $T_{ant}$ before taping | 13.0 | 11.1 | 10.2 |
| $T_{ant}$ after taping | 6.4 | 10.2 | Not measured |

# References

[1] W. A. Imbriale, M. S. Esquivel, and F. Manshadi, "Novel Solutions to Low-Frequency Problems with Geometrically Designed Beam-Waveguide Systems," *IEEE Transactions on Antennas and Propagation*, vol. 46, no. 12, pp. 1790–1796, December 1998.

[2] S. R. Rengarajan, V. Galindo-Israel, and W. A. Imbriale, "Amplitude and Phase-Shaping Effects in Beamwaveguides," *IEEE Transactions on Antennas and Propagation*, vol. 39, no. 5, pp. 687–690, May 1991.

[3] W. Imbriale and D. Hoppe, "Computational Techniques for Beam-waveguide Systems," *2000 IEEE Antennas Propagation Society International Symposium*, Salt Lake City, Utah, pp. 1894–1897, July 16–21, 2000.

[4] W. Imbriale and D. Hoppe, "Recent Trends in the Analysis of Quasioptical Systems," *Millennium Conference on Antennas and Propagation*, Davos, Switzerland, April 9–14, 2000.

[5] W. Veruttipong, J. C. Chen, and D. A. Bathker, "Gaussian Beam and Physical Optics Iteration Technique for Wideband Beam Waveguide Feed Design," *Telecommunications and Data Acquisition Progress Report 42-105*, vol. January–March 1991, http://tmo.jpl.nasa.gov/progress_report/issues.html Accessed October 2001.

[6] W. Veruttipong, "RF Design and Expected Performance of a 34-meter Multi-frequency Beam-Waveguide Antenna," JPL D-11853 (internal document), Jet Propulsion Laboratory, Pasadena, California, July 15, 1994.

[7] *DSS-24 RF Performance Test Plan*, DSN document 829-6 (internal document), Jet Propulsion Laboratory, Pasadena, California, January 1994.

[8] M. J. Britcliffe, *DSS-24 Antenna RF Performance Measurements*, JPL D-12277 (internal document), Jet Propulsion Laboratory, Pasadena, California, February 1, 1995.

[9] D. J. Rochblatt, P. M. Withington, and H. J. Jackson, "DSS-24 Microwave Holography Measurements," *Telecommunications and Data Acquisition Progress Report 42-121*, vol. January–March 1995, http://tmo.jpl.nasa.gov/progress_report/issues.html Accessed October 2001.

[10] T. Y. Otoshi, S. R. Stewart, and M. M. Franco, "A Portable X-Band Front-End Test Package for Beam Waveguide Antenna Performance Evaluation, Part I: Design and Ground Tests," *Telecommunications and Data Acquisition Progress Report 42-103*, vol. July–September 1990, http://tmo.jpl.nasa.gov/progress_report/issues.html Accessed October 2001.

[11] T. Y. Otoshi, S. R. Stewart, and M. M. Franco, "A Portable Ka-Band Front-End Test Package for Beam Waveguide Antenna Performance Eval-

uation, Part I: Design and Ground Tests," *Telecommunications and Data Acquisition Progress Report 42-106*, vol. April–June 1991, http://tmo.jpl.nasa.gov/progress_report/issues.html Accessed October 2001.

[12] W. Veruttipong, "S-band Efficiency and G/T Performance for Various Azimuth Positions," JPL Interoffice Memorandum 3327-94-113 (internal document), Jet Propulsion Laboratory, Pasadena, California, June 9, 1994.

[13] M. Gatti, "RF Characterization and Testing of the DSS-24 Feed Group S- and X-band Feeds," JPL Interoffice Memorandum 3320-94-101, Jet Propulsion Laboratory, Pasadena, California, March 2, 1994.

[14] W. Veruttipong, "RF Design and Expected Performances of a 34-Meter Multifrequency Beam-Waveguide Antenna," JPL D-11853 and DSN 890-261 (internal document), Jet Propulsion Laboratory, Pasadena, California, August 28, 1997.

[15] "DSS-25 Ka-band Downlink Optics Delta Detail Technical Review," internal document, Jet Propulsion Laboratory, Pasadena, California, May 8, 1997.

[16] J. C. Chen, P. H. Stanton, and H. R. Reilly, Jr., "A Prototype Ka-/Ka-Band Dichroic Plate with Stepped Rectangular Apertures," *Telecommunications and Data Acquisition Progress Report 42-124*, vol. October–December 1995, http://tmo.jpl.nasa.gov/progress_report/issues.html  Accessed October 2001.

[17] Y. H. Choung, K. R. Goudey, and L. G. Bryans, "Theory and Design of a Ku-Band TE21-Mode Coupler," *IEEE Transactions on Microwave Theory and Techniques*, vol. 30, no. 11, pp. 1862–1866, November 1982.

[18] M. A. Gudim, W. Gawronski, W. J. Hurd, P. R. Brown, and D. M. Strain, "Design and Performance of the Monopulse Pointing System of the DSN 34-Meter Beam-Waveguide Antennas," *Telecommunications and Mission Operations Progress Report 42-138*, vol. April–June 1999, http://tmo.jpl.nasa.gov/progress_report/issues.html Accessed October 2001.

[19] M. Franco and M. Britcliffe, "DSS-25 RF Performance Report," JPL Interoffice Memorandum 3336-97-003 (internal document), Jet Propulsion Laboratory, Pasadena, California, January 6, 1997.

[20] J. Prater, "System Noise Temperature Measurements of the Ka-band Feed at DSS-25," JPL Interoffice Memorandum 333-4-98-035 (internal document), Jet Propulsion Laboratory, Pasadena, California, August 4, 1998.

[21] K. Liewer, "A Problem with Two-Way Tracking at High Frequencies," JPL Interoffice Memorandum KML-91-66 (internal document), Jet Propulsion Laboratory, Pasadena, California, November 12, 1991.

[22] N. Rappaport, "Aberration Effects on Two-Way Tracking of a Spacecraft and Application to Cassini," JPL Interoffice Memorandum 312.F-97-037 (internal document) Jet Propulsion Laboratory, Pasadena, California, June 17, 1997.

[23] J. C. Chen, P. H. Stanton, and H. F. Reilly, "Performance of the X-/Ka-/KABLE-Band Dichroic Plate in the DSS-13 Beam-Waveguide Antenna," *Telecommunications and Data Acquisition Progress Report 42-115*, vol. July–September 1993, http://tmo.jpl.nasa.gov/progress_report/ issues.html  Accessed October 2001.

[24] P. H. Stanton, D. J. Hoppe, and H. Reilly, "Development of a 7.2-, 8.4-, and 32-Gigahertz (X-/X-/Ka-band) Three-Frequency Feed for the Deep Space Network," *Telecommunications and Mission Operations Progress Report 42-145*, vol. January–March 2001, http://tmo.jpl.nasa.gov/ progress_report/issues.html Accessed October 2001.

[25] "Preliminary Design Review (PDR) BWG Ka-band Upgrades, internal document," Jet Propulsion Laboratory, Pasadena, California, February 8, 2000.

# Chapter 9
# The Antenna Research System Task

The Antenna Research System Task (ARST) was a study and technology demonstration to explore the engineering aspects of combining transmitted pulses in position, time, and phase from independent radar systems [1], as well as to determine the applicability of using a beam-waveguide (BWG) system for high-power applications in large ground station antennas [2]. Toward this objective, the Jet Propulsion Laboratory (JPL) designed, constructed, and tested a Technology Demonstration Facility (TDF), consisting of two 34-m-diameter BWG antennas (see Fig. 9-1) and associated subsystems.

Fig. 9-1. The ARST antennas.

JPL's interest in uplink arraying and high-power technologies stems from its responsibilities in managing the Deep Space Network (DSN) for the National Aeronautics and Space Administration (NASA). Combining signals from two or more antennas in an uplink array has possible application in emergencies such as the loss of a high-power DSN transmitter or the loss of a high-gain antenna on a spacecraft. It would also be required to provide an uplink capability for an array of small reflectors.

In a BWG system, the feed horn and support equipment are placed in a stationary room below the antenna, and the energy is guided from the feed horn to the subreflector using a system of reflecting mirrors. Thus, there is no limit on the size of the transmitter system and it can be placed directly behind the feed, minimizing waveguide losses. One purpose of the TDF was to design a BWG system that would not degrade either the peak or average power handling capability of a large ground station antenna beyond the limitations imposed by either the feed or the dual-reflector configuration.

A 34-m main diameter was chosen for the study because JPL already had experience with conventional BWG antennas of the same size. The high-power design featured a transmit-only, four-port high-gain feed horn as input to a BWG system consisting of a single parabolic mirror and three flat plates. By using a single parabolic mirror, there is no further concentration of the field and, therefore, the highest field concentration is no greater than that caused by the feed horn itself. The feed horn is linearly polarized, and a duplexing reflector is used to reflect the orthogonal polarization into the receive feed. A rotatable dual polarizer provides for arbitrary transmit polarization. The dual-reflector system is shaped to provide uniform illumination over the main reflector and, therefore, maximum gain for the given size aperture. Unfortunately, due to funding limitations and unavailability of a suitable transmitter, the ARST antennas were never tested at high power. However, since significant effort went into the design and fabrication of the transmit feed horn, and the BWG demonstrated a design capable of maximum power, the details of the design and testing are included in this chapter.

Using a monopulse feed in the BWG receive path, along with a low-power transmitter, the ARST uplink arraying experiments were performed by tracking orbiting debris or satellites using the TDF's two antennas, each transmitting pulses to the objects. The TDF then measured the phase for each returned pulse and adjusted the phase and timing of one transmitter (the "slave") so that the next pulses would combine coherently at the object. These experiments permitted JPL to evaluate, characterize, and demonstrate the following:

- The ability of large antennas to track objects with sufficient accuracy to support timing and phase synchronization under various environmental conditions

- The ability to control phase and timing of pulses transmitted from two antennas to an object in order to maintain radio frequency (RF) carrier phase within calibration limits while tracking the object

- The significance of atmospheric effects

- The measurement accuracy of the average phase difference between signals reflected from complex-shaped objects as a function of signal-to-noise ratio (SNR) and object scintillation.

Upon completion of the uplink demonstration, the antennas were to be retrofitted with standard DSN equipment and used as additional resources for the DSN to track spacecraft. An S-band feed system and associated equipment were installed in one of the two antennas (Deep Space Station 27 [DSS-27]); the system is currently used to track high Earth-orbiting spacecraft.

## 9.1 Design of the Beam-Waveguide System

The more conventional BWG designs discussed in Chapters 7 and 8 use four mirrors to direct the feed-horn radiation around the two axes of rotation and to the subreflector. These designs use two curved mirrors, either ellipsoids or paraboloids, and two flat mirrors. This four-mirror system images the feed horn at a location near the apex of the main reflector. Such a configuration results in a dual reflector that is nearly a Cassegrain system, with the standard hyperbola-parabola configuration modified slightly (shaped) in order to provide essentially uniform illumination over the main reflector, thus providing nearly optimum gain.

For high power, the refocusing of the feed-system energy near the main reflector has two distinct disadvantages. First, since the peak field point is generally at or in front of the feed horn itself, the peak power point is imaged in front of the main reflector. If the peak field point can be contained within the BWG system, it is possible to fill the BWG tube with a gas to enhance the peak-power handling capability of the system. Also, when the energy is reflected from a surface such as a BWG mirror, there is a 6-dB enhancement of the power near the reflector, since the incident and reflected field add coherently near the reflection point. It is thus important to have the energy near the mirrors at least 6 dB below the peak point in order to prevent the mirrors from degrading the power handling capability of the system. This is difficult to do if the energy is refocused in the BWG system. For these reasons, an unconventional design [2,3] was chosen for the BWG optics. Figure 9-2 shows the detailed dimensions for the antennas.

For this design, only one curved mirror—a paraboloid—is used, along with three flat mirrors. The radiation from the feed horn is allowed to spread to the

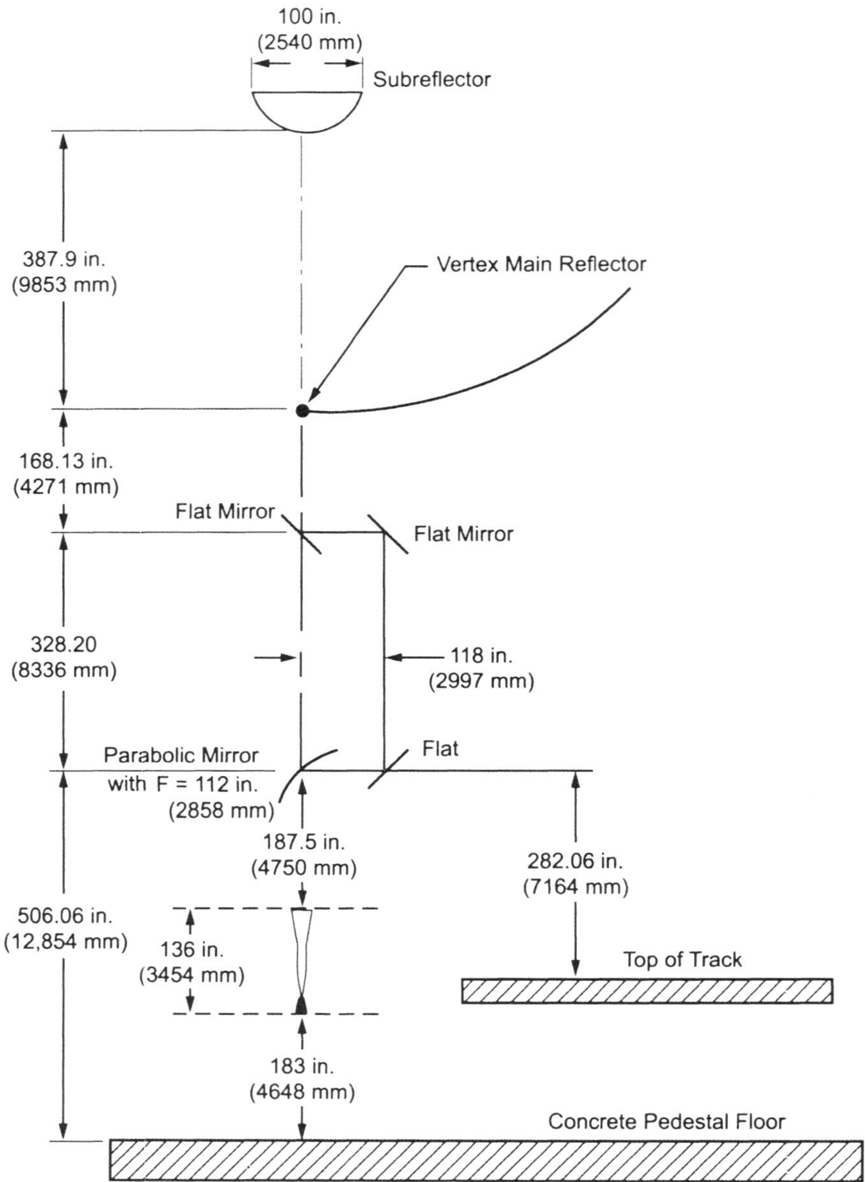



**Fig. 9-2. Geometry of the ARST antenna.**

paraboloid, where it is focused to a point at infinity. That is, after reflection, a collimated beam exists that is directed to the subreflector by the three flat reflectors. The energy is thus spread over the 2.743-m diameter of the BWG tube. Due to diffraction, the beam reflected by the paraboloid does not begin to spread significantly until it exits through the main reflector. At that point, additional spreading occurs in the region between the main reflector and the subreflector. Since a collimated beam exists beyond the first mirror, this antenna is closely related to a near-field Cassegrain design, where the feed system is defined to include both the feed horn and a parabolic mirror.

Both the main reflector and the subreflector are nominally paraboloids, with dual-reflector shaping used to increase the illumination efficiency on the main reflector by compensating for the amplitude taper of the feed radiation pattern. The design of the dual-shaped antenna is based upon geometrical optics, with the shape of the subreflector chosen to provide for uniform amplitude illumination of the main reflector, given the distribution of the radiation striking the subreflector. The curvature of the main reflector is then modified slightly from that of the parent paraboloid to compensate for any phase errors introduced by the subreflector shaping. In the present design, the deviation of the main reflector surface from that of the parent paraboloid is less than 2.5 cm at every point on the 34-m-diameter surface. Although the design of the surfaces is based on geometrical optics, the final design is then analyzed using physical optics, thus including diffraction effects, in order to predict the overall performance accurately.

Figure 9-3 shows the predicted far-field patterns of the overall antenna. The calculated gain (at the center frequency of 7.2 GHz) is 67.67 dB. This

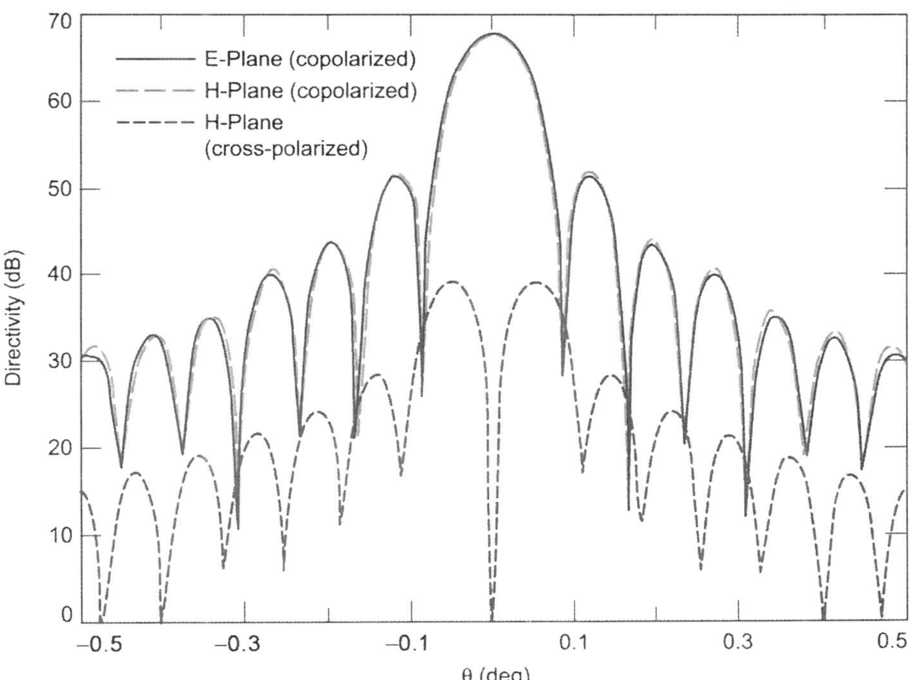


**Fig. 9-3. Far-field patterns of the overall antenna.**

number is based upon a purely theoretical calculation, and the following effects are neglected: polarizer and duplexing grid effects on the feed-horn radiation pattern, shroud effects, root-mean-square (rms) surface accuracy of all mirrors, subreflector and main reflector, quadripod blockage, and feed-horn loss. The specification for surface accuracy is 0.20-mm rms for the subreflector and BWG mirrors, and 0.89-mm rms for the main reflector at the rigging angle. If we define antenna efficiency as the actual gain for the antenna as designed divided by the maximum theoretical gain for a 34-m-diameter circular aperture at 7.2 GHz, we may estimate the following efficiency for the present design: the calculated gain quoted above represents an RF efficiency of 91 percent. The efficiency due to main reflector surface accuracy is calculated to be 93.1 percent and quadripod blockage gives an efficiency of 89.2 percent, for an overall efficiency of 75.6 percent. Including the other factors—BWG mirrors, polarizer, duplexing grid, ohmic loss from the mirrors, voltage standing-wave ratio (VSWR) effects, and BWG mirror alignments—gives an efficiency referenced to the feed input of 73 percent. The antenna was never actually measured from the transmit feed port, but rather by using a monopulse feed in the receive system path [4]. The feed efficiency budget is shown in Table 9-1 and shows a predicted efficiency of 68 percent. The measured efficiency referenced to the feed input was 65 percent for the east and 64 percent for the west antenna. The noise temperature budget, shown in Table 9-2, was 26 K. The measured noise temperature was 21 K.

## 9.2   Design of the Transmit Feed Horn

The basic requirements of the primary transmit feed horn were

- Sufficient gain to minimize spillover past the first parabola in the BWG system

- The peak field point positioned inside the feed horn so that a vacuum system for the feed would result in maximum power handling capability

- Equal E- and H-plane patterns to maximize efficiency in the dual-reflector system

- Arbitrary polarization capability

- Placement far enough from the first mirror so that the energy level at the first mirror was 6 dB below the peak level between the feed and the mirror

- A window design for the feed system mechanically capable of holding a vacuum with minimum radio frequency (RF) loss

- Multiple input ports to enable combining a number of transmitters.

**Table 9-1.  The ARST 34-m BWG antenna receive (monopulse) feed efficiency budget.**

| Element | Sum Pattern at 7.2 GHz | Notes |
|---|---|---|
| Main reflector | | |
| Ohmic loss | 0.99957 | Calculated |
| Panel leak | 0.99992 | Calculated from DSS-13 model |
| Gap leak | 0.9982 | Calculated from DSS-13 model |
| (main reflector + subreflector) rms | 0.94509 | Calculated (rms = 0.79 mm) |
| Subreflector ohmic loss | 0.99957 | Calculated |
| 4 BWG mirrors | | |
| Ohmic loss | 0.99829 | Calculated |
| rms | 0.99251 | Calculated (rms = 0.20 mm) |
| Hyperboloid mirror | | |
| Ohmic loss | 0.99957 | Calculated |
| rms | 0.99625 | Calculated (rms = 0.20 mm) |
| BWG/Cassegrain VSWR | 0.999 | Estimated |
| Polarizer | | |
| Ohmic loss | 0.998 | Estimated |
| Reflection | 0.998 | Estimated |
| Duplexing grid | | |
| Ohmic loss | 0.999 | Estimated |
| Reflection | 0.995 | Estimated |
| Feed-support blockage | 0.8946 | Estimated |
| Pointing squint | 0.9696 | 0.008-deg pointing loss |
| BWG mirror alignments | 0.9994 | Estimated |
| Subtotal | 0.79732 | |
| Physical optics efficiency | 0.85114<br>67.48 dB | Physical optics analysis |
| Total at feed aperture | 0.6786<br>68.18 dB = 100% | |

All of the above requirements are satisfied using a four-input port, linearly polarized, square, multiflare feed [5] with a double output window, as shown in Fig. 9-4. Arbitrary polarization is accomplished using dual-rotary vane polarizers described in Section 9.4. The various flare angles in the design are used to achieve equal E- and H-plane patterns and a tapered amplitude distribution at the output, which puts the peak energy point inside the feed. This is demon-

**Table 9-2. The ARST 34-m BWG antenna receive (monopulse) feed noise-temperature budget.**

| Element | Sum Pattern at 7.2 GHz | Notes |
|---|---|---|
| Cosmic background | 2.50 | Effective blackbody |
| Main reflector | | |
|    Ohmic loss | 0.13 | Calculated |
|    Panel leak | 0.02 | DSS-13 model |
|    Gap leak | 0.50 | DSS-13 model |
|    Rear spill | 0.55 | $T_{\mathit{eff}} = 240$ K |
| Subreflector ohmic loss | 0.13 | Calculated |
| 4 BWG mirrors | | |
|    Ohmic loss | 0.50 | Calculated |
| Hyperboloid mirror | | |
|    Ohmic loss | 0.13 | Calculated |
|    Spill | 6.62 | $T_{\mathit{eff}} = 240$ K |
| Polarizer | 1.16 | Estimated |
| Duplexing grid | 1.74 | Estimated |
| Atmosphere | 2.17 | Goldstone (average clear) |
| $F_1$ fields/forward spill/ground | 0.02 | $T_{\mathit{eff}} = 5$ K |
| Spillover from 4 upper BWG mirrors | 7.15 | $T_{\mathit{eff}} = 240$ K |
| Quadripod scatter | 2.70 | Estimated (DSS-24) |
| Subtotal (K) | 26.02 | Noise at feed aperture |

strated in Fig. 9-5, which plots the relative power density on axis as a function of distance from the feed horn. Figure 9-5 also shows that the energy level at the distance of the first mirror (187 in. [4.75 m]) is 12 dB below the peak value between the feed and mirror.

The design of the window system for this feed horn is a key issue. A dual-quartz window is used. Figure 9-6 shows the geometry of the window system sitting on top of the ARST primary feed aperture. Each window is approximately one-half wavelength thick at the center frequency of operation. The spacing between the windows is chosen to minimize reflections for some band of frequencies. The calculated reflection properties for the window section are shown in Fig. 9-7, along with results for a single half-wave-thick quartz window. Specifications for typical transmitters indicate that the VSWR for the entire feed assembly should be less than 1.1 over a ± 5 percent frequency band. As can be seen from Fig. 9-7, a single half-wave-thick quartz window fails to meet the requirement, particularly since additional factors such as combiner

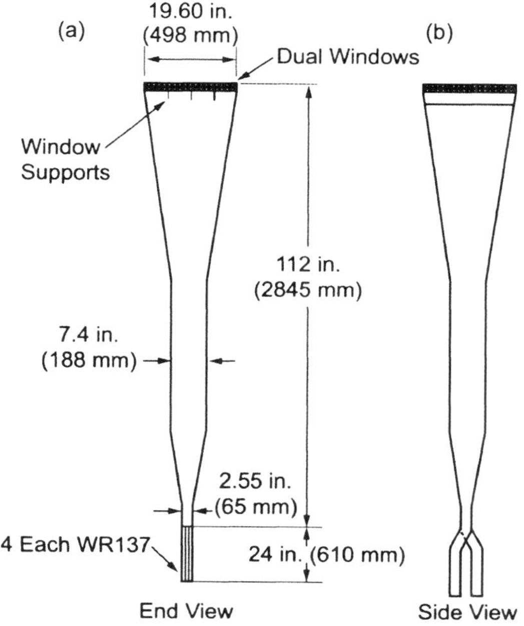



**Fig. 9-4. Important dimensions of the feed horn:
(a) end view and (b) side view.**

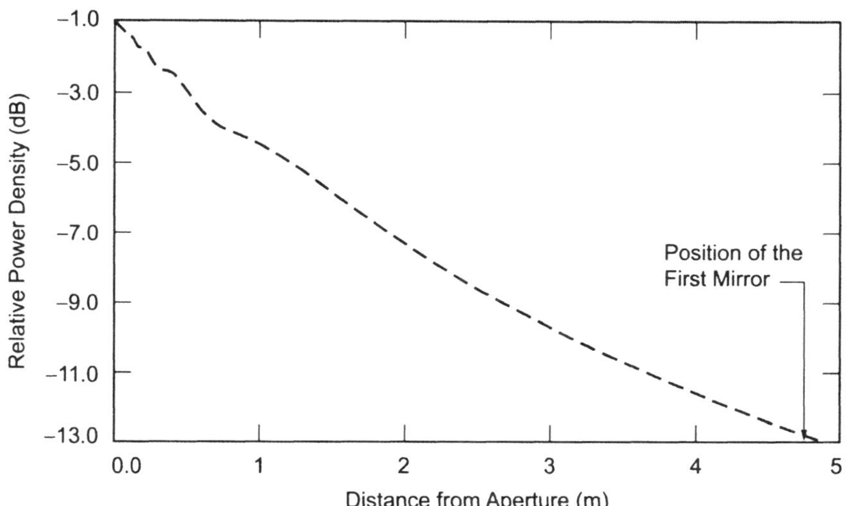



**Fig. 9-5. Relative power density on axis as a function of the
distance from the feed horn.**

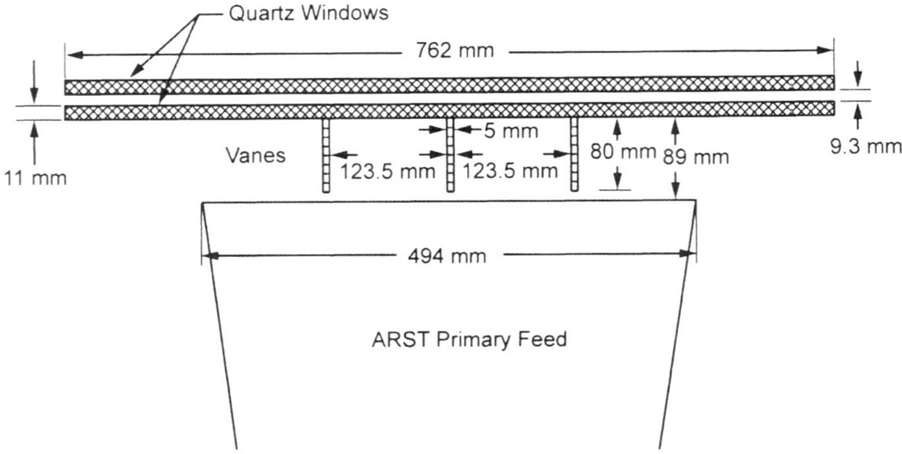



**Fig. 9-6. Geometry of the window system.**

design is able to meet the requirement. This is accomplished by choosing the window thickness so that both windows are perfectly matched at one frequency, and choosing the window spacing so that reflections from the two windows cancel at a second frequency. For frequencies between these values a small reflection will exist, as is shown in the figure, but sufficient bandwidth is available from the dual-quartz window design to meet the VSWR requirement.

Another important property of the window is its mechanical integrity; a 20-in. (508-mm)-diameter window would not be capable of supporting a vacuum. The solution to the window problem was found by using the fact that the radiation inside the feed horn is linearly polarized. This allows the placement of

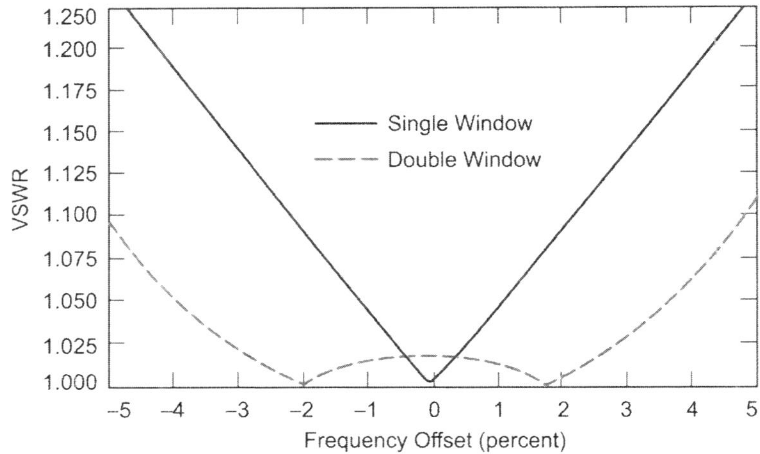



**Fig. 9-7. Calculated reflection from single and double windows as a function of frequency.**

support fins under the window that are orthogonal to the electric field. With the help of a structural engineer, a solution giving minimum blockage was found consisting of three support fins. A finite element analysis of the entire structure was performed to verify the structural integrity of the window, and RF calculations indicate that the VSWR requirements placed on the feed horn by the transmitter can still be met.

This design combines high efficiency with a realizable window design and is, therefore, the design of choice. All of the antenna optics have been designed around this feed horn, and RF testing of a scale model was completed. Figure 9-8 shows a comparison of the calculated and measured radiation patterns for the scale-model feed horn. A picture of the scale-model feed horn is shown in Fig. 9-9 showing the fins that support the window. The measured pattern was for a scale-model feed horn at 31.4 GHz, and the theoretical pattern was computed for the design shown in Fig. 9-4 at 7.2 GHz. The transmit feed horn has four input ports so that four transmitters could be combined in the feed horn. A picture of the four-port coupler is shown in Fig. 9-10. Details of the full-scale primary feed development can be found in [6].

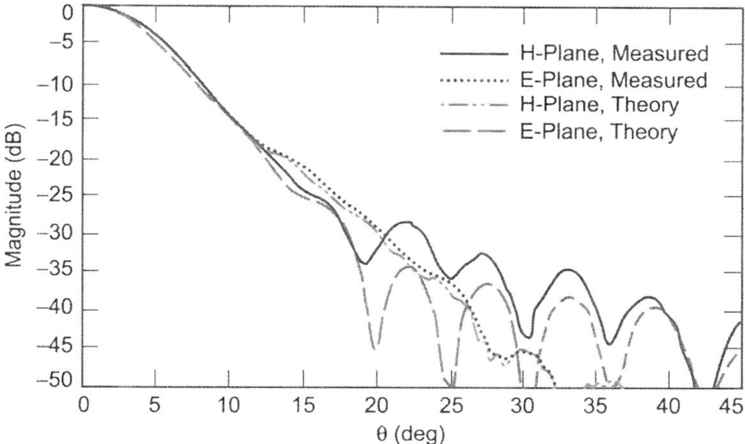


**Fig. 9-8. Comparison of calculated and measured radiation patterns for the scale-model feed horn.**

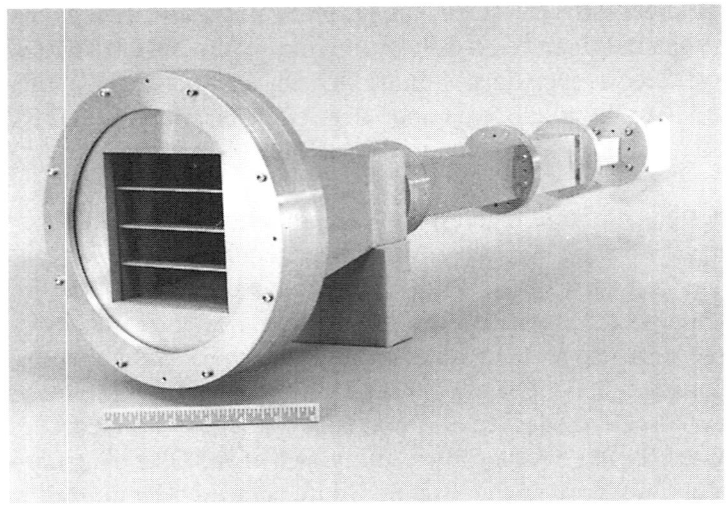

**Fig. 9-9. Scale model of the ARST primary feed.**

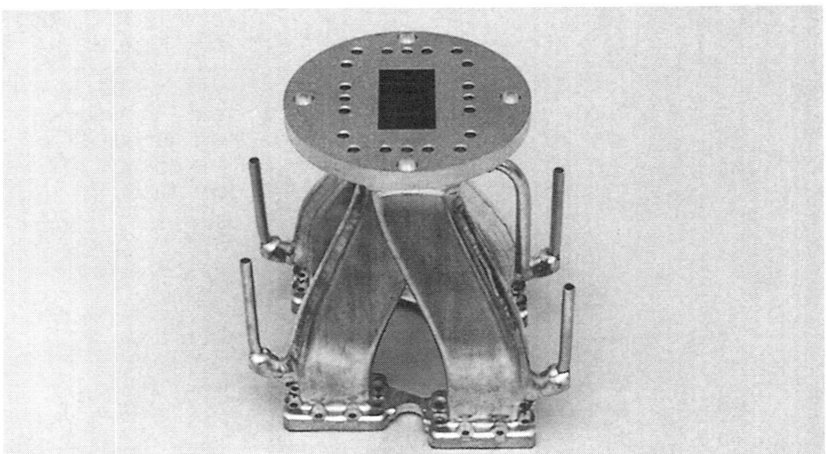

**Fig. 9-10. Primary feed-horn combiner.**

## 9.3  Receive-System Design

The receive system design is shown in Fig. 9-11. It consists of a duplexing grid that reflects the orthogonal polarization towards a curved mirror that is designed to be used with a monopulse receive feed horn. The curved mirror is a hyperbola with focal points at the receive feed horn and the reflected image of the parabola focal point. The monopulse receive feed horn is a scaled version of

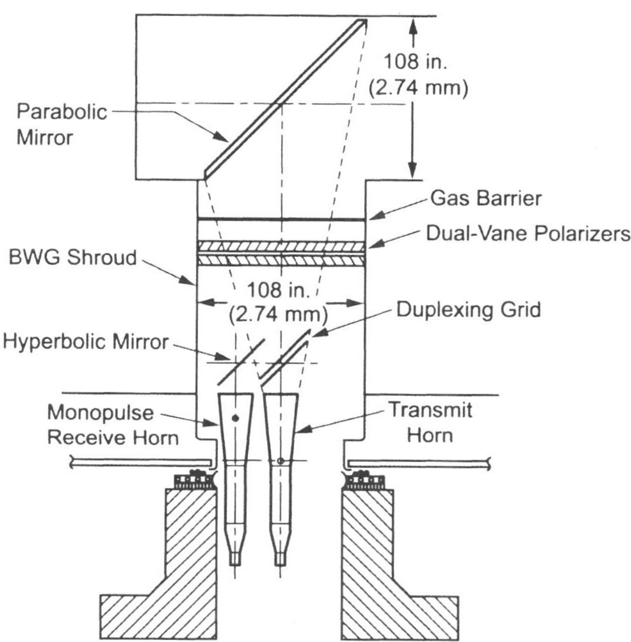



**Fig. 9-11. Schematic diagram of the total feed system.**

the Haystack feed horn [7]. Preliminary calculations were carried out in order to verify that acceptable monopulse performance—that is, tracking to within 1/20 of a beamwidth—could be obtained on the 34-m antenna.

The duplexing grid is a plane reflector that is formed by a series of thin metal strips. If the strips are spaced closer than one-half wavelength, a signal that is polarized parallel to the strips (in this case the received signal) is reflected, while one perpendicular to the strips (in this case the signal from the transmit feed horn) is transmitted. In order to achieve minimal reflection in the perpendicular mode, the dimensions of the strips must be adjusted for the frequency band of interest. Figure 9-12 shows the computed reflection from the grid in the perpendicular mode over the band of interest. It can be seen that less than 0.1 percent of the incident energy is reflected over the band of interest. It should be noted that this reflected energy does not reflect back into the feed horn, but rather is reflected off the grid to the right (see Fig. 9-11), where it will either be absorbed or reflected again. Therefore, the grid should have minimum impact on the VSWR seen by the transmit feed horn, and essentially no effect on the transmitted energy. A Ka-band scale model of the ARST primary feed window system and duplexing grid were built and RF measurements made in the JPL BWG test facility [8]. The setup is shown in Figs. 9-13 and 9-14. Notice the duplexing grid attached to the top of the feed. The measured patterns (including window and grid) in Fig. 9-15 were compared to the theoretical pat-

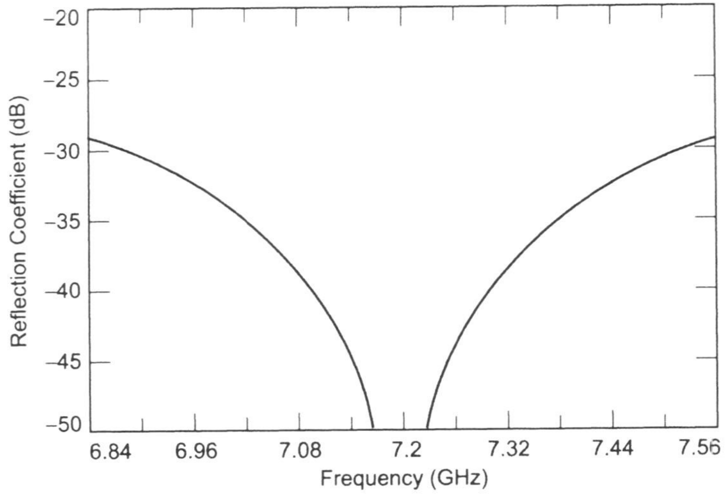

Fig. 9-12.  Calculated reflection from the duplexing
grid in the perpendicular mode.

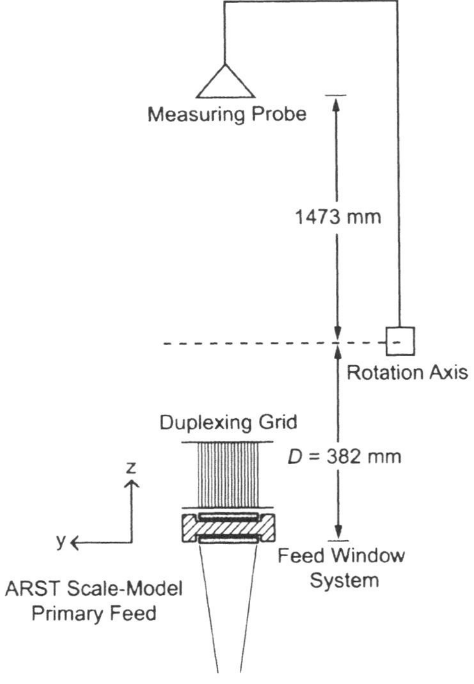


Fig. 9-13.  Measurement setup of the scale model.

Fig. 9-14. RF testing of the scale-model feed.

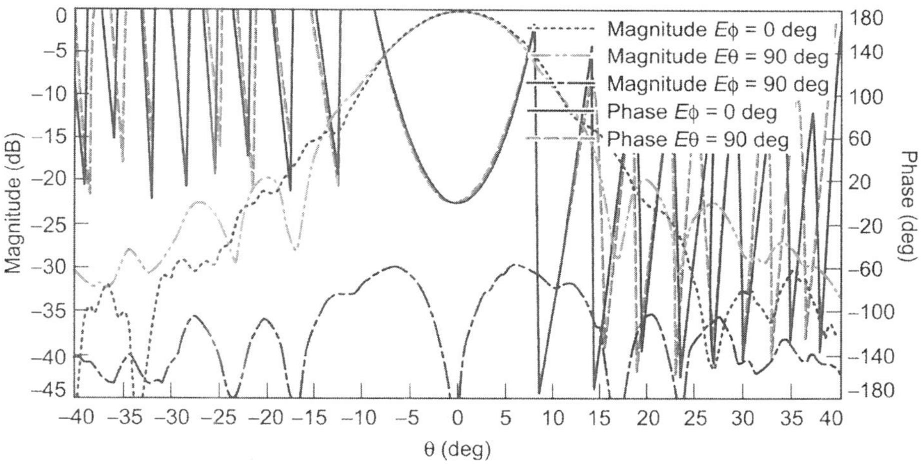



Fig. 9-15. Measured pattern of the scale model at Ka-band.

terns of Fig. 9-16, which did not include the effects of the window or grid. The difference between the measured and predicted radiation patterns is shown in Fig. 9-17.

The agreement of the measured and computed patterns in the range [−10 deg, +10 deg] is within ±1 dB for the amplitude and within ± 20 deg for the phase. This means that the effects of the window system and the duplexing grid on the ARST primary-feed radiation pattern at the center frequency of 7.2 GHz are minimal.

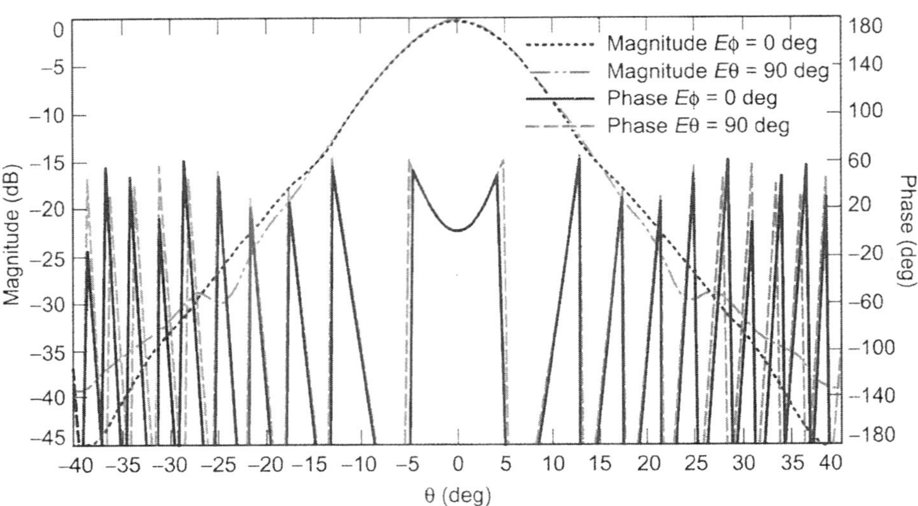

Fig. 9-16. Theoretical radiation patterns at X-band.

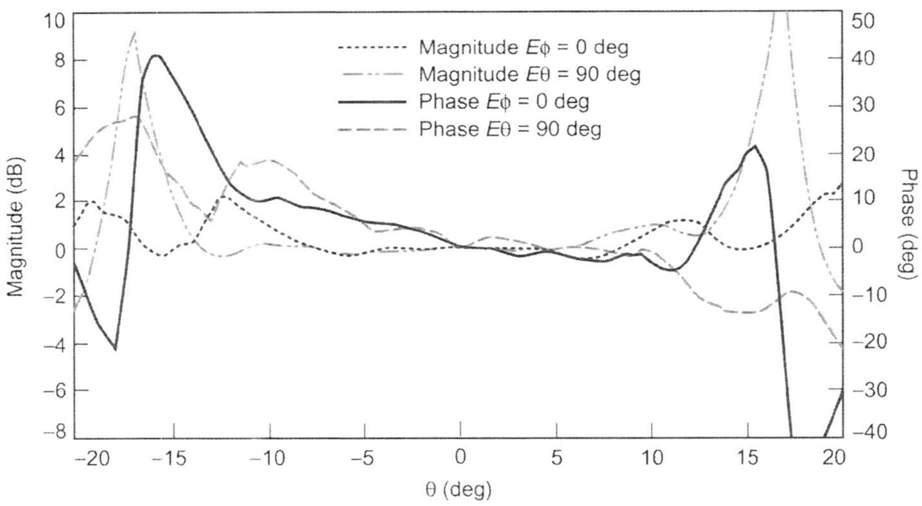

Fig. 9-17. Difference between measured and computed
radiation patterns.

## 9.4   Dual-Vane Polarizers

A broadband venetian blind dual-vane polarizer was designed to be used in the ARST 34-m BWG antenna [9]. At first, a single-vane polarizer design was considered. However, when a prototype was built and tested in the JPL antenna range, it was discovered that the radiation pattern was distorted for the E-perpendicular mode in the cut made perpendicular to the vanes. Upon further investigation, it was determined that the cause of the problem was the extreme sensitivity of the single-vane polarizer to off-axis illumination, which created a high reflection at the center design frequency.

It was at that point that the polarizer was broken up into two sets of vanes separated by some distance. The space between the sets of vanes was selected so that the reflections from the second set would cancel those from the first one. Also, each set of vanes was designed to create a 45-deg phase shift in order to make the overall change equal to 90 deg. Figure 9-18 shows the design dimensions of the venetian blind dual-vane polarizer.

The calculated return loss for the E-parallel and E-perpendicular modes of the dual-vane polarizer is shown in Fig. 9-19. This figure shows that the worst reflection from this polarizer is close to –25 dB at the edge of the 5 percent frequency band.

Figure 9-20 shows a comparison of two measured radiation patterns of a test feed horn. One pattern is of the feed horn by itself and the other one is with the polarizer. The patterns are those for the E-field mode in the cut perpendicu-

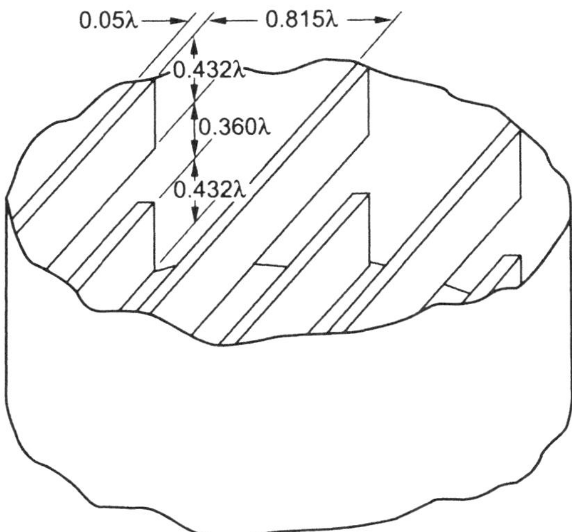


**Fig. 9-18.  Venetian blind dual-vane polarizer.**

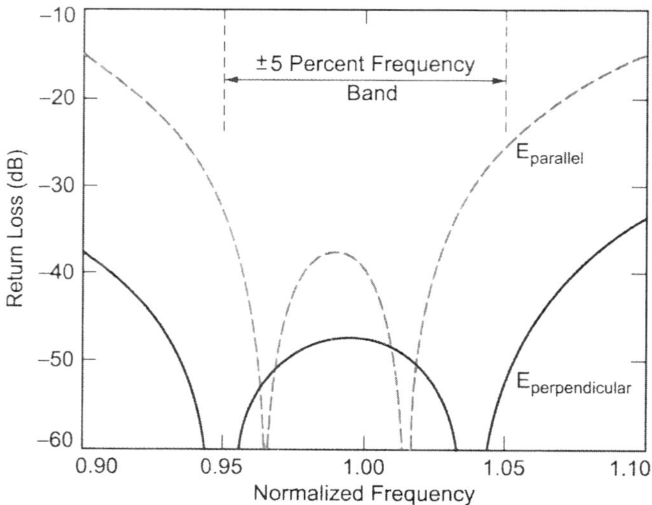

**Fig. 9-19.  Return loss of the dual-vane polarizer.**

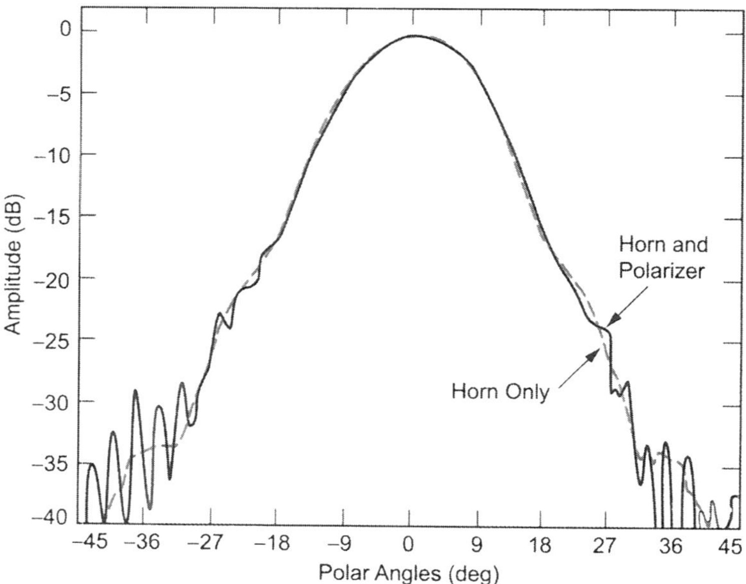

**Fig. 9-20.  Effects of the dual-vane polarizer on the
horn pattern.**

lar to the vanes. It was this mode that showed noticeable pattern deformation
when a single-vane polarizer was used. As can be seen, the dual-vane polarizer
has very little effect on the transmit signal of the illuminating feed horn.

Figure 9-21 details the three main modes of operation of the dual-vane polarizer system. In the first two modes, the first polarizer is set so the vanes are perpendicular to the E-field, and the second polarizer is oriented to produce right-hand circular polarization (RCP) for the first mode or left-hand circular polarization (LCP) for the second mode. In the third mode, the first polarizer is set at 45 deg to the E-field from the feed horn, thus producing RCP between the two polarizers. The second polarizer accepts this RCP and outputs a linear signal at 45 deg to the orientation of the vanes. In the figure, this is shown as being in the y-direction, but linear polarization can be produced at any angle by rotation of the second polarizer.

## 9.5  Uplink Arraying

An uplink arraying experiment was performed using a monopulse feed and a low-power transmitter in the receive system path and is well documented in [1]. Since the uplink array experiment is primarily a system demonstration, and the main focus of this book is primarily on the RF design and performance of the antennas, only the abstract of [1] is presented here. The report describes the

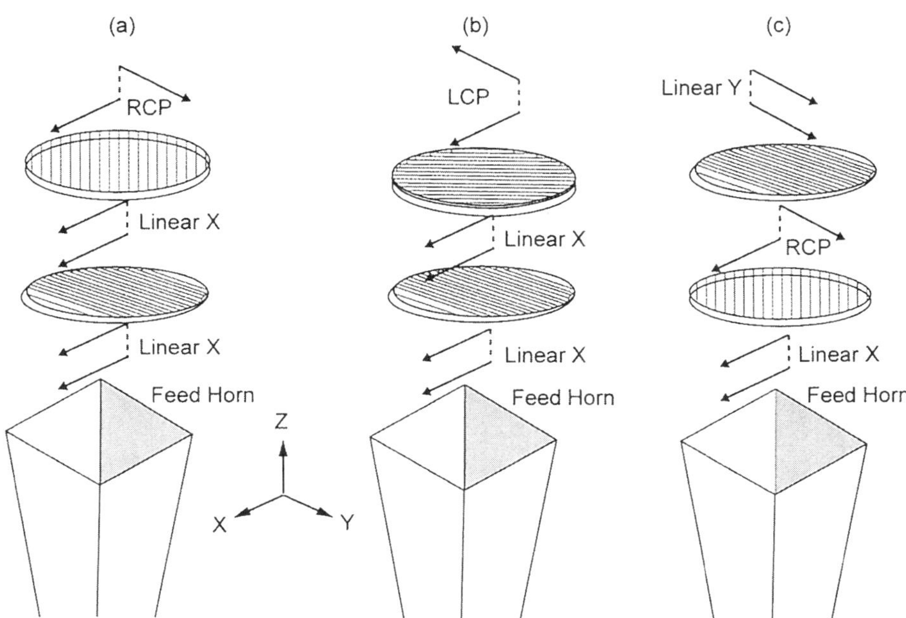


**Fig. 9-21. Modes of operation of the dual-polarizer system: (a) right-hand circular polarization, (b) left-hand circular polarization, and (c) linear y.**

design, development, testing, and performance of a phase-locked, two-element uplink array of 34-m-diameter BWG azimuth-elevation (az–el) radars operating at 7.19 GHz. The rms phase error of two coherent 100-ms, 50-pps (pulses per second), 5-kW peak pulses reflected from low Earth-orbiting debris with SNR > 23 dB is less than 4 deg, with monopulse pointing errors of less than 11-mdeg rms in winds up to 8.9 m/s (32.2 km/h). Each element has a radar SNR performance of 20.8 dB on a 0-dBsm target at 1000-km range. The elements are separated by 204 m on an east-west line at an altitude of approximately 1000 m at the Goldstone Deep Space Communications Complex in the Mojave Desert of California. Signal combination efficiencies approaching 98 percent have been achieved for tracks from 10-deg elevation at signal rise to 4-deg elevation at signal set. Acquisition and tracking are based upon orbital elements and the Simplified General Perturbations Model 4 (SGPM4), along with pointing calibrations based upon radio star and Defense Satellite Communications System (DSCS)-III racks. Both the 15-cm- (6-in.) and 10-cm- (4-in.) diameter spheres placed in orbit by the Space Transportation System (STS)-60 flight have been tracked.

## 9.6   Deep Space Station 27

When the ARST technology demonstration was completed, the antennas were turned over to the DSN to be used to track spacecraft. The first intended use was to track the SOHO (Solar and Heliospheric Observatory) spacecraft, which required one of the antennas—DSS-27—to be retrofitted with an S-band (2.2–2.3 GHz) feed system. Since the BWG optics was designed for 7.2 GHz, it was necessary to use a focal plane analysis method [10,11] to design the feed horn. The first step is to carry out a focal plane analysis of the antenna to compute a theoretical horn pattern (THP). The THP was then approximated by radiation patterns from corrugated horns. It was difficult to match the THP with corrugated feed-horn patterns, so several different cases were analyzed [12,13]. Ultimately, however, the aperture size was used that maximized gain and enabled the feed horn to be attached at the plate position that was built for the ARST primary feed system. The feed horn had a 24.8-dB gain at 2.295 GHz and was placed as shown in Fig. 9-22. The predicted S-band efficiency and noise-temperature performance are given in Tables 9-3 and 9-4. From Table 9-4 the predicted efficiency at 2.295 GHz referenced to the feed input is 49 percent.

The antenna efficiency was measured using the conventional three-point boresight technique on celestial radio sources [14]. The aperture efficiency at 2.29 GHz is 47 ± 2 percent at all elevation angles. These measurements are referenced to the low-noise-amplifier (LNA) input. The estimated uncertainty is

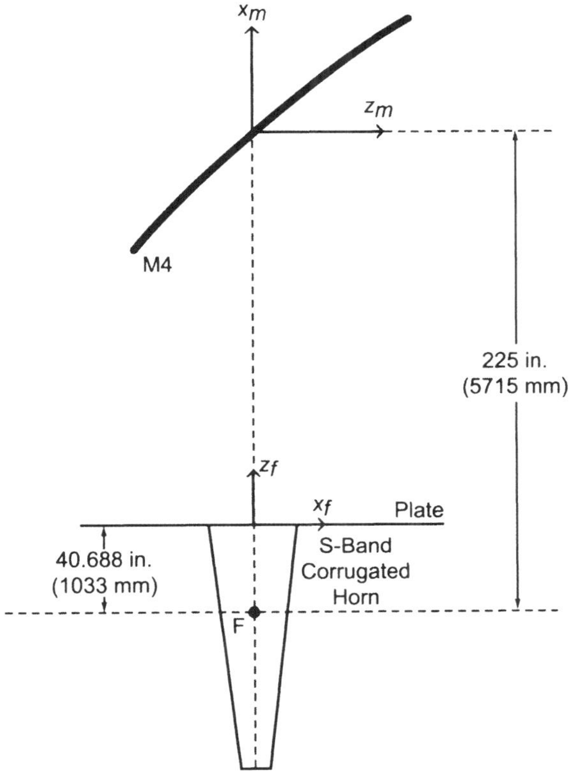



**Fig. 9-22. S-band geometry.**

2 percent. The data is corrected for atmospheric effects using a Goldstone standard atmosphere at 2.295 GHz of 0.03 dB.

The feed and waveguide loss must be subtracted from the efficiency measurements above to quantify the aperture efficiency of the antenna mechanical subsystem defined at the feed input. The feed and waveguide loss is estimated by converting the feed/waveguide noise temperature to loss at a physical temperature of 290 K.

The noise contribution of the feed horn (12.5 K) corresponds to a loss of 0.18 dB or 4.3 percent. The aperture efficiency referenced to the feed input is obtained by multiplying the efficiency referenced to the LNA input by this factor. This results in an efficiency referenced to the feed input of 49 ± 2 percent.

The individual components of the measured total system noise temperature at 2.290 GHz are shown in Table 9-5.

## Table 9-3. S-band efficiency budget.

| Element | Frequency (GHz) | | | Notes |
|---|---|---|---|---|
| | 2.035 | 2.200 | 2.295 | |
| Main reflector | | | | |
|   Ohmic loss | 0.99976 | 0.99976 | 0.99976 | Calculated |
|   Panel leak | 1 | 1 | 1 | Calculated from DSS-24 model |
|   Gap leak | 1 | 1 | 1 | Calculated from DSS-24 model |
|   (Main reflector + subreflector) rms | 0.99428 | 0.99428 | 0.99428 | Calculated (rms = 0.20 mm) |
| Subreflector | | | | |
|   Ohmic loss | 0.99976 | 0.99976 | 0.99976 | Calculated |
| 4 BWG Mirrors | | | | |
|   Ohmic loss | 0.99903 | 0.99903 | 0.99903 | Calculated |
|   rms | 0.99924 | 0.99924 | 0.99924 | Calculated (rms = 0.20 mm) |
| BWG/Cassegrain VSWR | 0.995 | 0.995 | 0.995 | Estimated |
| Polarizer | | | | |
|   Ohmic loss | 1 | 1 | 1 | Removed from system |
|   Reflection | 1 | 1 | 1 | Removed from system |
| Duplexing grid | | | | |
|   Ohmic loss | 1 | 1 | 1 | Removed from system |
|   Reflection | 1 | 1 | 1 | Removed from system |
| Feed support blockage | 0.9 | 0.9 | 0.9 | DSS-24 model |
| Pointing squint | 1 | 1 | 1 | Removed from system |
| BWG mirror alignments | 1 | 1 | 1 | Unknown |
| Measurement uncertainty | 0.9333 | 0.9333 | 0.9333 | −0.3 dB |
| Subtotal | 0.8291 | 0.8291 | 0.8291 | Not applicable |
| Physical optics efficiency | 0.5433 | 0.5722 | 0.5929 | Physical optics analysis |
| Total at feed aperture | 0.4505[a] | 0.4744[b] | 0.4916[c] | Not applicable |

[a] 57.21 dB = 100 percent; 53.75 dB = 45 percent.
[b] 57.88 dB = 100 percent; 54.64 dB = 47.4 percent.
[c] 58.25 dB = 100 percent; 55.17 dB = 49.1 percent.

### Table 9-4. S-band noise-temperature budget.

| Element | Frequency (GHz) | | | Notes |
|---|---|---|---|---|
| | 2.035 | 2.200 | 2.295 | |
| Main reflector | | | | |
|     Ohmic loss | 0.07 | 0.07 | 0.07 | Calculated |
|     Panel leak | 0.00 | 0.00 | 0.00 | DSS-24 model |
|     Gap leak | 0.50 | 0.50 | 0.50 | DSS-24 model |
|     Rear spill | 24.10 | 26.35 | 25.03 | $\alpha = 240$ K |
| Subreflector | | | | |
|     Ohmic loss | 0.07 | 0.07 | 0.07 | Calculated |
| Polarizer | 0.00 | 0.00 | 0.00 | Removed from system |
| Duplexing grid | 0.00 | 0.00 | 0.00 | Removed from system |
| Cosmic background | 2.70 | 2.70 | 2.70 | Effective blackbody |
| Atmosphere | 1.80 | 1.80 | 1.80 | Goldstone (average clear) |
| $F_1$ fields/forward spill/ ground | 1.14 | 1.04 | 0.99 | $\alpha = 5$ K |
| 4 upper BWG mirror spill | 24.17 | 21.10 | 17.53 | $\alpha = 180$ K |
| 4 BWG mirror ohmic loss | 0.28 | 0.28 | 0.28 | Calculated |
| Quadripod scatter | 2.50 | 2.50 | 2.50 | Estimated (DSS-24) |
| Subtotal (K) | 61.33 | 56.41 | 51.48 | Noise at feed aperture |

### Table 9-5. Measured noise temperature.

| Noise-Temperature Component | T (K) | Notes |
|---|---|---|
| Sky | 4.5 | 1.8-K atmosphere, 2.7-K cosmic background |
| Receiver | 1.0 | Measured |
| Low-noise amplifier | 45.0 | Measured |
| Feed/waveguide | 12.5 | From measurements of microwave system on ground |
| Antenna | 40.1 | Calculated by subtracting all the above from the total operating temperature |
| Total operating temperature | 103.1 | Total system-noise temperature |

The noise contribution of the microwave subsystem is the sum of the LNA and the feed/waveguide contribution, or 57.5 K. The antenna mechanical subsystem contribution is 39.1 K. Observe that this temperature is lower than the 51.48 K given in Table 9-4. This is because the analysis used to compute the noise temperature did not include the effect of the BWG tube. A more accurate analysis based upon the techniques given in [15–17] yielded a more accurate estimate of 42.9 K.

The second antenna, DSS-28, has not been upgraded to include DSN frequencies and, for that reason, is currently not in use.

# References


[1]  H. Cooper, B. Conroy, J. Craft, R. Dickinson, D. Losh, and C. Yamamoto, *Antenna Research System Task Final Report*, JPL D-21208 (internal document), Jet Propulsion Laboratory, Pasadena, California, June 30, 1994.

[2]  W. A. Imbriale, D. J. Hoppe, M. S. Esquivel, and B. L. Conroy, "A Beam-waveguide Design for High-Power Applications," *Intense Microwave and Particle Beams III*, proceedings of the SPIE conference, vol. 1629, Los Angeles, California, pp. 310–318, January 20–24, 1992.

[3]  M. S. Esquivel, D. J. Hoppe, and W. A. Imbriale, *RF Design and Analysis of the ARST Antenna*, JPL D-11259 (internal document), Jet Propulsion Laboratory, Pasadena, California, November 1, 1993.

[4]  M. Britcliffe, L. Alvarez, and M. Franco, "ARST Antenna RF Performance Measurements," JPL Interoffice Memorandum 3328-17-94 (internal document), Jet Propulsion Laboratory, Pasadena, California, March 4, 1994.

[5]  S. B. Cohn, "Flare-Angle Changes in a Horn as a Means of Pattern Control," *The Microwave Journal*, pp. 41–46, October 1970.

[6]  M. S. Esquivel, R. Frazer, D. J. Hoppe, F. Manshadi, R. Manvi, and H. Reilly, *ARST Primary Feed System Development*, JPL D-21209 (internal document), Jet Propulsion Laboratory, Pasadena, California, June 1994.

[7]  K. R. Goudey and A. F. Sciambi, Jr., "High Power X-band Monopulse Tracking Feed for the Lincoln Laboratory Long-Range Imaging Radar," *IEEE Transactions on Microwave Theory and Techniques*, vol. MTT-26, no. 5, May 1978.

[8]  J. R. Withington, W. A. Imbriale, and P. Withington, "The JPL Beam-waveguide Test Facility," *IEEE Antennas and Propagation Society Inter-*

*national Symposium*, vol. 2, Ontario, Canada, pp. 1194–1197, June 24–28, 1991.

[9] B. L. Conroy, D. J. Hoppe, and W. A. Imbriale, "Broadband Venetian Blind Polarizer with Dual Vanes," *International Journal of Infrared and Millimeter Waves*, vol. 14, no. 5, pp. 897–996, May 1993.

[10] W. A. Imbriale, "Design Techniques for Beam Waveguide Systems," *Proceedings of the International Conference on Millimeter and Submillimeter Waves and Applications III*, vol. 2842, Denver, Colorado, pp. 192–200, August 5–7, 1996.

[11] W. A. Imbriale, M. S. Esquivel, and F. Manshadi, "Novel Solutions to Low-Frequency Problems with Geometrically Designed Beam-Waveguide Systems," *IEEE Transactions on Antennas and Propagation*, vol. 46, no. 12, pp. 1790–1796, December 1998.

[12] M. S. Esquivel, "ARST 34-m BWG Antenna PO Analysis at S-band and X-band," JPL Interoffice Memorandum MSE-93-920 (internal document), Jet Propulsion Laboratory, Pasadena, California, October 14, 1993.

[13] M. S. Esquivel, "More S-band PO Analysis of DSS-27," JPL Interoffice Memorandum MSE-94-003 (internal document), Jet Propulsion Laboratory, Pasadena, California, January 2, 1994.

[14] M. J. Britcliffe, *DSS-27 Antenna RF Performance Measurements*, JPL D-12543 (internal document), Jet Propulsion Laboratory, Pasadena, California, March 15, 1995.

[15] A. G. Cha and W. A. Imbriale, "A New Analysis of Beam Waveguide Antennas Considering the Presence of the Metal Enclosure," *IEEE Transactions on Antennas and Propagation*, vol. 40, no. 9, pp. 1041–1046, September 1992.

[16] W. A. Imbriale, T. Y. Otoshi, and C. Yeh, "Power Loss for Multimode Waveguides and Its Application to Beam-Waveguide System," *IEEE Transactions on Microwave Theory and Techniques*, vol. 46, no. 5, pp. 523–529, May 1998.

[17] W. A. Imbriale, "On the Calculation of Noise Temperature in Beam Waveguide Systems," proceedings of the International Symposium on Antennas and Propagation, Chiba, Japan, pp. 77–80, September 24–27, 1996.

# Chapter 10
# The Next-Generation Deep Space Network

This monograph has presented a comprehensive summary of all the large ground antennas in the Deep Space Network (DSN), including the evolutionary paths that have led to the current operational configuration. In this chapter, future scenarios are considered for both short-term and long-term needs.

A key element of the DSN, the 70-m antennas were built starting in the mid-1960s. They first became operational as 64-m antennas starting in 1966 and continued until 1973. They were extended to 70 m in support of the Voyager spacecraft encounter with Neptune in 1989. The original 64-m antenna had a design service life of 10 years for mechanical components, based on 25 percent use. Though life-extension efforts continue, recent failures and repairs indicate that extended outages can be expected. For example, on August 14, 2001, the radial bearing failed on the 70-m antenna at Goldstone, causing the antenna to be out of service for 4 days. Future catastrophic failures are possible, but not predictable.

The 70-m antennas are nonresilient, single points of failure. In 1991, the National Aeronautics and Space Administration (NASA) established a 70-m backup plan that uses four 34-m beam-waveguide (BWG) antennas per DSN complex to replace the downlink capability of a 70-m antenna [1]. This plan was only partially completed due to lack of funding.

In 1999, a 3-year study was commissioned to (a) investigate and estimate the remaining service life of the existing 70-m antennas and (b) investigate alternatives for backup and eventual replacement of the 70-m subnetwork capability. Some preliminary results from the study are discussed below.

However, there needs to be a look far beyond just the replacement of the 70-m antennas. The support of NASA's four major space science themes (Structure and Evolution of the Universe, Origins, Solar System Exploration, and the Sun-Earth Connection) demands a much greater DSN capability to support high-data-rate science instruments such as synthetic-aperture radar and multispectral and hyperspectral imagers. In addition, video or high-definition television (HDTV) can provide greater data return for public outreach from Mars missions.

Three complementary technologies need to be explored to fulfill these needs: (a) significantly larger (10 to 100 times) antenna aperture size in the DSN, (b) optical communications, and (c) relay satellites orbiting nearby planets.

## 10.1 The Study to Replace 70-Meter Antennas

The DSN 70-m antennas are crucial for deep-space critical events, both for planned operations and anomalous unplanned operations. Examples of critical planned operations are encounters, entry-descent-landing (Mars missions), limited-life vehicles (Solar Probe [to the Sun] and Europa [to Jupiter] missions), and limited data-return span (Cassini [to Saturn] high-activity periods and Near Earth Asteroid Rendezvous [NEAR] Eros descent). Examples of anomalous unplanned operations are spacecraft emergency recovery (Solar and Heliospheric Observatory [SOHO], NEAR, and Voyager [tour of the solar system]) and saving damaged missions [Galileo, to Jupiter]. Compared to 34-m antennas, the DSN 70-m antennas can significantly improve the data return of space science missions.

The 70-m antennas, if baselined, can produce a fourfold increase in mission data return (at any given frequency) or reduce mission power, mass, volume, and associated cost. But the 70-m antenna can only be baselined for a mission if it is reliable and backed up—hence, the need for the study to replace the 70-m antennas.

The replacement study considered the following options [2]:

- Extending the life of the existing 70-m antennas

- Designing a new 70-m single-aperture antenna

- Arraying four 34-m aperture antennas

- Arraying small reflector antennas

- Arraying flat-plate antennas

- Implementing a spherical pair of high-efficiency reflecting elements (SPHERE) antenna concept.

### 10.1.1   Extending the Life of the Existing 70-Meter Antennas

Two options were studied: life extension (a) with and (b) without a Ka-band upgrade.

A complete antenna structure model was updated for analysis of both strength and fatigue life, using the current antenna-use factor. The weakest sections were identified, and proposals for retrofitting the suspected areas were investigated. Items to be overhauled consisted of the subreflector positioner, azimuth drives, azimuth tangential links, azimuth bull gear, hydrostatic bearing, elevation bearing assembly, elevation bull gear, elevation drives, and the radial bearing assembly. Costs and tasks for both a 10-year and 25-year extension were identified.

New technology for adding Ka-band consisted of a new deformable subreflector with actuators; a much simpler, less expensive yet more accurate pointing instrument (replacement for the existing Master Equatorial precision pointing instrument); and a new X-/X-/Ka-band feed. If this option is selected, the X-/X-/Ka-band feed will replace the existing X-band feed and provide a Ka-band receive capability. The deformable subreflector will be used to compensate for gravity distortion at Ka-band.

### 10.1.2   Designing a New 70-Meter Single-Aperture Antenna

The new antenna design option includes state-of-the-art technology for gain recovery through gravity compensation and precision pointing at frequencies up to 50 GHz. This frequency range is sufficient to support the future Ka-band communication requirements and NASA's Human Exploration and Development of Space (HEDS) program. This design concept provides an increase in performance of the existing 70-m antennas. The primary configuration is for Ka-band (32 GHz) downlink and X-band downlink and uplink operations, with allowance for future expansion to include Ka-band uplink and HEDS radio frequency (RF) equipment. In addition, this antenna design would accommodate the existing 70-m antenna RF equipment, to the maximum extent possible, should the existing 70-m antennas become inoperable.

Key features of the new design are

- A 70-m-diameter main reflector
- Dual-shaped RF optics
- A center-fed BWG
- Feeds and front-end electronics located in alidade enclosures that rotate in azimuth
- Double elevation wheel and counterweights
- An electric drive for the azimuth wheels

- An actuated main reflector surface for gravity compensation
- A low-profile concrete foundation.

In addition, to provide precision blind pointing and focus corrections, the following technologies should be incorporated:

- Insulated and ventilated backup structure
- Thermal correction of pointing error and subreflector focus error
- Wind-sensor correction of pointing error and subreflector focus error
- A metrology-controlled subreflector.

### 10.1.3   Arraying Four 34-Meter Aperture Antennas

This configuration is an array of four 34-m BWG antennas to provide a 70-m equivalent aperture. This option is considered the most well understood, as the antenna cost is readily available from recent 34-m antenna construction. Throughout the 1980s, as part of the Voyager mission, development of downlink array technology was completed and tested for telemetry and tracking at the DSN facilities in Goldstone, California.

A problem was discovered when a high-power uplink (equivalent to 20 kW on the 70-m antenna) was considered. If this uplink effective isotropic radiated power (EIRP) was to be produced with a single 34-m antenna, a transmitter power of 80 kW would be required, and the power density in the near-field beam would exceed the aircraft safety standard of 10 mW/cm$^2$. The EIRP could be achieved with a lower power density if more than one antenna had a transmitter and the array was properly phased; the required EIRP could be produced by installing 5-kW transmitters on each of the four antennas. The proper phasing could be obtained by combining calibration of transmitter phase, using a spacecraft power monitor with knowledge of the geometric change in path length to each antenna as the pointing is changed. Atmospheric effects are small at X-band but are appreciable at Ka-band. Demonstration of X-band uplink phasing is being planned in the near future.

Future work was proposed to improve antenna performance for support of HEDS by providing

- A 100 percent solid reflector surface
- Lower reflector root-mean-square (rms) surface error
- Higher drive capacity to handle the additional wind load due to solid panel utilization
- Better antenna pointing requirements
- More reliable antenna design.

Developing and demonstrating an uplink arraying technique was also proposed.

### 10.1.4  Arraying Small Antennas

An array of three-hundred and fifty 6.1-m antennas is being constructed in Hat Creek, California, by the SETI (Search for Extraterrestrial Intelligence) Institute and the University of California, Berkeley [3]. The array is known as the Allen Telescope Array (after benefactor Paul Allen), is planned for completion in 2005, and will have the equivalent area of a 114-m-diameter single telescope. The antenna will utilize a log-periodic feed and low-noise amplifier (LNA) covering the entire 0.5- to 11.0-GHz frequency range.

A similar array, with one-hundred and forty 8-m antennas and higher-performance narrowband receivers at 2.2, 8.4, and 32 GHz, was considered for the 70-m replacement study. There are a number of advantages to this approach.

The first advantage is that, since the cost of an antenna varies with diameter (usually taken to be $D^{2.7}$), the cost per unit total area will be less with an array of small antennas. For example, if a 70-m antenna costs \$100 million, the cost of one hundred 7-m antennas will be $100 \leftrightarrow (0.1)^{2.7} = \$20$ million. However, the cost of the electronics for the array will be higher. It can be shown [2] that the minimum total cost for a given total area will be achieved with an antenna cost that is 2.86 times the electronics cost. This ratio is dependent only upon the antenna cost exponent, here assumed to be 2.7.

The second advantage is that higher data throughput can be obtained by virtue of the flexibility and multiplicity of digital beam forming. Specifically, having multiple beams within the main beam of the small antenna allows simultaneous communication with multiple spacecraft orbiting a planet. Furthermore, subdividing the array into subarrays allows communication with a number of spacecraft located in different parts of the sky.

The third advantage is that the high reliability and availability are guaranteed by virtue of eliminating single-point failures. The array is sized to give the $G/T$ and EIRP performance of a 70-m antenna, even when only 10 percent of the antennas are calibrated or maintained. And maintenance can be performed during a 40-hour workweek.

The fourth advantage is that very high angular resolution is ensured. Indeed, the proposed array has a beamwidth 14 times sharper than that of a 70-m antenna. The improved resolution allows new paradigms for determining spacecraft position. The beamwidth can be further sharpened by adding outrigger antenna elements.

The fifth advantage is extended frequency range. Small, stamped solid aluminum antennas can operate at short wavelengths much more easily than large structures. In fact, very wide bandwidth communication is feasible at frequencies as high as 100 GHz.

The required number of antennas to produce a 62.4-dB/K $G/T$ at 8.4 GHz (70-m equivalent) as a function of diameter is shown in Table 10-1.

**Table 10-1.  Required number of antennas as function of diameter.**

| Minimum Performance Requirements | Antenna Diameter (m) | | |
|---|---|---|---|
| | 5 | 8 | 12 |
| Elements required for 62.4-dB $G/T$ at 8.4 GHz | 328 | 128 | 58 |
| Allowance for continuous calibration, 6 percent | 20 | 8 | 4 |
| Allowance for maintenance, 3 percent | 10 | 4 | 2 |
| Total required array elements | 358 | 140 | 64 |

A number of primary enabling technologies are required to make a low-cost array feasible:

* Inexpensive, mass-produced, stamped parabolic dishes
* Multiple-frequency or decade bandwidth feeds
* Low-noise indium phosphide (InP) high-electron-mobility transistors (HEMT) amplifiers
* Commercially available 80-K cryogenics
* Wideband fiber-optic links
* Satellite transmitter and timing calibration
* Low-cost solid-state high-power amplifiers.

Whereas the technology for downlink arraying is quite well understood, the techniques for uplink arraying need to be developed.

## 10.1.5   Arraying Flat-Plate Antennas

This type of array consists of thousands of low-cost, flat-plate antennas arranged on the ground to enable signal detection from any direction within the hemisphere. The array orientation is optimized to maximize the signal-detection capability.

To provide a 70-m capability with active planar phased arrays, millions of elements are required. Many alternative phased arrays—including planar horizontal, hybrid mechanical/electronically steered, mechanically steered reflectors, multifaceted planar, planar reflect-arrays, and phased array-fed lens antennas—were compared and their viability assessed.

Although they have many advantages, including higher reliability and near-instantaneous beam switching or steering capability, the arrays are currently prohibitively expensive, and it has been concluded that the only viable array options at present are the arrays of modest-sized reflector antennas.

### 10.1.6 Implementing a Spherical Pair of High-Efficiency Reflecting Elements Antenna Concept

The original spherical pair of high-efficiency reflecting elements (SPHERE) concept consists of two 100-m nontipping spherical reflectors pointed at 30- and 70-deg elevation. The antennas are fully rotatable in azimuth, and switching between antennas is required as a spacecraft crosses 50-deg elevation. The 100-m-diameter provides a 70-m spot for all scans and, thus, nearly equivalent performance to a 70-m parabolic reflector. The concept utilizes an Arecibo Observatory (Puerto Rico)-style [4] Gregorian feed system, with a linear motion for elevation scan that covers a ± 20-deg elevation range. This concept has major cost advantages over more conventional structures:

* No tipping of the large structure
* No counterweight
* No gravity effects (thus permitting high-frequency operation)
* Simple alignment procedure
* Simple backup structure
* Identical, low-cost panels.

A spherical design was implemented in 1997 for the McDonald Observatory (Texas) 10-m optical Hobby–Eberly Telescope (HET) [5]. The cost of HET is significantly lower than that of an equivalently sized tippable structure such as the W. M. Keck Observatory (Hawaii) 10-m optical telescope.

The Chinese are also considering a spherical antenna [6]. They are proposing a 500-m sphere with adjustable main reflector panels and a simple focal-point feed. The main reflector panels are actuated to form a parabola for each direction of scan.

Additional work at the Jet Propulsion Laboratory (JPL) on the SPHERE concept [7] produced two significant improvements. First, the number of antennas was reduced from two to one by enlarging the aperture in the scan plane to 135 m and having the feed travel extended to the elevation angular range of 90 deg so the entire sky can be covered with just the one antenna. Second, the feed system was simplified from a two-mirror Arecibo-style system to a single-mirror system that is a phase-correcting subreflector.

A spherical antenna can potentially provide a large aperture at a much-reduced cost compared to that of conventional tipping antennas, and construction of such an antenna should, therefore, be vigorously pursued.

## 10.2 Towards the Interplanetary Network

For future exploration of Mars, Edwards [8] has proposed a communications and navigation infrastructure that includes orbiting relay satellites (see Fig. 10-1). The use of orbiting satellites around a planet to communicate with assets on the surface of that planet and to rely data back to Earth from the orbiter is a key element in an interplanetary network that may eventually include relays around the majority of planets.There are already two infrastructure assets (Mars Global Surveyor and 2001 Mars Odyssey) that can provide for some relay communications between the surface of Mars and Earth. There are also several proposals for additional assets to provide near-continuous contact between the surface of Mars and Earth. They include the Mars Areostationary Relay Satellite (MARSat) and the 2007 (Agenzia Spaziale Italiana) ASI/NASA Marconi mission.

Abraham [2] has made a compelling case for maintaining equivalent 70-m long-term capability. He analyzed aggregate future mission demographics and identified four key trends:

- Growth of proximity links and consequent "trunk-line" demand

- Migration of NASA's Space Science Enterprise mission set into deep space

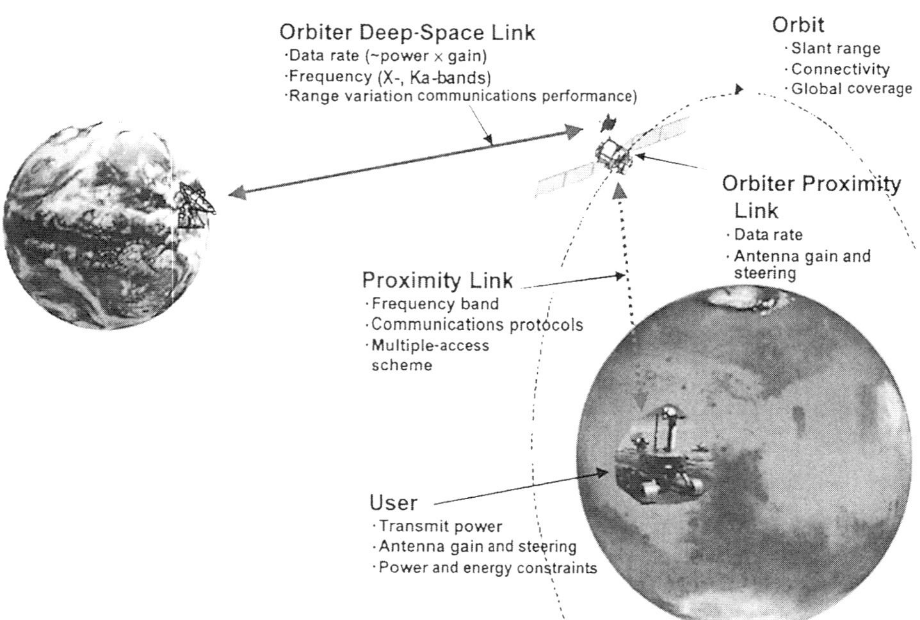


**Fig. 10-1. Key considerations for telecommunications relay-system design.**

- Mission plan reliance on large-aperture ground stations
- Evolution toward more data-intensive instruments and media.

Nonetheless, the present DSN capability places severe constraints on deep-space science due to data-rate limits, spacecraft antenna and power requirements, and cost. The DSN's current facilities are aging and costly to maintain, while demand is increasing for more complex and data-rich missions. Optical technology, which has limitations due to pointing requirements and high quantum-limited noise, is also being developed. It is not an either/or proposition, however. There is likely to be a continued need for an RF link from the ground for the foreseeable future.

It is exciting to speculate what a 100-fold increase in aperture size could do for space science or spacecraft design, including possibly enabling missions not yet envisioned. However, the only way to accomplish such an increase is through the use of arrays of low-cost, modest-sized reflector antennas. The radio astronomy community believes that

> a revolutionary new instrument at radio wavelengths is [possible], one with an effective collecting area more than 30 times greater than the largest telescope ever built. Such a telescope could reveal the dawn of galaxy formation, as well as a plethora of other new discoveries in all fields of astronomy. Vigorous technological developments in computing and radio frequency devices make it possible for such a telescope to be built within the next decade, and the international radio astronomical community is proposing that such a telescope, with 1,000,000 $m^2$ of collecting area, be the next major radio telescope to be built [9].

Completion is currently scheduled for some time between 2010 and 2020. The project is called the Square Kilometre Array (SKA).

## 10.3 Final Thoughts

Radio frequency design of the large antennas of the Deep Space Network has been a great success story. They have seen a significant evolution in performance over the last four decades. Sizes have increased from 26 to 70 m, efficiencies have been enhanced through dual-reflector shaping, and operating frequencies have evolved from L-band to S- and X-bands, to a plan for Ka-band. There has also been a growth from single-frequency to multiple-frequency operation. In addition, operability and maintainability have been enhanced through the use of beam-waveguide designs. However, designs have developed and matured to a point where only very small improvements in performance are possible, especially at S-band and X-band. In fact, virtually the only option available to improve performance at S- and X-band would be to use a clear-aperture antenna such as the Green Bank Telescope (West Virginia) [10]. However, this is a very expensive and mechanically complicated alternative that increases gain by only a fraction of a decibel (or possibly by 1 dB in gain over temperature, or $G/T$). Consequently, it would be far more cost effective to just make the blocked aperture larger. On the other hand, at Ka-band, there is room for improvement in surface accuracy and in gravity-deformation performance.

If the DSN would like a ten- to one-hundredfold improvement in $G/T$ performance, it will have to follow the lead of the designers of the next generation of large radio telescopes, who all plan to use large numbers of smaller antennas.

# References


[1] S. Brunstein, "Comparison of Implementation Costs Between a 70 m BWG DSS with an Array of Four 34m BWG Antennas," JPL Interoffice Memorandum 3330-90-116 (internal document), Jet Propulsion Laboratory, Pasadena, California, September 4, 1990.

[2] D. Abraham, "User/Mission Requirements (August 4, 2000)," (JPL internal website), http://eis.jpl.nasa.gov/antrep70m   Accessed February 2002.

[3] J. Dreher, "The Allen Telescope Array," presented at SKA: Defining the Future, Berkeley, California, July 9–12, 2001, http://www.skatelescope.org/skaberkeley/html/presentations/pdfbytopic.htm  Accessed February 2002.

[4] P. S. Kildal, L. Baker, and J. Hagfors, "Development of a Dual-Reflector Feed for the Arecibo Radio Telescope: An Overview," *Antennas and Propagation Magazine*, vol. 33, pp. 12–18, October 1991.

[5] V. L. Krabbendam, T. A. Sebring, F. B. Ray, and J. R. Fowler, "Development and Performance of Hobby Eberly Telescope 11 meter Segmental Mirror," *SPIE Conference on Advanced Technology Optical/IR Telescopes VI*, Kona, Hawaii, March 1998.

[6] B. Peng, R. Nan, R. G. Strom, B. Duan, G. Ren, J. Zhai, Y. Qiu, S. Wu, Y. Su, L Zhu, and C. Jin, "The Technical Scheme for FAST," *Perspectives on Radio Astronomy-Technologies for Large Antenna Arrays*, A. B. Smolders and M. P. van Haarlem (eds.), Netherlands Foundation for Research in Astronomy, 1999.

[7] W A. Imbriale, S. Weinreb, V. Jamnejad, and J. Cucchissi, "Exploring the Next Generation Deep Space Network," 2002 IEEE Aerospace Conference, Big Sky, Montana, March 9–16, 2002.

[8]  C. D. Edwards, J. T. Adams, D. J. Bell, R. Cesarone, R. DePaula, J. F. Durning, T. A. Ely, R. Y. Leung, C. A. McGraw, S. N. Rosell, "Strategies for Telecommunications and Navigation in Support of Mars Exploration," *Acta Astronautica*, vol. 48, no. 5–12, pp. 661–668, 2001.

[9]  The Square Kilometre Array Website, http://www.skatelescope.org/ Accessed November 2001.

[10] 100-meter Green Bank Telescope [National Radio Astronomy Website], http://info.gb.nrao.edu/GBT/GBT.html  Accessed February 2002.

# Acronyms and Abbreviations

| | |
|---|---|
| **AAS** | Advanced Antenna System |
| **AFCS** | array-feed compensation system |
| **AGARD** | Advisory Group for Aerospace Research and Development |
| **AIAA** | American Institute of Aeronautics and Astronautics |
| **AP** | antennas and propagation |
| **APC** | antenna-pointing computer |
| **ARISE** | Advanced Interferometry between Space and Earth |
| **ARST** | Antenna Research System Task |
| **ASAP** | Ariane structure for auxiliary payloads |
| **AZ** | azimuth |
| **BWG** | beam waveguide |
| **CL** | centerline |
| **CONSCAN** | conical scanning |
| **COSMIC** | University of Georgia software distribution center [formerly under contract with NASA] |
| **CP** | circular polarization |
| **DFP** | deformable flat plate |
| **DFT** | discrete Fourier transform |
| **DOY** | day of year |
| **DS1** | Deep Space 1 |

| | |
|---|---|
| **DSCS** | Defense Satellite Communications System |
| **DSN** | Deep Space Network |
| **DSPA** | digital-signal-processing assembly |
| **DSS** | deep space station |
| **EDM** | electrical discharge machining |
| **EIRP** | effective radiative power |
| **EL** | elevation |
| **FDSPA** | digital signal processing assembly |
| **FEM** | finite-element method |
| **FET** | field-effect transistor |
| **FFT** | fast Fourier transform |
| **FM** | figure of merit |
| **FSS** | frequency-selective surface |
| **GAVRT** | Goldstone–Apple Valley Radio Telescope |
| **GO** | geometrical optics |
| **HDTV** | high-definition television |
| **HEF** | high efficiency |
| **HEMT** | high-electron-mobility transistor |
| **HET** | Hobby–Eberly Telescope |
| **HPLE** | half-path-length error |
| **IECE** | International Electronics and Communications Exhibition |
| **IEEE** | Institute of Electrical and Electronics Engineers |
| **IF** | intermediate frequency |
| **IRE** | Institute of Radio Engineers |
| **IRG** | integral ring girder |
| **ISAP** | International Symposium on Antennas and Propagation |
| **JPL** | Jet Propulsion Laboratory |
| **LCER** | Lewis Center for Educational Research |
| **LCP** | left-hand circular polarization |
| **LHCP** | left-hand circular polarization |
| **LNA** | low-noise amplifier |

| | |
|---|---|
| **MAHST** | Microwave Antenna Holography System |
| **MARSat** | Mars Aerostationary Relay Satellite |
| **NASA** | National Aeronautics and Space Administration |
| **NEAR** | Near Earth Asteroid Rendezvous [mission] |
| **OPS** | operations, operational |
| **PLL** | phase-locked loop |
| **PNL** | panel |
| **PO** | physical optics |
| **pps** | pulses per second |
| **R&D** | research and development |
| **RA/DEC** | right-ascension/declination |
| **RCP** | right-hand circular polarization |
| **RF** | radio frequency |
| **RFI** | radio frequency interference |
| **RH** | relative humidity |
| **RHCP** | right-hand circular polarization |
| **rms** | root-mean-square |
| **RTLT** | round-trip light time |
| **SCM** | S-band Cassegrain monopulse |
| **SETI** | Search for Extraterrestrial Intelligence |
| **SKA** | Square Kilometer Array |
| **SNR** | signal-to-noise ratio |
| **SOHO** | Solar and Heliospheric Observatory |
| **SOHO** | Solar and Heliospheric Observatory [spacecraft] |
| **SPC** | signal processing center |
| **SPHERE** | spherical pair of high-efficiency reflecting elements |
| **SPIE** | Society for Optical Engineering |
| **SWE** | spherical wave expansion |
| **TDF** | Technology Demonstration Facility |
| **TE** | transverse-electric |
| **THP** | theoretical horn pattern |

| | |
|---|---|
| **TM** | transverse-magnetic |
| **TMOD** | Telecommunications and Mission Operations Directorate |
| **TPR** | total-power radiometry |
| **TWM** | traveling-wave master |
| **ULNA** | ultralow noise amplifier |
| **USNC URSI** | The United States National Committee for the International Union of Radio Science |
| **UWV** | microwave |
| **VSWR** | voltage standing-wave radio |
| **XCE** | X-band Cassegrain experimental feed cone |
| **XEL** | cross-elevation |
| **XRO** | X-band receive-only feed cone |

# Index